NOVEL PACKAGING SYSTEMS FOR FRUITS AND VEGETABLES

Postharvest Biology and Technology Series

NOVEL PACKAGING SYSTEMS FOR FRUITS AND VEGETABLES

Edited by

Kirtiraj K. Gaikwad, PhD

Suman Singh, PhD

AAP APPLE ACADEMIC PRESS

First edition published 2025

Apple Academic Press Inc.
1265 Goldenrod Circle, NE,
Palm Bay, FL 32905 USA

760 Laurentian Drive, Unit 19,
Burlington, ON L7N 0A4, CANADA

CRC Press
2385 NW Executive Center Drive,
Suite 320, Boca Raton FL 33431

4 Park Square, Milton Park,
Abingdon, Oxon, OX14 4RN UK

Library and Archives Canada Cataloguing in Publication

Title: Novel packaging systems for fruits and vegetables / edited by Kirtiraj K. Gaikwad, PhD, Suman Singh, PhD.
Names: Gaikwad, Kirtiraj K., editor. | Singh, Suman, editor.
Description: First edition. | Series statement: Postharvest biology and technology series | Includes bibliographical references and index.
Identifiers: Canadiana (print) 20230598587 | Canadiana (ebook) 20230598617 | ISBN 9781774914960 (hardcover) | ISBN 9781774914977 (softcover) | ISBN 9781003415701 (ebook)
Subjects: LCSH: Vegetables—Packaging. | LCSH: Fruit—Packaging.
Classification: LCC TP440 .N68 2024 | DDC 664/.809—dc23

ISBN: 978-1-77491-496-0 (hbk)
ISBN: 978-1-77491-497-7 (pbk)
ISBN: 978-1-00341-570-1 (ebk)

ABOUT THE BOOK SERIES: POSTHARVEST BIOLOGY AND TECHNOLOGY

Editor-in-Chief

Mohammed Wasim Siddiqui, PhD

Scientist-cum-Assistant Professor, Department of Food Science and Technology, Bihar Agricultural University, Sabour, Bhagalpur, Bihar, India
AAP Acquisitions Editor | Horticultural Science
Founding/Managing Editor | *Journal of Postharvest Technology*

As we know, preserving the quality of fresh produce has long been a challenging task. In the past, several approaches were in use for the postharvest management of fresh produce, but due to continuous advancement in technology, the increased health consciousness of consumers, and environmental concerns, these approaches have been modified and enhanced to address these issues and concerns.

The Postharvest Biology and Technology series features edited books that address several aspects related to postharvest technology of fresh produce. The series presents existing and novel management systems that are in use today or that have great potential to maintain the postharvest quality of fresh produce in terms of microbiological safety, nutrition, and sensory quality.

The books are aimed at professionals, postharvest scientists, academicians researching postharvest problems, and graduate-level students. This series is intended to be a comprehensive venture that provides up-to-date scientific and technical information focusing on postharvest management for fresh produce.

Books in the series address the following themes

- Nutritional composition and antioxidant properties of fresh produce
- Postharvest physiology and biochemistry
- Biotic and abiotic factors affecting maturity and quality
- Preharvest treatments affecting postharvest quality

- Maturity and harvesting issues
- Nondestructive quality assessment
- Physiological and biochemical changes during ripening
- Postharvest treatments and their effects on shelf life and quality
- Postharvest operations such as sorting, grading, ripening, de-greening, curing etc.
- Storage and shelf-life studies
- Packaging, transportation, and marketingVase life improvement of flowers and foliage
- Postharvest management of spice, medicinal, and plantation crops
- Fruit and vegetable processing waste/byproducts: management and utilization
- Postharvest diseases and physiological disorders
- Minimal processing of fruits and vegetables
- Quarantine and phytosanitory treatments for fresh produce
- Conventional and modern breeding approaches to improve the postharvest quality
- Biotechnological approaches to improve postharvest quality of horticultural crops

BOOKS IN THE SERIES

Postharvest Biology and Technology of Horticultural Crops: Principles and Practices for Quality Maintenance
Editor: Mohammed Wasim Siddiqui, PhD

Insect Pests of Stored Grain: Biology, Behavior, and Management Strategies
Ranjeet Kumar, PhD

Postharvest Management of Horticultural Crops: Practices for Quality Preservation
Editors: Mohammed Wasim Siddiqui, PhD, and Asgar Ali, PhD

Innovative Packaging of Fruits and Vegetables: Strategies for Safety and Quality Maintenance
Editors: Mohammed Wasim Siddiqui, PhD,
Mohammad Shafiur Rahman, PhD, and Ali Abas Wani, PhD

Advances in Postharvest Technologies of Vegetable Crops

Editors: Bijendra Singh, PhD, Sudhir Singh, PhD, and Tanmay K. Koley, PhD

Plant Food By-Products: Industrial Relevance for Food Additives and Nutraceuticals
Editors: J. Fernando Ayala-Zavala, PhD, Gustavo González-Aguilar, PhD, and Mohammed Wasim Siddiqui, PhD

Emerging Technologies for Shelf-Life Enhancement of Fruits
Editors: Basharat Nabi Dar, PhD, and Shabir Ahmad Mir, PhD

Sensor-Based Quality Assessment Systems for Fruits and Vegetables
Editors: Bambang Kuswandi, PhD, and Mohammed Wasim Siddiqui, PhD

Emerging Postharvest Treatment of Fruits and Vegetables
Editors: Kalyan Barman, PhD, Swati Sharma, PhD, and Mohammed Wasim Siddiqui, PhD

Packaging and Storage of Fruits and Vegetables: Emerging Trends
Editor: Tanweer Alam, PhD

Postharvest Physiological Disorders of Fruits
Editors: Ghulam Khaliq, PhD, and Sajid Ali, PhD

Bioengineered Fruit and Vegetables: Trends in Postharvest Quality Improvement
Editors: Mohammad Irfan, PhD, and Pankaj Kumar, PhD

Novel Packaging Systems for Fruits and Vegetables
Editors: Kirtiraj K. Gaikwad, PhD, and Suman Singh, PhD

Ozone Applications in the Food Industry: Trends and Future Prospects
Editors: Ravi Pandiselvam, PhD, Anjineyulu Kothakota, PhD, and Kaavya Rathnakumar, PhD

Nonthermal Food Processing Technologies: Impact on Color Profile
Editors: Kaavya Rathnakumar, PhD, and R. Pandiselvam, PhD

Ready-to-Eat Snacks: Emerging Technologies for Production and Safety
Editors: R. Pandiselvam, PhD and Gülşah Çalışkan Koç, PhD

ABOUT THE SERIES EDITOR

Mohammed Wasim Siddiqui, PhD, is an Assistant Professor and Scientist in the Department of Food Science and Post-Harvest Technology, Bihar Agricultural University, Sabour, India, and author or co-author of peer-reviewed research articles, book chapters, two manuals, and conference papers. He has over 20 edited and an authored books to his credit, published by Elsevier, CRC Press, Springer, and Apple Academic Press. Dr. Siddiqui has established the international peer-reviewed *Journal of Postharvest Technology*. He is Editor-in-Chief of two book series (*Postharvest Biology and Technology and Innovations in Horticultural Science*), published by Apple Academic Press. Dr. Siddiqui is also a Senior Acquisitions Editor for Apple Academic Press, USA, for Horticultural Science. He has been serving as an editorial board member and active reviewer of several international journals.

Dr. Siddiqui has received several grants and respected awards for his research work by a number of organizations. He is an active member of the organizing committees of several national and international seminars, conferences, and summits. He is one of key members in establishing the World Food Preservation Center (WFPC), LLC, USA, and is currently an active associate and supporter.

Dr. Siddiqui acquired a BSc (Agriculture) degree from Jawaharlal Nehru Krishi Vishwa Vidyalaya, Jabalpur, India. He received his MSc (Horticulture) and PhD (Horticulture) degrees from Bidhan Chandra Krishi Viswavidyalaya, Mohanpur, Nadia, India, with specialization in Postharvest Technology. He is a member of Core Research Group at the Bihar Agricultural University (BAU), which provides appropriate direction and assistance in prioritizing research.

ABOUT THE EDITORS

Kirtiraj K. Gaikwad, PhD
Assistant Professor, Department of Paper Technology, Indian Institute of Technology Roorkee, Roorkee, Uttarakhand, India

Kirtiraj Kundlik Gaikwad, PhD, is an Assistant Professor in the Department of Paper Technology, IIT Roorkee, India. He is a fellow of the Linnean Society of London, UK. Before joining IIT Roorkee, he worked as a postdoctoral fellow at Ecole Polytechnic de Montreal, Canada. He has mainly worked in the field of food packaging technology. Dr. Gaikwad received several national and international awards, including a FSSAI Research Award 2021, DST INSPIRE Faculty Award by DST Government of India; Young Scientist Award by the International Association for Food Protection (IAFP), USA; and A J Bank Research Award by the Society of Chemical Industries (SCI) London, UK. Dr. Gaikwad has published 70 research papers in SCI-indexed journals, two international patents, and three book chapters, and has co-authored one book titled, *Food Packaging System.* He has received several grants from DST and SERB, UCOST-funding agencies to carry out his research projects. He is dynamically involved in teaching (graduate and doctorate students) and research, and he has proved himself as an active scientist in the area of packaging technology.

He received his PhD in Packaging Technology in 2018 from Yonsei University, Seoul, South Korea; his MS in Packaging Technology from Michigan State University, Lansing, USA in 2013; his MTech in Food Technology in 2011 from SHUATS, Allahabad, India; and BTech in Food Science from Dr. Panjabrao Deshmukh Agriculture University in 2009.

Suman Singh, PhD
Assistant Professor, Department of Food Engineering, Institute of Food Science & Technology, Veer Chandra Singh Garhwali, Uttarakhand; University of Horticulture & Forestry, Mazri Grant, Dehradun, Uttarakhand, India

Suman Singh, PhD, is an Assistant Professor at VCSG Uttarakhand University of Horticulture and Forestry, Dehradun, India. Dr. Singh worked as a postdoctoral fellow at Yonsei University, South Korea, and mainly worked on active packaging materials (antimicrobial, scavenge, and temperature-sensitivity materials) for the delivery of high-quality and safe fresh produce and prolonging the shelf life of fresh produce using packaging. Dr. Singh has received several national and international awards, including aincluding a FSSAI Research award 2021, University Gold Medal in the MTech program in 2011. She was a recipient of the prestigious INSPIRE Fellowship for her PhD program in 2011, and was a recipient of a Young Researcher Award 2017 at the IFT Annual meeting, Las Vegas, USA. Dr. Suman has published 50 research papers in SCI-indexed journals and three book chapters, and has co-authored the book, *Food Packaging System.*

She received her PhD in Food Process and Engineering from the G. B. Pant University of Agriculture and Technology Pantnagar, India, in 2014; MTech in Food Process and Engineering from Sam Higginbottom University of Agriculture, Technology & Sciences, Allahabad, India, in 2011; and BTech, in Agriculture Engineering from G. B. Pant University of Agriculture and Technology Pantnagar, India, in 2009.

CONTENTS

Contributors .... *xiii*

Abbreviations .... *xv*

Foreword by Prof. (Dr.) Jay Singh .... *xix*

Preface .... *xxi*

1. **Biopolymers as Packaging Materials for Fresh Produce Applications: Preparation and Properties** **1**
Vidhi Gupta and Punyadarshini Punam Tripathy

2. **Physical and Mechanical Properties of Biopolymers Used for Packaging Agricultural Produce** **33**
Ajit Kumar Singh and Youn Suk Lee

3. **Recent Trends in Films and Gases for Modified Atmosphere Packaging of Fresh Produce** **67**
Shefali Tripathi, Lokman Hakim, Deepika Gupta, Ram Kumar Deshmukh, and Kirtiraj K. Gaikwad

4. **Natural Antimicrobial Agents for Active Packaging of Fruits and Vegetables** **95**
Saurabh Kumar Kardam and Dharm Dutt

5. **Natural Oxygen Scavengers for Maintaining the Quality of Fruits and Vegetables** **115**
Dakuri Ramakanth, Kirtiraj K. Gaikwad, and Pradip K. Maji

6. **Natural Clay-Based Ethylene Scavengers for Active Packaging of Fruits and Vegetables** **145**
Afreen Sultana, Anushikha, and Sampat Singh Bhati

7. **Natural Antioxidants for Active Packaging of Fresh-Cut Fruits** **169**
Ram Kumar Deshmukh and Kirtiraj K. Gaikwad

8. **New Smart Packaging Technologies for Monitoring Quality of Agriculture Produce** **195**
Konala Akhila, Suman Singh, and Kirtiraj K. Gaikwad

9. Active Edible Coatings and Films for the Preservation of the Quality of Agricultural Produce 219
Lokesh Kumar, Suman Singh, and Kirtiraj K. Gaikwad

10. Protective Packaging for Distribution and Storage of Fruits and Vegetables 247
Aaditya Pandey, Abhijit Maiti, and Kirtiraj K. Gaikwad

11. Package Designing Aspects for Marketing of Fruits and Vegetables 265
Pragya Srivastava, Dakuri Ramakanth, Lokman Hakim, and Kirtiraj K. Gaikwad

12. Advance Packaging for Safety and Security of Agricultural Produce in Supply Chain 289
Shiva Singh, Kaushik Ghosh, and Pradip K. Maji

Index 319

CONTRIBUTORS

Konala Akhila
Department of Paper Technology, Indian Institute of Technology Roorkee, Roorkee, Uttarakhand, India

Anushikha
Department of Paper Technology, Indian Institute of Technology Roorkee, Roorkee, Uttarakhand, India

Sampat Singh Bhati
Department of Paper Technology, Indian Institute of Technology Roorkee, Roorkee, Uttarakhand, India

Ram Kumar Deshmukh
Department of Paper Technology, Indian Institute of Technology Roorkee, Roorkee, Uttarakhand, India

Dharm Dutt
Department of Paper Technology, Indian Institute of Technology Roorkee, Roorkee, Uttarakhand, India

Kirtiraj K. Gaikwad
Department of Paper Technology, Indian Institute of Technology Roorkee, Roorkee, Uttarakhand, India

Kaushik Ghosh
Department of Chemistry, Indian Institute of Technology Roorkee, Roorkee, Uttarakhand, India

Deepika Gupta
Department of Paper Technology, Indian Institute of Technology Roorkee, Roorkee, Uttarakhand, India

Vidhi Gupta
Department of Agricultural and Food Engineering, Indian Institute of Technology Kharagpur, Kharagpur, West Bengal, India

Lokman Hakim
Department of Paper Technology, Indian Institute of Technology Roorkee, Roorkee, Uttarakhand, India

Saurabh Kumar Kardam
Department of Paper Technology, Indian Institute of Technology Roorkee, Roorkee, Uttarakhand, India

Lokesh Kumar
Department of Paper Technology, Indian Institute of Technology Roorkee, Roorkee, Uttarakhand, India

Youn Suk Lee
Department of Packaging & Logistics, Yonsei University Mirae Campus, Wonju, South Korea

Abhijit Maiti
Department of Polymer and Process Engineering, Indian Institute of Technology Roorkee, Saharanpur Campus, Saharanpur, Uttar Pradesh, India

Pradip K. Maji
Department of Polymer and Process Engineering, Indian Institute of Technology Roorkee, Saharanpur Campus, Uttar Pradesh, India

Aaditya Pandey
Department of Polymer and Process Engineering, Indian Institute of Technology Roorkee, Saharanpur Campus, Saharanpur, Uttar Pradesh, India

Dakuri Ramakanth
Department of Polymer and Process Engineering, Indian Institute of Technology Roorkee, Saharanpur Campus, Saharanpur, Uttar Pradesh, India

Ajit Kumar Singh
Department of Packaging & Logistics, Yonsei University Mirae Campus, Wonju, South Korea

Shiva Singh
Department of Polymer and Process Engineering, Indian Institute of Technology Roorkee, Saharanpur Campus, Uttar Pradesh, India

Suman Singh
Department of Food Engineering, Institute of Food Science & Technology, Veer Chandra Singh Garhwali Uttarakhand University of Horticulture & Forestry, Mazri Grant, Dehradun, Uttarakhand, India

Pragya Srivastava
Department of Paper Technology, Indian Institute of Technology Roorkee, Roorkee, Uttarakhand, India

Afreen Sultana
Department of Paper Technology, Indian Institute of Technology Roorkee, Roorkee, Uttarakhand, India

Punyadarshini Punam Tripathy
Department of Agricultural and Food Engineering, Indian Institute of Technology Kharagpur, Kharagpur, West Bengal, India

Shefali Tripathi
Department of Paper Technology, Indian Institute of Technology Roorkee, Roorkee, Uttarakhand, India

ABBREVIATIONS

AA	ascorbic acid
AP	active packaging
BC	bacterial cellulose
BCG	Boston Consulting Group
BCT	box compressive strength
BHA	butylated anisole
BHT	butylated hydroxytoluene
BOPP	biaxially-oriented polypropylene
CAP	controlled atmosphere packaging
Cat	catalase
CFB	corrugated fiber board
CFR	code of federal regulations
CH	chitosan
CIN	cinnamaldehyde
CMAP	controlled modified atmospheric packaging
CMC	carboxymethyl cellulose
CO	carbon monoxide
DBD	dielectric barrier discharge
DE	degree of esterification
DHAA	dehydroascorbic acid
DNA	deoxyribonucleic acid
EAB	elongation at break
EC	edible coatings
EF	edible films
EFSA	European Food Safety Authority
EO	essential oils
FAP	food additive petition
FAO	Food and Agriculture Organization
FCM	food contact materials
FCS	food contact substance
FC&V	fresh-cut fruits and vegetables
FDA	food loss and waste
FFS	film-forming solution

FI	freshness indicators
GA	gallic acid
GMP	glycomacropeptide
GOx	glucose oxidase
GRAS	Generally Recognized As Safe
HDPE	high-density polyethylene
HEC	hydroxyethyl cellulose
HMC	hydroxymethyl cellulose
HNT	halloysite nanotubes
ILAC	International Laboratory Accreditation Cooperative
IP	intelligent packaging
IUPAC	International Union of Pure and Applied Chemistry
KGM	konjac glucomannan
LAB	lactic acid bacteria
LDPE	low-density polyethylene
LLDPE	linear low-density polyethylene
LSE	litchi shell extract
MAP	modified atmosphere packaging
MC	methyl cellulose
Micro-EDM	micro-electric discharge method
MMT	materials montmorillonite
MOF	metallic organic frame work
MRP	mangosteen rind powder
MT	metric tonnes
MW	molecular structure
MWCNT	multiwalled carbon nanotubes
NPs	nanoparticles
ORAC	capacity for radical oxygen absorption
OTG	on the go
OTR	oxygen transmission rate
PAM	pressure-assisted micro-syringe printing
PBAT	polybutylene adipate terephthalate
PBS	polybutylene succinate
PdNP	palladium nanoparticles
PE	polyethylene
PET	polyethylene terephthalate
PG	propyl gallate
PHAs	polyhydroxyalkanoates

PHB	poly(3-hydroxybutyrate)
PHBV	poly(3-hydroxybutyrate-co-3-hydroxyvalerate)
PLA	polylactic acid
PP	polypropylene
PPO	polyphenol oxidase
PRS	pattern reorganization system with algorithms
PS	polystyrene
PVA	polyvinyl alcohol
PVC	polyvinylchloride
PVDC	polyvinylidene chloride
QR	quick response
RFID	radio-frequency identification
RO	residual oxygen
ROS	reactive oxygen species
RSS	reduced space symbology
RTC	ready-to-cook
RTE	ready-to-eat food
SC	sodium carbonate
SOPs	standard operating procedures
SO_2/MA	sulfur dioxide/modified atmosphere
SP	special packaging
SPC	soy protein concentrate
SPI	soy protein isolate
TPS	thermoplastic starch
TS	tensile strength
TSS	total soluble solids
TTI	time-temperature indicators
UGVs	unmanned ground vehicles
WHO	World Health Organization
WP	whey proteins
WPC	whey protein concentrates
WPI	whey protein isolates
WVP	water vapor permeability
ZnONP	zinc oxide nanoparticles

FOREWORD

Fruits and vegetables contain protein, fatty acids, vitamins, minerals, dietary fibers, and antioxidants. A diet rich in vegetables and fruits can prevent or reduce several serious health issues such as heart attacks, some types of cancer, digestive issues, and blood sugar. With the constantly evolving international trade, fresh fruits and vegetables are increasingly becoming one of the most important categories in global supermarkets.

Due to its restrictive shelf life, fresh produce typically requires expedited movement from farm to retail. While the use of optimal postharvest treatments helps maintain quality of the produce—packaging methods, processes, and technology can enhance shelf life and quality of the fresh produce through its distribution. An optimal rate of respiration, transpiration, ethylene production, and other metabolic processes are essential factors in choosing suitable packaging for fresh produce. With the target of maintaining or enhancing their shelf life, novel packaging techniques such as active packaging, intelligent packaging, and bioactive packaging involves an intentional interaction between the produce, packaging, and the internal/external environment.

This book on packaging technologies for quality preservation of fresh fruits and vegetables brings together a unique insight into the various novel packaging systems, packaging materials, and design elements. Some of the most innovative information on various aspects of packaging to effectively preserve the quality of fruits and vegetables can be found in this book. It is hoped that students currently enrolled in graduate and post-graduate programs and teaching professionals in packaging technology, food science and technology, and horticultural science will find this book useful. Benefits will also accrue to academic scientific research community members and the horticulture and packaging industries.

—Prof. (Dr.) Jay Singh
Professor & Director, Packaging Program
Orfalea College of Business
Cal Poly State University
San Luis Obispo, California, USA

PREFACE

Due to high demand for fresh fruits and vegetables, the market for packaging materials has experienced notable growth over the last 10 years, as per the US Department of Agriculture. This growth is expected to continue in the coming years. The growth is the result of immense knowledge on packaging technology learned through research and development work and material science and processing technology advances. For the past 10 years, there have been several patents, published research articles, and research projects undertaken by the food and packaging industries, academics, and scientists. The influence of sustainability agendas, carbon footprinting, and a keen interest in biodegradable and natural materials have further driven, and will continue to boost, the growth and interest in packaging materials. In addition, polymeric packaging material and novel technologies are adding value to the agricultural and food processing industries. With such a background, this book on Novel Packaging Systems for Fruits and Vegetables was structured. This book brings together packaging materials and novel techniques from different scientific disciplines from the academic and research institutes and the food industry (fruits and vegetables processing industries).

Novel packaging techniques considerably increase the shelf life of fresh produce. For example, freshness and quality indicators can be impregnated into the packaging to check the quality and safety of packaged fresh produce in smart packaging.

This book is composed of 12 chapters contributed by packaging technology experts. The book starts with important biopolymeric films for fresh produce packaging and a historical and common overview, followed by two chapters on the mechanical and physical and permeability properties and recent developments in investigative techniques of biopolymers as well as their applications in modified atmosphere packaging used in fresh produce packaging. These two chapters also comprise a detailed description of polymers' fabrication and engineering properties utilizing biomaterials. The next four chapters discuss detailed applications of natural/organic active agents, including oxygen scavenger, ethylene scavenger, antioxidants, antimicrobial agents, etc., for the fabrication of active packaging for maintaining the quality of fresh produce during storage and transportation. The next chapter deals with active (antimicrobial, antioxidant) edible films and coating used to preserve the quality of fresh produce. The following three chapters in the book deal with protective packaging, package designing aspects, and safety and security packaging for agricultural produce in the supply chain.

This book discusses the commercially available packaging technologies for fresh produce. Each author has constructed their chapter comprehensively so that each chapter could stand on its own. This book was constructed to aid a beginner in packaging technology in addition to those already engaged in the area, with the intention that the topics covered in the edited book will prompt future novel ideas and processes.

Our honest and immense gratitude goes to all authors for their contributions in writing the book chapters and to all the professor/scientists/authors who have worked hard to keep the interest and continued advancement of packaging technology. A special thanks goes to Dr. Mohammed Wasim Siddiqui, editor-in-chief of the Postharvest Biology and Technology book series, and Apple Academic Press, Inc. for their motivation, patience, and wholehearted support for the development of this book.

CHAPTER 1

BIOPOLYMERS AS PACKAGING MATERIALS FOR FRESH PRODUCE APPLICATIONS: PREPARATION AND PROPERTIES

VIDHI GUPTA and PUNYADARSHINI PUNAM TRIPATHY

Department of Agricultural and Food Engineering, Indian Institute of Technology Kharagpur, West Bengal, India

ABSTRACT

Food packaging plays a crucial role in storing, protecting, preserving, handling, and transporting a food product. Synthetic plastics are nonbiodegradable and their usage over the century has created serious damage to ecosystem. There is urgent need of substituting them with an eco-friendly alternative. In this context, biopolymers are the novel materials that can be explored for their potential in packaging sector. They can be classified into three classes: extracted from biomass, polymers derived from bio-based monomers, and the polymers produced by microorganisms. This chapter discusses different types of biopolymers for packaging of fresh produce, their key properties, and different methods employed to produce films and coatings from these biopolymers. These biopolymers can be an effective tool to prevent postharvest losses of fruits and vegetables.

Novel Packaging Systems for Fruits and Vegetables. Kirtiraj K. Gaikwad & Suman Singh (Eds.)

1.1 INTRODUCTION

Oil-derived plastics have been an integral part of mankind as they are cheap, easily producible, flexible, lightweight and possess superior properties. These plastics are derived from nonrenewable sources and account for 45% of the packaging sector with a global production of 360 metric tonnes (MT) in 2018. One of the limitations of these petrochemical plastics is that they are nonbiodegradable in nature and consequently leads to the generation of enormous amount of waste. Packaging sector is one of the dominant sources of plastic waste in the world. In 2015, approximately 407 MT of plastics were manufactured and three quarters of it ended up as waste. It is evaluated that out of the 6300 MT plastic waste produced, only 9% goes for recycling, while the rest of it gets accumulated as a waste in the landfill sites or seabed (Ritchie and Roser, 2020). Thus, to mitigate the negative ecological impact of plastics, there is an urgent need to shift from nonbiodegradable plastic packaging to natural biodegradable polymer-based packaging materials.

Biopolymers are natural, biodegradable and can be a promising alternative to the synthetic plastics. They are generally produced from food sources like polysaccharides, proteins, lipids or synthesized by natural or genetically modified microorganisms. They are developed with the primary goal of reducing plastic waste to promote sustainability and circular economy. In 2019, the world's production of biopolymer-based packaging was 2.11 MT with a market value of USD 4.65 billion, and it is estimated to increase up to 2.43 MT by the end of 2025 with a CAGR of 17.04% (Shaikh et al,, 2021). The use of biopolymer packaging materials offers many advantages over the conventional ones such as nontoxicity, excellent recyclability, easy processability, and good film formability properties. However, these biopolymers lack some of the properties like poor barrier and mechanical properties, along with high cost of the packaging material. These limitations have hindered the large-scale commercialization of the biopolymer packaging films. Although these issues have been addressed through the incorporation of plasticizers, fillers etc., still much attention is required to explore their full potential as packaging materials.

Fresh fruits and vegetables are consumed on a huge scale globally owing to their high nutritional profile (mineral, vitamins, antioxidants). The perishable nature of fresh produce makes them prone to postharvest deterioration and invasion by microorganisms. This results in shorter shelf life and huge loss. In addition to this, any mechanical injury in the supply chain may trigger biochemical changes accelerating senescence in fruits and vegetables. Biopolymer packaging is a cheap, effective post-harvest technique that can be applied to improve the storability characteristics of fresh produce. This chapter provides an outline of different types of biopolymers which can be utilized to develop packaging materials for fruits and vegetable packaging. Different methods like casting, extrusion, coating, and electrospinning which can be utilized for the production of these packaging materials are discussed in detail. Diverse applications of the biodegradable packaging for elevating the safety, freshness, and shelf-life of different climacteric and non-climacteric fruits and vegetables are also reviewed.

1.2 BIOPOLYMERS FOR PACKAGING OF FRESH PRODUCE

Polymers that are biodegradable and derived from natural renewable sources can be classified as biopolymers. Biopolymers used for the preservation of different horticultural commodities are mainly divided into three categories: biopolymers extracted from natural sources (polysaccharides and proteins), biopolymers synthesized via chemical methods (polylactic acid, polyvinyl alcohol, polycaprolactones), and biopolymers produced by microorganisms (polyhydroxyalkanoates (PHAs) or bacterial cellulose), as illustrated in Figure 1.1. The detailed description of these biopolymers is given below.

1.2.1 NATURALLY DERIVED BIOPOLYMERS

1.2.1.1 PLANT SOURCE-BASED BIOPOLYMERS

1.2.1.1.1 Cellulose and Its Derivatives

Cellulose is the naturally occurring biopolymer found in ample amount. It is a polysaccharide comprising unbranched chain of D-glucopyranosyl units linked with β-1,4 bonds and forms an important constituent of plants cell wall. Due to high crystallinity, cellulose is insoluble in almost all the solvents. As a result, it is not possible to cast or extrude pure cellulose into films. Hence, it is modified into cellulose derivatives which are utilized for the films development with better structural properties. These include methyl and acetate derivatives like carboxymethyl cellulose

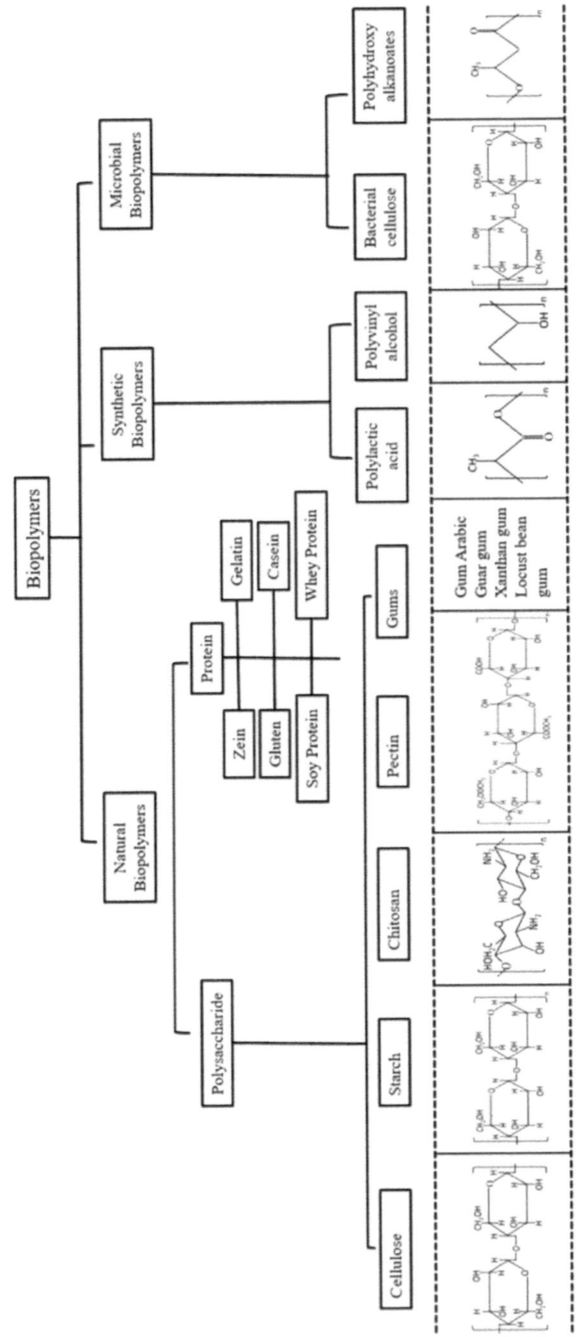

FIGURE 1.1 Classification of biopolymers.

(CMC), hydroxymethyl cellulose (HMC), methyl cellulose (MC), and cellulose acetate. These compounds are water soluble and used to develop cellulose-based films and coatings (Cazón et al., 2017). The incorporation of glutaraldehyde has shown a positive impact on decreasing the water vapor permeability of methylcellulose films. Similarly, water resistance capacity of CMC films can be enhanced by chemical cross-linkers and photo-cross-linkers for food packaging applications (Deng et al., 2022). Cellulose acetate is derived through the acetylation process of cellulose. These films are more popularly used than the films from other cellulose derivatives due to their good barrier properties. It forms clear films which can be used to pack fresh produce. The composite films produced by blending cellulose with chitosan display very good functional properties as they yield homogeneous film solution due to similar structures. The degradation of cellulose in soil is complex as it is conjugated with hemicellulose and lignin, but pure cellulose degrades quickly.

1.2.1.1.2 Starch

It is an abundantly available renewable biopolymer found in nature and can be found in many sources like wheat, rice, corn, millets, tapioca, cassava, yam, potato, and many more. It is a polysaccharide comprising glucose monomers and main source of energy for plants. It is composed of two subunits: 20–25% amylose (D-glucose linked by α-1,4 glycosidic bonds) along with 75–80% amylopectin (α-1,4- and α-1,6-D-glucopyranoside). Starches are not solubilized in cold water due to the presence of strong hydrogen bonding. Upon heating, these interactions get weakened leading to starch solubilization (gelatinization). Gelatinization process converts starch into thermoplastic starch (TPS) which is important for the production of starch-based films. This starch suspension is casted onto a Teflon substrate and dried to form the film. During drying, recrystallization occurs leading to the formation of new bonds between polymer chains (retrogradation) (Adeyeye et al., 2019).

Films and coatings made from starch without any additives are brittle, fragile and lack tensile strength. Hence, plasticizers like sorbitol, polyethylene glycol, and glycerol are added to increase the films flexibility. Films made from starch offer several advantages such as biodegradability, nontoxicity, odorless, tasteless, and good oxygen barrier properties (Sameen et al., 2021). However, due to hydrophilic nature of starch, they possess poor water barrier capacity. These technical issues can be overcome by employing certain physical, chemical, enzymatic, and thermal post treatments that helps to modify the starch structure and alter film characteristics. In one study high amylose starch was first modified using the extrusion process by exposing it to high temperature and pressure conditions and then formed into thin films using the casting method. The developed films displayed improved barrier properties and helped to preserve the quality attributes of mango (Calderón-Castro et al., 2018). The characteristics of starch films are related to many factors such as plasticizer concentration, starch source, amylose/amylopectin ratio, and method of processing. Depending upon the type and amount of plasticizer used, or the method like injection molding, compression molding or extrusion or casting can have a significant effect on the functional and structural properties of the films (Hejna et al., 2019).

Over the past few decades, bioactive components like antioxidants and antimicrobial agents are being added to the starch-based films to develop active packaging solutions for different fruits and vegetables. In relation to the biodegradation rate of starch, it is being reported that major part of starch starts to degrade within 3 days due to the cleavage of glycosidic bonds by the action of glycoside hydrolases. Thus, biodegradable starch films could be a potential packaging material to prevent postharvest losses in fruits and vegetables.

1.2.1.1.3 Gums

Gums or hydrocolloids are also polysaccharides. They can be derived from various sources like seed endosperm (guar gum), plant exudates (tragacanth gum), bacteria (xanthan gum), algae (agar), and trees (gum arabic) (Tahir et al., 2019). Gums-based edible coatings provide some benefits such as delaying ripening, extending shelf life, decreasing respiration rate, and maintaining postharvest quality of fruits and vegetables. The gums are very hydrophilic as the number of hydroxyl groups present in their structure is large. This leads to the formation of strong inter- and intramolecular hydrogen bonds which is a requisite for film formation. Polysaccharide gum coatings act as a semipermeable barrier and help to control the exchange of gases and moisture between the product and its external environment. Films produced from locust bean gum and tara gum show higher value of tensile strength and %EB (elongation at break) than the films made from guar gum (Liu et al., 2020). Gums-based films/coatings offer many advantages over the synthetic ones like sustainability, eco-friendly, biodegradability, biosafety, and they are approved as GRAS by FAO. Main disadvantages of gums-based edible films are high solubility and low strength. However, optimizing film formulations such as type of gum, amount of the gum used, plasticizers, cross-linkers, and mixing gums with other biopolymers can produce films with desirable mechanical, physical, and thermal properties (Zibaei et al., 2021). These materials can be considered prospective replacers of synthetic plastic polymers for packaging of fresh produce.

1.2.1.1.4 Pectin

Pectin is a polysaccharide present in the cell wall and middle lamellae of ripe fruits. It comprises linear chain of D-galacturonic acid units connected to each other by α-1,4 glycosidic bonds. Depending upon the degree of esterification (DE), it is of two types: high methoxyl pectin where DE is greater than 50% and low methoxyl pectin in which DE is less than 50%. It can be extracted from a variety of biomass like peels, rind, and skins of different fruits and vegetable. The gelling behavior of the pectin can be utilized to develop pectin-based coatings or films for packaging of perishable fruits and vegetables (Mellinas et al., 2020). Production of neat pectin films through melt technology is not feasible due to low thermal stability of pectin and hence, the solvent casting method is the commonly used method to formulate pectin-based films. It has been demonstrated by many studies that edible coatings made from pectin pose excellent barrier to oxygen and oil, help to lock in aroma, and display good mechanical properties. Nevertheless, high hydrophilicity and low mechanical strength of pectin-based films compared to polypropylene (PP) and high-density polyethylene (HDPE) limits its industrial applications.

1.2.1.1.5 Plant Proteins

Owing to their abundant availability, outstanding film-forming capability, excellent barrier properties, and biodegradability, plant-protein-based films have received much attention in comparison to its counterparts (polysaccharides and lipids). The main plant proteins are sourced from cereals (wheat, rice, corn, barley), pseudo-cereal (amaranth, quinoa, chia), oilseeds (sesame, hemp, sunflower, cotton), legumes (soy, pea, mung bean, lentils), and green leaf (algae, beet, alfalfa) which have the tendency to form adhesive film. Soy protein, derived from soybeans, is widely utilized in packaging applications in two forms: soy protein concentrate (SPC) and soy protein isolate (SPI). Good resistance against oxygen and oil makes soy protein isolate a promising material to fabricate films and coatings for a variety of F&V. The choice of method used for producing films based on SPC and SPI depends on two factors: concentration of protein and pH of biopolymer solution. When the protein concentration is low (~5%), the solvent casting method is used and if the concentration is about 80%, the extrusion method is employed (Assad et al., 2020). Wheat gluten is also an appealing biopolymer used for preparing packaging films with remarkable oxygen barrier properties. During film preparation, glutenin gets cross-linked to gliadin via thiol-disulphide exchange reaction leading to the development of three-dimensional gluten network. This contributes to the excellent barrier properties against atmospheric gases of wheat gluten-based films (Chen et al., 2022). The tensile strength (TS) and elongation at break (%EB) of films from soft wheat gluten are higher than the hard wheat gluten films.

Another popular and widely explored protein is zein, found in corn. It is hydrophobic in nature and has remarkable film formability. High thermal stability and viscoelastic behavior above T_g aids to process zein-based films in the presence of a suitable plasticizer through the extrusion technique which is a widely used method for obtaining plastic polymers on a commercial scale. It is a hydrophobic material and has an excellent film-forming ability. Therefore, zein has been extensively used to produce biodegradable packaging films and coatings to protect fresh produce from spoilage. Various studies have demonstrated that zein films possess lower gas permeability and better strength than other biopolymer films (Martins et al., 2021).

1.2.1.2 ANIMAL SOURCE-BASED BIOPOLYMERS

1.2.1.2.1 Chitosan

Chitin is an acetylated polysaccharide comprising N-acetyl-d-glucosamine monomers bonded by β-1,4 glycosidic linkages. It is a naturally occurring biopolymer, present in exoskeletons of insects or crustaceans shells. Chitosan is a polysaccharide which is obtained by carrying out the deacetylation process of chitin. The linear structure of chitosan comprises N-acetyl-glucosamine and D-glucosamine. The former is an acetylated unit while the latter is a deacetylated unit. It is insoluble in water but soluble in acids like acetic acid. It is very well known for its good film-forming capability. Chitosan-based films and coatings are selectively permeable to the atmospheric gases like oxygen and carbon dioxide (O_2 and CO_2). Moreover, they also possess excellent mechanical properties (Y. Liu et al., 2020). But they have high permeability to water vapor, which limits their applications for packaging of minimally processed fruits and vegetables. Different approaches can be undertaken to tackle this issue. Some of them include cross linking, incorporation of fillers, blending with other biopolymers, and enzyme treatment. Blending of chitosan with other biopolymers such as gelatin, cellulose, gums, pectin, and many more has been explored extensively to improve the physical, mechanical, barrier, and thermal properties of films and coatings made from chitosan (Haghighi et al., 2020). Recently, in a study chitosan was blended with the starch derived from white turmeric rhizome and films were fabricated by the solvent casting technique. The prepared films not only exhibited excellent antioxidant and antimicrobial characteristics, but also water resistance capacity and mechanical properties were improved significantly (Hiremani et al., 2021). Furthermore, degree of acetylation and molecular weight of chitosan are two important factors which help to determine the different properties of chitosan film-forming solution, particularly mechanical properties (González-Espinosa et al., 2019).

The antimicrobial activity of chitosan against a broad spectrum of microbes is widely known and explored for various applications in food. The remarkable antimicrobial activity could be due to a number of mechanisms. One such mechanism involves the disruption of cellular membrane of microorganisms due to interaction between negatively charged microbes and positive amine groups of chitosan. This results in leakage of bacterial constituents to outside and eventually leads to the death of microbes. Another possible reason for the antimicrobial property could be related to the low molecular weight of chitosan which enables it to enter microbial cell and disrupt the synthesis of DNA (Nair et al., 2020). Chitosan films/coatings are biocompatible, biodegradable, eco-friendly, and nontoxic.

1.2.1.2.2 Gelatin

Gelatin is an odorless, tasteless, and water-soluble protein. Collagen is a fibrous protein found in abundant amount in the connective tissues, skin, hides, and bones of the animals and acid/alkali hydrolysis of collagen yield gelatin. The sources from which gelatin can be derived include mammalian sources (pig, bovine), poultry sources (fish), and marine sources (chicken, duck). Depending upon the method used for gelatin extraction, it is of two types: Type A gelatin extracted from acid-treated collagen of pig skin, isoelectric point at pH 8-9, and Type B gelatin obtained from alkali-treated collagen of beef skin or cattle hides with isoelectric point at pH 4-5. Gelatin is a mixture of three polypeptide chains—α, β, and γ chains (Ahmad et al., 2017). It is used in food industries in many applications like thickening agent, gelling agent, stabilizer etc.

With the increase in demand of bio-based polymers for food packaging, gelatin has also been explored as a promising biopolymer for fabricating films and coatings because of low cost, excellent film-forming ability, and abundant availability as byproduct of agro-food industry. Generally, films made from gelatin display good oxygen barrier properties against oxygen at low or intermediate relative humidity along with excellent mechanical properties. The triple helical structure of the gelatin provides physical strength to the films, while the presence of different amino acids like proline and glycine helps to block UV radiation to pass through the packaging films. This aids in safeguarding packaged fruit and vegetables from oxidative damage. Several studies have reported the tensile strength for mammalian gelatin-based film within the range of 2.40–63.25 MPa. This is comparable to conventional plastics like high-density polyethylene (HDPE) and low-density polyethylene (LDPE) having tensile strength of 17.3–44.8 MPa and 8.2–31.4 MPa, respectively (Said et al., 2021). Despite these advantages, there are certain limitations in the application of gelatin-based films. Standalone gelatin films are highly sensitive to moisture and provide very poor protection from the humid environment. They tend to swell and disintegrate in contact with moisture (Andreuccetti et al., 2017). Hence, it becomes crucial to modify gelatin using different methods to address these constraints. Functional properties of gelatin films/coatings can be altered by the use of additives such as plant extracts (Riahi et al., 2021), cross-linkers, plasticizers (Suderman et al., 2018) or strengthening biopolymers (Soo and Sarbon, 2018).

1.2.1.2.3 Casein

Milk primarily consists of two proteins: casein (~80–85%) and whey proteins (~15–20%). Casein is the most important protein and comprises four subfractions: αS1-, αS2-, β-, and κ-caseins. Casein proteins are valuable source for the production of biopolymeric films and coatings with excellent gas barrier attributes, suitable thermal stability, nontoxicity, and protection against flavor loss (Mihalca et al., 2021). Casein-based films can be developed from caseinate solutions by solubilizing them in a suitable solvent, casting onto a nonreactive substrate and then drying it adequately under room conditions. The ability of sodium caseinates to form composite films can be attributed to its unsystematic bobbin structure which helps to form intermolecular interactions between the chains, resulting in a coherent polymer matrix. Calcium caseinate films have higher tensile strength and thermal stability than those made from sodium caseinate as calcium ions lead to cross-linking between the polypeptide chains giving a tougher film structure (Karaca et al., 2019).

However, the major drawbacks of milk protein-based packaging systems like brittleness, poor resistance to moisture, and low elasticity, etc. limit their industrial applications and hence should be addressed to replace synthetic polymers with biodegradable ones. These limitations of films prepared from caseinates can be modified by incorporating chemical/enzymatic cross-linkers, plasticizers, bioactive compounds, and nanofillers into the film matrix. Wu et al. (2020) explored the potential of dielectric barrier discharge (DBD) cold plasma as a modification technique to improve packaging performance of casein-based film. Mechanical (61% increment in tensile strength) and barrier properties (61% decrement in WVP (water vapor permeability)) of the formed films were notably improved due to casein aggregation. Similarly, addition of tannic acid (polyphenolic compound) as a cross linking agent in casein films helped to improve the thermal stability along with different physico-chemical properties of films (Picchio et al., 2018). Recently, several research studies have developed different approaches to create functional films of casein using additives like nanoparticles, antimicrobial agents, and antioxidants. Films obtained by blending polyvinyl alcohol with casein hydrolysates were utilized as a protective carrier of biocontrol agents (lactic acid bacteria) (Settier-Ramírez et al., 2021) which can be used for potential active packaging applications. A casein-based nanocomposite coating incorporated with

ZnO nanoparticles was formulated through double polymerization. The films not only demonstrated antibacterial properties but also the water barrier property increased markedly (Ma et al., 2019).

1.2.1.2.4 Whey Protein

Whey is a by-product of dairy industry obtained during manufacturing of cheese. Whey proteins (WP) account for approximately 15% of the total proteins present in milk and comprise five major globular proteins: 50% β-lactoglobulin, 20% α-lactalbumin, 15% glymacropepetide (GMP), 8% bovine serum albumin, and 10% immunoglobulin.

Whey protein is an important bio-based polymer that can be used to produce edible and biodegradable packaging films for fresh fruits and vegetables. Whey-based films can be produced from two forms of whey protein namely whey protein concentrates (WPC) and isolates (WPI). WPC are obtained by the ultrafiltration process of whey, and it comprises 40–80% protein with very less amount of fats and cholesterol. On the other side, further processing of WPC through ion exchange chromatography yields WPI having 90% protein content with zero lactose (Carter et al., 2021). Thermal denaturation is the commonly used technique to develop coherent films of whey protein. The film-making process of whey protein-based films can be separated into different steps. Firstly, at pH 6 and temperature >40°C, β–lactoglobulin converts from dimeric to monomeric form. Furthermore, when the temperature reaches above 60°C unfolding of β–lactoglobulin molecule occurs, exposing the functional and hydrophobic groups of whey protein. The exposed groups form new interactions like intermolecular disulfide bonding, or hydrogen bonding, van der Waals interaction, and hydrophobic bonding. This leads to irreversible denaturation of WP resulting in the formation of a 3D network. Hence, the films formed have uniform structure (Schmid and Müller, 2019).

Whey proteins-based films and coatings are transparent, biodegradable, flexible, odorless, colorless, and have great barrier and mechanical properties compared to the films made from polysaccharides and other proteins (Tsai and Weng, 2019). Compared to the films made from other proteins, films made from WPC and WPI possess better oxygen permeability, high WVP, and low tensile strength (Kandasamy et al., 2021). They also exhibit excellent oil, oxygen, and aroma barrier properties (KS Silva 2018). Films/coatings based on whey protein have also been utilized as carriers of active ingredients like probiotics, antioxidants, antimicrobials, etc., maintaining the desired properties of packaging films (Fernandes et al., 2020).

1.2.2 SYNTHETICALLY PRODUCED BIOPOLYMERS

1.2.2.1 POLYLACTIC ACID

Polylactic acid is a bio-plastic and popularly known for its availability, biodegradability, biocompatibility, and properties. It belongs to the family of aliphatic polyesters and consists of building blocks of lactic acid, synthesized by the fermentation or chemical method. PLA possesses good mechanical, physical, optical, and barrier properties compared to synthetic plastic based packaging material. For example, the permeability coefficients of different atmospheric gases like N_2, O_2, CO_2, and water vapor are lower for PLA than polystyrene (PS), but the values are higher in case of polyethylene terephthalate (PET). In terms of the mechanical properties, polylactic acid exhibits good strength and stiffness, but it is brittle and fragile in nature. Tensile strength of PLA is higher than oriented polystyrene, high density polyethylene (HDPE), and polypropylene (PP). Similarly, thermal stability of PLA is the same as that of polyvinylchloride (PVC), but it is notably lower than PS, PP, PE, and PET (Nilsen-Nygaard et al., 2021). Brittleness, poor barrier properties (water vapor, oxygen), and high price of PLA often limit its commercialization. Overall, the properties of polylactic acid are on par with that of plastics currently used in food packaging. Hence, PLA could be a suitable replacement for the nonbiodegradable fossil-based polymers.

1.2.2.2 POLYVINYL ALCOHOL

Polyvinyl alcohol (PVA) is a thermoplastic, synthetic, biodegradable, and water-soluble biopolymer, obtained via hydrolysis of poly(vinylacetate). Based upon the extent of hydrolysis, PVA can be categorized into two types: fully hydrolyzed and partially hydrolyzed polyvinyl alcohol; and the latter being commonly used for versatile applications in the food processing industry. It exhibits many remarkable physical and chemical properties like film formability, good dispersing and emulsifying power, tensile strength, and flexibility (Aslam et al., 2018). It not only exhibits good mechanical and thermal properties but also is capable of producing transparent films having good resistance capacity against oxygen permeation. Therefore, PVA can be used to produce polymer matrix for different films and coatings. However, it has low degradation rate, high cost, and poor barrier properties against moisture as a large number of hydroxyl groups are present in its structure making it highly hydrophilic in nature. This limits its scalability on an industrial level. Blending of polyvinyl alcohol with starch results in the improvement of biodegradability and reduction of capital.

1.2.3 BIOPOLYMERS PRODUCED BY MICROORGANISMS

1.2.3.1 POLYHYDROXYALKANOATES (PHAs)

PHAs belong to the family of aliphatic polyesters and are obtained by the fermentation of sugars and lipids by microorganisms under unbalanced growth conditions. They consist of two copolymers namely poly(3-hydroxybutyrate) (PHB) and poly(3-hydroxybutyrate-co-3-hydroxyvalerate) (PHBV) that are highly biodegradable in nature and used in different packaging applications (Samui and Kanai, 2019). PHAs could be produced with properties close to conventional petrochemical polymers such as polyethylene terephthalate, polyethylene, or polypropylene. Under aerobic conditions, both PHB and PHBV degrade completely into carbon dioxide and water and hence are capable of replacing nonbiodegradable plastic packaging films completely (Meereboer et al., 2020). Moreover, due to their hydrophobicity, films prepared from PHAs display low WVP coefficient, similar to that of low-density polyethylene (LDPE). In addition, these films display barrier properties for gases like O_2 and CO_2, making them an appealing bio-based packaging material for minimally process fruits and vegetables.

1.2.3.2 BACTERIAL CELLULOSE

The bacterial cellulose (BC) is a linear polysaccharide composed of β-1,4-D-glucopyranose monomers. It is pure cellulose free from hemicellulose, lignin, or pectin. The BC is excreted as exopolysaccharide by aerobic bacteria, such as *Achromobacter, Aerobacter, Agrobacterium, Alcaligenes, Azotobacter, Dickeya, Komagataeibacter, Pseudomonas, Rhodobacter,*

Rhizobium, and *Sarcina* (Lin et al., 2020). High thermal resistance, excellent mechanical properties, and barrier to oxygen, moisture, and UV radiation of BC make it a promising alternative to nonbiodegradable petroleum-based packaging materials. Moreover, high surface area and porosity of BC help in establishing physical interaction with antimicrobials and other active compounds (Azeredo et al., 2019). Thus, it can be a suitable material for the development of smart packaging films for fruits and vegetables. Due to extensive intermolecular interactions, bacterial cellulose films are brittle in nature with low elasticity, and this makes processing and handling of the films very difficult. But the functional properties of BC can be modified through the *in-situ* and *ex-situ* methods. Addition of plasticizer or blending BC with other polymers having better elasticity (like PVA or chitosan) helps to reinforce BC composite films and improve flexibility of the films/coatings (Cazon and Vazquez, 2021).

1.3 METHODS OF PREPARATION

Different methods are employed to develop biopolymer-based packaging materials. This includes solvent casting, extrusion, coating, and electrospinning discussed below. A schematic illustration of different approaches applied for the preparation of biopolymer-based films and coatings is shown in Figure 1.2

1.3.1 CASTING

Solvent casting is the simplest and commonly used method for fabricating biopolymer-based films. It is a wet method which mainly involves three steps: preparation of the film-forming solution (FFS) through solubilization of biopolymer in a suitable solvent, casting the FFS onto a mold/substrate followed by drying step, and finally peeling off the film from the substrate as shown in Figure 1.2a. Drying is the most important step in the film formation. Several intramolecular interactions occur during the drying process which helps to give a desirable microstructure to the film. Due to this reason, slow drying of the films at room temperature is carried out as quick drying methods have shown a negative effect on the structural and physical properties of the films (Velaga et al., 2018). Apart from the biopolymer and solvent, functional additives such as plasticizers, fillers, antioxidants, and antimicrobial agents are also added to the FFS (Wang et al., 2022). Biopolymers like polysaccharides and proteins are stiff in nature due to strong bonding between their polymeric chains. It becomes very difficult to process the films made up of biopolymers. Hence, to address this issue, plasticizers like sorbitol, glycerol, polyethylene glycol, sucrose, xylitol, and mannitol are used. They are low molecular weight compounds which reduce the glass transition temperature (T_g). They help to increase the flexibility of the films and ease the processability by hindering the interactions between polymeric chains (Tan et al., 2022). Biopolymer films have gained much attention owing to their high biodegradability and abundant availability of the raw materials. However, the films made of biopolymers lack mechanical strength. In the context to this, fillers such as metal oxides, nanoclays, nanoparticles, and nanocrystals are added to the matrix to reinforce and improve the mechanical strength of the biodegradable packaging films (Gupta et al., 2021).

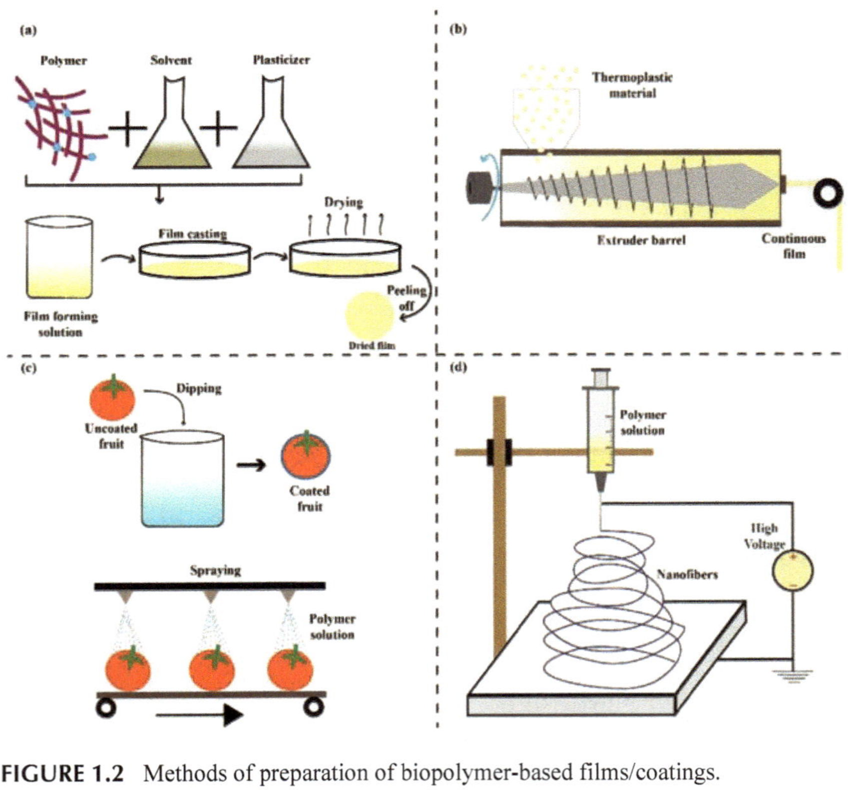

FIGURE 1.2 Methods of preparation of biopolymer-based films/coatings.

The solvent casting method offers numerous advantages such as simplicity, better dispersion of active agents into the polymer matrix, and uniform film thickness. Due to the absence of high temperatures, degradation of heat-sensitive active components is avoided, and therefore it is a widely accepted method for formulating functional packaging films like antioxidant and antimicrobial films. One of the major disadvantages of the solvent casting technique is that it is time consuming and hence cannot be scaled up for mass production.

1.3.2 EXTRUSION

Extrusion is a continuous process used for the production of films with uniform shape and density. It is a dry method in which the biopolymeric material is heated above its melting temperature under pressurized conditions as a result of which it starts to flow, gets mixed thoroughly and extruded out through a tiny nozzle called die. Generally, single-screw and twin-screw extruders are utilized for film production (Figure 1.2b). The extrusion can be divided into three zones: feeding, kneading, and heating zone where unit operations such as mixing, compression, heating, shearing, cooling, and shaping take place (Prabha et al., 2021). In the feeding zone, biopolymer(s) and plasticizer are added where they are mixed thoroughly to form a uniform and homogeneous mixture. The rotating action of the screw helps to convey the material from feed section toward the end of the barrel. They are then kneaded under high temperature–pressure combination followed by heating above the glass transition temperature (T_g). The polymer mix gets converted into a melt form and extruded out in the form of thin film (Wang et al., 2022).

Several factors like pressure, temperature, screw speed, and shear rate influence the final quality of the developed films. During the fabrication of protein films by extrusion technology, disruption of interactions between polypeptide chains occurs due to which the three-dimensional structure of proteins gets altered and new intermolecular bonds are formed. This leads to improved polymeric network resulting in better physical properties of the films (Lim, 2022). For example, transparent egg white protein films were developed using a twin-screw extruder maintained at the temperature 40–50–60–70–75°C from feeder to die section. The obtained extrudates were then calendered with the help of double-drum dryer to produce a continuous and thin film. The resulting film displayed better thermal stability, increased tensile strength, and good resistant capability to oxygen when compared to polylactic acid (PLA), a commonly used bioplastic (Pranata et al., 2019). Compared to the solvent casting method, the extrusion method offers many advantages such as less time consumption, large-scale production, no use of solvents, and low consumption of energy. However, it requires complicated machinery and hence, it is a costly technique for the production of biopolymeric films (da Silva Rios et al., 2022).

1.3.3 COATING

Coatings have been applied for the preservation of horticultural products since time immemorial. The first use of coating can be dated back to the 12th century where wax was coated over oranges in China to improve its aesthetic value (Ulusoy et al., 2018). Coating is a thin protective covering of biopolymer applied on the outer surface of the fresh F&V which helps to protect them from the external factors like microbial attack, temperature, and relative humidity. They act as a protective layer against the invasion of pathogenic microorganisms, gases, and moisture during storage, transportation, and handling of F and V. This coating is generally edible and consumed along with the whole fruit. Therefore, it is essential that the biopolymers used to develop edible coatings should be in GRAS (Generally Recognized as Safe) category. Edible coatings help to retard the microbial spoilage, reduce transpiration rate, delay senescence, enhance organoleptic/nutritional attributes, and improve the post-harvest quality of fresh commodities for a longer time.

The biopolymers used for edible coatings belong to three classes: polysaccharides, proteins, and lipids. Polysaccharides like cellulose, cellulose derivatives, starch, chitosan, gums, alginate, carrageenan, seaweed extracts, and pectin are used for making coatings. Polysaccharides-based coatings have good aroma barrier and oxygen barrier properties. Excellent mechanical strength of these films is related to the packed structure. But due to high hydrophilicity of polysaccharides, these coatings have poor resistance to water vapor (Sharma et al., 2019). Proteins of plant origin such as zein, soy protein, wheat protein, and animal proteins like gelatin, whey protein, collagen, casein, egg protein are generally used for developing packaging films for fruits and vegetables. Protein-based coatings have excellent barrier properties against atmospheric gases. They also exhibit good water vapor permeability (Hadidi et al., 2022). On the other hand, lipids like waxes are hydrophobic in nature and are used to increase the resistance power of the coatings against detrimental factors. They are applied for the preservation effect.

Different techniques are utilized to apply edible coatings to the F&V, *viz.*, dipping, spraying, panning, fluidized bed method, electrospraying, and brushing (Figure 1.2c). Dipping is the most common technique for coating fresh produce. It involves immersing the fruit into coating solution/emulsion for 5–30 seconds at a constant rate followed by the deposition of the coating onto the product surface. At last, excess solution drains off and the solvent is then evaporated either by drying in ambient conditions or employing some drying techniques (Oduro, 2021). In the spraying method, the coating solution is sprayed over the product surface. Firstly, the coating mixture is atomized into tiny droplets and then sprinkled over fruits and vegetable through the nozzle. Atomization of the liquid solution can be achieved by two methods: air pressure atomization (high velocity air assisted) and pressure atomization (high pressure assisted). This technique is likely used for less viscous biopolymer solutions. The panning method is used to coat round-shaped products in a batch mode. The fruit or vegetable to be coated is kept inside a rotating pan, and the coating solution is sprayed with the help of a spray gun. The tumbling action of the pan results in even coating of the solution (Suhag et al., 2020).

Electrospraying is a novel and more efficient technique for coating fresh produce. In this, the coating suspension is subjected to high voltage which results in the formation of tiny droplets. Then these micro and nano-sized droplets uniformly spread out on the product. This method results in better and uniform coverage of coating material over the fruits surface (Cakmak et al., 2018). The selection of the coating method relies on the fruit's physical properties and coating polymer along with the desirable final attributes. The surface roughness, shape, and size of the fruits and parameters such as viscosity, density, and surface tension of the coating solution play a vital role in determining the efficacy of the coating method. Furthermore, the interfacial interaction between the coating suspension and the fruit epicarp is important for designing coatings. This is influenced by the physicochemical properties like wettability and surface free energy. The wettability of the coating formulations can be characterized by estimating contact angle on the surface by the sessile drop method. If the value of contact angle comes equal to 0°, it denotes total wettability and the value between 0° and 180° shows partial wettability. On the other hand, contact angle of 180° relates to zero wettability which means that the surface is hydrophobic. This helps to understand the relationship between the coating and product and modify the properties of the coatings according to the requirements (Osorio et al., 2017).

1.3.4 ELECTROSPINNING

Electrospinning is a top-down spinning technique used for the fabrication of nano- and/or micro-sized polymeric nonwoven fibers. It involves pumping of a viscoelastic biopolymer solution through a capillary tube, application of high-voltage electrostatic field to produce electrically charged jet, ejection of jet from the Taylor cone resulting in the elongation of jet into fibers. This is followed by spinning and shearing of polymer solution, solvent evaporation, and deposition of intertwined nanofibers onto the grounded metal collector (Janssen 2019), as shown in Figure 1.2d. Electrospun fibers offer several advantages over the convention fibers. This may include highly specific surface area, low cost of production, high porosity, and interconnectivity, tunable thickness (Aman Mohammadi et al., 2020). Different forms of the fibers can be achieved by altering electrospinning processing parameters (applied voltage, flow rate, collection distance), biopolymer solution characteristics (surface tension, viscosity, molecular weight, electrical conductivity, solvent used), and ambient conditions (air velocity, temperature, and relative humidity in the electrospinning chamber) (Vidal et al., 2022).

Electrospun nanofibers have been widely explored for their applications in food packaging, especially in smart packaging (active/intelligent) of various fruits and vegetables. Natural (polysaccharides and proteins) and synthetic (polylactic acid) biopolymers are often utilized to produce nanofibers because of their biocompatibility and biodegradable nature. Some of the natural biopolymers used in electrospinning techniques include chitosan, starch, cellulose, pullulan, gums, gelatin, zein, soy protein, whey protein, and many more (Kumar et al., 2019). In food packaging applications, the electrospinning technique is used to enhance the postharvest quality or encapsulate various bioactive components for active packaging of fruits and vegetables. Fruits like strawberry, peach, mango, and grapes are highly sensitive to the fungal decay and hence to preserve their post-harvest quality, electrospun packaging materials are often combined with the refrigeration. Yue et al. (2018) studied the fresh keeping effects of carboxymethyl chitosan/polyoxyethylene oxide nanofiber membranes on strawberries. They showed a remarkable effect in extending the shelf-life of strawberries owing to their excellent breathability and antibacterial activity.

Over the past few decades, electrospinning has also been used for the encapsulation of active ingredients such as antioxidants, and antimicrobial agents to produce fiber mats for smart packaging applications. For encapsulating purposes, the active ingredient can be incorporated into the nanofibers by three different ways: (i) coaxial electrospinning, (ii) blend electrospinning, and (iii) emulsion electrospinning. In the first process, the active ingredient is mixed with the biopolymer solution prior to electrospinning while in blend type multiple feed ducts are provided to simultaneously spin two solutions. This leads to the production of core-shell nanofibers. Emulsion electrospinning is used to electrospin W/O (water-in-oil) or O/W (oil-in-water) emulsions stabilized by an emulsifier (Hemmati et al., 2021). For example, chitosan/poly(vinyl alcohol) eletrospun nanofibrous mats loaded with fish protein hydrolysate were fabricated using a high voltage of 15 kV. The developed mats exhibited high encapsulation efficiency (>94%) along with sustained release of the active component (Hosseini 2019). Similarly, in another study, gelatin nanofibers impregnated with peppermint and cinnamon essential oils exhibited excellent antioxidant and antibacterial activities against two common pathogens *Staphylococcus aureus* and *Escherichia coli*. These nanofibers could be promisingly utilized as edible packaging material and help to store fruits and vegetables for a longer time (Tang et al., 2019).

Furthermore, nanofibrous mats can be converted into continuous films by subjecting them to a thermal posttreatment called annealing. It is carried out at the temperature below the melting point of the biopolymer; as a result of which nanofibers aggregate together to form continuous and homogenous films. These films have the potential application as adhesive interlayers, active, and barrier layers. The films obtained after electrospinning and annealing of poly(3-hydroxybutyrate) at 160°C exhibited enhanced optical properties and tensile strength, while water vapor permeability was found to be similar to the conventionally extruded films (Cherpinski et al., 2017).

1.4 APPLICATIONS OF VARIOUS BIOPOLYMERS IN FRUITS AND VEGETABLES INDUSTRY

Fruits and vegetables are highly perishable commodities. Approximately, 20% of F&V are lost globally due to microbial spoilage and ineffective methods to prolong the shelf life. Two phenomena namely respiration and transpiration are responsible for fruit shrinkage, early onset of decay, and senescence of the fresh produce, especially in case of climacteric fruits. These issues can be addressed through utilization of bio-based smart packaging materials. Some of the recent applications of biopolymers for the preservation of different F&V are outlined below.

1.4.1 ANTIOXIDANT PACKAGING

Oxidation reactions can cause browning reactions in minimally processed F&V turning their surface into brown color. This diminishes the aesthetic value of sliced and cut pieces of fruits, making them unacceptable by the consumers. Degradation of fresh produce can also be initiated by the production of reactive oxygen species (ROS) due to respiration. This promotes early decay and loss of organoleptic and nutritional quality. Antioxidants are the compounds that scavenge these reactive species. Various antioxidants from different sources such as plant extracts (lemongrass, basil leaf, mulberry leaf, grape seed, green tea, etc.), herbs (rosemary, oregano, basil etc.), and spices (clove, cinnamon, ginger, turmeric etc.) can be successfully incorporated into the biopolymer matrix to develop antioxidant packaging films and coatings. These films have shown promising results in preserving the freshness, maintaining the postharvest quality, and enhancing the shelf-life of sliced/diced F&V. Majority of the antioxidant compounds such as polyphenols and flavonoids prevent autoxidation reactions by hindering the formation and propagation of free radical's chain reaction. Different types of fruits and vegetables like plum, melon, kiwi, blackberry, pear, strawberry, cucumber, tomatoes, eggplant etc. have successfully shown prolonged shelf-life and improved storage stability after being coated or packed with antioxidant films (Khan et al., 2021). The detailed description of the different types of antioxidant and applications of antioxidant packaging for the protection of fruits and vegetables have been discussed in Chapter 7.

1.4.2 ANTIMICROBIAL PACKAGING

Fruits and vegetables are full of nutrients, and hence they are highly vulnerable to the microbial spoilage. Antimicrobial packaging can be defined as an active packaging which consists of antimicrobial agents as a functional ingredient. They are either incorporated into the polymer matrix during film forming or coated onto the surface on the polymer or added in the form of sachet/pads. These agents are effective against inhibiting or killing the microbes responsible for the contamination of fresh produce. The hydroxyl group of the polyphenols interact with the microorganisms cell membrane leading to the disruption of its cellular structure. The ability of an antimicrobial agent to modify cell permeability results in leakage of cell contents, damage to DNA, and denaturation of protein, ultimately leading to the cell death (Pandey et al., 2022). This helps to prevent foodborne pathogens from multiplying. There are many antimicrobial agents explored worldwide for the development of antimicrobial packaging systems for the protection of F&V. The further details on antimicrobial packaging are given in Chapter 4 of this book.

1.4.3 EDIBLE FILMS AND COATINGS

Edible films (EF) and edible coatings (EC) are a very fast emerging and eco-friendly packaging alternative to conventional synthetic polymers. Natural biopolymers such as polysaccharides, proteins, and lipids have been proven to be some potential candidates for the development of edible packaging. The latest developments in this packaging include the extensive use of proteins, polysaccharides, and lipids as the polymer matrix of the edible coatings to which different functional agents are added, like antioxidants extracts, antimicrobial agents, and probiotics (Kumar et al., 2021). The role of edible packaging is not only limited to act as a protective layer but also to enhance its functional properties. An edible coating of modified sweet potato starch impregnated with cumin essential oil was formulated to delay the postharvest decay and enhance the storability of early crisp. The coating significantly reduced the deterioration of the fruit and exhibited antifungal activity against *Alternaria alternate* (Oyom et al., 2022). EF and EC also help in controlling microbial spoilage, antioxidant properties, delaying senescence, preserving firmness, improving nutritional quality and better organoleptic quality of the fruits and vegetables, resulting in the extension of shelf-life and better consumer acceptance. The retardation of weight losses, decay percentage rate, and the prolonged shelf life have a huge impact on the reduction of global food wastages and, consequently, help to ensure food security.

1.4.4 INTELLIGENT PACKAGING

Owing to the increase in demand of more nutritious yet safer food, many innovations have been done in food industry including food packaging. Consumers are not only curious about the quality of food but also want to know about the food safety during transport and handling. One such is the development of intelligent packaging. The safety of the food can be compromised intentionally or unintentionally. Deliberate counterfeiting the content/quality or temperature fluctuations in cold chain may pose a serious threat to the food safety and consequently, the consumers. Intelligent packaging is a smart packaging system where smart functions are added to the packaging material which help to detect, sense, record, monitor, and track the condition of packed fruits and vegetables and tell about their quality. These include pH indicators, sensors, oxygen indicators, freshness indicators, and time-temperature indicators (Halonen et al., 2020). This can guarantee food quality, provide information about food safety, and give warning about spoilage. Anthocyanins, curcumin, and carotenoids are some functional compounds which can be added to biopolymer matrix to produce pH-sensitive films. Presence of microorganisms on any fruit or vegetable surface leads to the production of metabolites. As a result of this, pH changes and a visual color change is detected. This is correlated with the information related to freshness and microbiological quality of F&V.

1.5 LEGISLATIONS AND SAFETY ISSUES

Migration of monomeric compounds from the polymer matrix to the food product may pose serious health risk to the consumers. Furthermore, to reinforce the functional and barrier properties of the biodegradable polymers, certain additives like plasticizers, nanofillers, antioxidants, antimicrobial agents, cross-linkers etc. are incorporated into them. As these compounds are not chemically bonded to the polymer matrix, there are chances that they might migrate to the fruits or vegetables and induce toxicity. Green and sustainable approaches of fabrication should be adopted in order to maximize the safety of the bioplastics. Ingestion of fruits and vegetable coated with biopolymers containing nanoparticles (NPs) may cause damage to cells; hence, the permissible limit for the addition of NPs in the packaging films/coatings should be implemented strictly.

The edible coatings for fruits and vegetables are mainly made from the proteins and polysaccharides extracted from wheat, soy, milk, and peanut. These products are well-known allergens and may cause allergy to people. Thus, it is crucial to properly label these coatings for allergen information and make consumers aware of it. Allergen information should be present on the label even if the quantity used is very small. Furthermore, the polymers used in edible coatings and films should be FDA approved and hold GRAS status.

1.6 FUTURE RECOMMENDATIONS

Currently, a wide variety of biopolymer-based packaging material have been developed, characterized, analyzed for their different properties, modified to enhance their inherent properties and utilized for the packaging of different F&V to prolong their shelf-life and postharvest quality. However, many limitations need to be addressed to successfully commercialize the applications of different biopolymers in food processing. First, the materials and equipment required for manufacturing biodegradable materials should be cheap, available in surplus quantity, and capable of mass production. But the processes employed in the scientific literature to obtain these biopolymers are time consuming, inefficient, and not feasible to be adopted at industrial scale. Next, the performance of these materials during storage, handling, transportation and in real environmental conditions needs to be carefully monitored and thoroughly examined in order to partially or completely replace nonbiodegradable plastics from the market shelves. Despite many efforts of the scientific community to undertake extensive research in bio-based packaging, these materials are still not at par with the conventional petroleum-based polymers in terms of their market size and practical applications. Government regulations, policies, and standard operating procedures (SOPs) should be developed and implemented properly to ensure efficient commercialization of natural and biodegradable polymers as packaging materials. Moreover, consumer perception toward the acceptance of biopolymer-based packaging material should be taken into consideration.

1.7 CONCLUSIONS

Till date, there is a huge demand of plastics as primary packaging materials in almost every industry. The nonbiodegradable nature and disposal problem of plastics has resulted in environmental problems. The willingness to focus toward sustainability and minimize the negative impact of plastics on environment has necessitated to develop biopolymers as alternative packaging materials to the synthetic ones. The biopolymers like cellulose, starch, gelatin, chitosan, PLA, PHB etc. are sustainable, biodegradable and can be applied as primary packaging material for the preservation of fruits and vegetables. These materials help to prevent spoilage, delay senescence, and prolong the keeping quality of a variety of horticultural products, thereby helping to combat global food wastage problem. Use of different methods like addition of plasticizers, bioactive agents, or blending of two or more biopolymers in order to achieve desirable characteristics has been proposed to tailor the barrier (moisture resistance capacity) and mechanical properties (tensile strength) of biopolymer films/coatings. Even after lot of research has been done on the development of natural polymers, the application of these polymers on a commercial level is still very scarce. Thus, it is of utmost importance that the biopolymer packaging materials should be tested rigorously to bridge the gap between natural and fossil-based polymers. In nutshell, substituting plastics with biodegradable-based packaging materials will help to contribute toward sustainable development environment, and promote circular economy.

KEYWORDS

- **biopolymers**
- **fruits and vegetables**
- **biodegradable**
- **food packaging**
- **shelf-life**

REFERENCES

Adeyeye, O. A.; Sadiku, E. R.; Babu Reddy, A.; Ndamase, A. S.; Makgatho, G.; Sellamuthu, P. S. et al. The Use of Biopolymers in Food Packaging. In *Green Biopolymers and their Nanocomposites*. Springer, 2019; pp 137–158.

Ahmad, T.; Ismail, A.; Ahmad, S. A.; Khalil, K. A.; Kumar, Y.; Adeyemi, K. D.; Sazili, A. Q. Recent Advances on the Role of Process Variables Affecting Gelatin Yield and Characteristics with Special Reference to Enzymatic Extraction: A Review. *Food Hydrocoll.* **2017,** *63,* 85–96.

Aman Mohammadi, M.; Hosseini, S. M.; Yousefi, M. Application of Electrospinning Technique in Development of Intelligent Food Packaging: A Short Review of Recent Trends. *Food Sci. Nutr.* **2020,** *8* (9), 4656–4665.

Andreuccetti, C.; Galicia-García, T.; González-Nuñez, R.; Martínez-Bustos, F.; Grosso, C. R. F. Native and Modified Gelatin Films Produced by Casting, Extrusion, and Blowing Extrusion Processes. *Polym. Renew. Res.* **2017,** *8* (1), 11–26.

Aslam, M.; Kalyar, M. A.; Raza, Z. A. Polyvinyl Alcohol: A Review of Research Status and Use of Polyvinyl Alcohol Based Nanocomposites. *Polym. Eng.. Sci.* **2018,** *58* (12), 2119–2132.

Assad, I.; Bhat, S. U.; Gani, A.; Shah, A. Protein Based Packaging of Plant Origin: Fabrication, Properties, Recent Advances and Future Perspectives. *Int. J. Biol. Macromol.* **2020,** *164*, 707–716.

Azeredo, H. M. C.; Barud, H.; Farinas, C. S.; Vasconcellos, V. M.; Claro, A. M. Bacterial Cellulose as a Raw Material for Food and Food Packaging Applications. *Front. Sustain. Food Syst.* **2019,** *3,* 7.

Cakmak, H.; Kumcuoglu, S.; Tavman, S. Production of Edible Coatings with Twin-Nozzle Electrospraying Equipment and the Effects on Shelf-Life Stability of Fresh-Cut Apple Slices. *J. Food Process Eng.* **2018,** *41* (1), e12627.

Calderón-Castro, A.; Vega-García, M. O.; de Jesús Zazueta-Morales, J.; Fitch-Vargas, P. R.; Carrillo-López, A.; Gutiérrez-Dorado, R.; … Aguilar-Palazuelos, E. Effect of Extrusion Process on the Functional Properties of High Amylose Corn Starch Edible Films and Its Application in Mango (Mangifera indica L.) cv. Tommy Atkins. *J. Food Sci. Technol.* **2018,** *55* (3), 905–914.

Carter, B. G.; Cheng, N.; Kapoor, R.; Meletharayil, G. H.; Drake, M. A. Invited Review: Microfiltration-Derived Casein and Whey Proteins from Milk. *J. Dairy Sci.* **2021,** *104* (3), 2465–2479.

Cazon, P.;Vazquez, M. Improving Bacterial Cellulose Films by Ex-Situ and In-Situ Modifications: A Review. *Food Hydrocoll.* **2021,** *113,* 106514.

Cazón, P.; Velazquez, G.; Ramírez, J. A.; Vázquez, M. Polysaccharide-Based Films and Coatings for Food Packaging: A Review. *Food Hydrocoll.* **2017,** *68,* 136–148.

Chen, Y.; Li, Y.; Qin, S.; Han, S.; Qi, H. Antimicrobial, UV Blocking, Water-Resistant and Degradable Coatings and Packaging Films Based on Wheat Gluten and Lignocellulose for Food Preservation. *Compos. B. Eng.* **2022,** 109868.

Cherpinski, A.; Torres-Giner, S.; Cabedo, L.; Lagaron, J. M. Post-Processing Optimization of Electrospun Submicron Poly(3-Hydroxybutyrate) Fibers to Obtain Continuous Films of Interest in Food Packaging Applications. *Food Addit. Contam. Part A Chem. Anal. Control. Expo. Risk. Assess.* **2017,** *34* (10), 1817–1830.

da Silva Rios, D. A.; Nakamoto, M. M.; Braga, A. R. C.;da Silva, E. M. C. Food Coating Using Vegetable Sources: Importance and Industrial Potential, Gaps of Knowledge, Current Application, and Future Trends. *Appl. Food Res.* **2022,** 100073.

Deng, J.; Zhu, E.-Q.; Xu, G.-F.; Naik, N.; Murugadoss, V.; Ma, M.-G.; ... Shi, Z.-J. Overview of Renewable Polysaccharide-Based Composites for Biodegradable Food Packaging Applications. *Green Chem.* **2022,** *24* (2), 480–492.

Fernandes, L. M.; Guimarães, J. T.; Pimentel, T. C.; Esmerino, E. A.; Freitas, M. Q.; Carvalho, C. W. P.; ... Silva, M. C. Edible Whey Protein Films and Coatings Added with Prebiotic Ingredients. In *Agri-Food Industry Strategies for Healthy Diets and Sustainability*. Elsevier, 2020; pp 177–193.

González-Espinosa, Y.; Sabagh, B.; Moldenhauer, E.; Clarke, P.;Goycoolea, F. M. Characterisation of Chitosan Molecular Weight Distribution by Multi-Detection Asymmetric Flow-Field Flow Fractionation (AF4) and SEC. *Int. J. Biol. Macromol.* **2019,** *136,* 911–919.

Gupta, V.; Yadav, R.; Tanwar, R.;Gaikwad, K. K. κ-Carrageenan-Based Bio-Nanocomposite Film Reinforced with Cellulose Nanocrystals Derived from Amla Pomace for Food Packaging. *Biomass Convers. Biorefin.* **2021,** 1–10.

Hadidi, M.; Jafarzadeh, S.; Forough, M.; Garavand, F.; Alizadeh, S.; Salehabadi, A.; ... Jafari, S. M. Plant Protein-Based Food Packaging Films; Recent Advances in Fabrication, Characterization, and Applications. *Trends Food Sci. Technol.* **2022**.

Haghighi, H.; Licciardello, F.; Fava, P.; Siesler, H. W.;Pulvirenti, A. Recent Advances on Chitosan-Based Films for Sustainable Food Packaging Applications. *Food Packag. Shelf Life* **2020,** *26,* 100551.

Halonen, N.; Pálvölgyi, P. S.; Bassani, A.; Fiorentini, C.; Nair, R.; Spigno, G.;Kordas, K. Bio-Based Smart Materials for Food Packaging and Sensors—A Review. *Front. Mater.* **2020,** *7*, 82.

Hejna, A.; Lenża, J.; Formela, K.;Korol, J. Studies on the Combined Impact of Starch Source and Multiple Processing on Selected Properties of Thermoplastic Starch/ Ethylene-Vinyl Acetate Blends. *J. Polym. Environ.* **2019,** *27* (5), 1112–1126.

Hemmati, F.; Bahrami, A.; Esfanjani, A. F.; Hosseini, H.; McClements, D. J.;Williams, L. Electrospun Antimicrobial Materials: Advanced Packaging Materials for Food Applications. *Trends Food Sci. Technol.* **2021,** *111,* 520–533.

Hiremani, V. D.; Khanapure, S.; Gasti, T.; Goudar, N.; Vootla, S. K.; Masti, S. P.; ... Chougale, R. B. Preparation and Physicochemical Assessment of Bioactive Films Based on Chitosan and Starchy Powder of White Turmeric Rhizomes (Curcuma Zedoaria) for Green Packaging Applications. *Int. J. Biol. Macromol.* **2021,** *193,* 2192–2201.

Kandasamy, S.; Yoo, J.; Yun, J.; Kang, H.-B.; Seol, K.-H.; Kim, H.-W.;Ham, J.-S. Application of Whey Protein-Based Edible Films and Coatings in Food Industries: An Updated Overview. *Coatings* **2021,** *11* (9), 1056.

Karaca, O. B.; Oluk, C. A.; Taşpinar, T.;Güven, M. New Concept in Packaging: Milk Protein Edible Films. In *Health and Safety Aspects of Food Processing Technologies*. Springer, 2019; pp 537–565.

Khan, M. R.; Di Giuseppe, F. A.; Torrieri, E.;Sadiq, M. B. Recent Advances in Biopolymeric Antioxidant Films and Coatings for Preservation of Nutritional Quality of Minimally Processed Fruits and Vegetables. *Food Packag. Shelf Life* **2021,** *30,* 100752.

Kumar, L.; Ramakanth, D.; Akhila, K.;Gaikwad, K. K. Edible Films and Coatings for Food Packaging Applications: A Review. *Environ. Chem. Lett.* **2021,** 1–26.

Kumar, T. S. M.; Kumar, K. S.; Rajini, N.; Siengchin, S.; Ayrilmis, N.; Rajulu, A. V. A Comprehensive Review of Electrospun Nanofibers: Food and Packaging Perspective. *Compos. B. Eng.* **2019,** *175,* 107074.

Lim, L.-T. Plant-Based Protein Films and Coatings. *Plant Protein Foods* **2022,** 271–311.

Lin, D.; Liu, Z.; Shen, R.; Chen, S.;Yang, X. Bacterial Cellulose in Food Industry: Current Research and Future Prospects. *Int. J. Biol. Macromol.* **2020,** *158,* 1007–1019.

Liu, F.; Chang, W.; Chen, M.; Xu, F.; Ma, J.;Zhong, F. Film-Forming Properties of Guar Gum, Tara Gum and Locust Bean Gum. *Food Hydrocoll.* **2020,** *98,* 105007.

Liu, Y.; Yuan, Y.; Duan, S.; Li, C.; Hu, B.; Liu, A.; ... others. Preparation and Characterization of Chitosan Films with Three Kinds of Molecular Weight for Food Packaging. *Int. J. Biol. Macromol.* **2020,** *155,* 249–259.

Ma, J.; An, W.; Xu, Q.; Fan, Q.;Wang, Y. Antibacterial Casein-Based ZnO Nanocomposite Coatings with Improved Water Resistance Crafted via Double in Situ Route. *Prog. Org. Coat.* **2019,** *134,* 40–47.

Martins, V. G.; Romani, V. P.; Martins, P. C.; Nogueira, D. Protein-Based Materials for Packaging Applications. In *Bio-Based Packaging: Material, Environmental and Economic Aspects*, 2021; pp 27–49.

Meereboer, K. W.; Misra, M.;Mohanty, A. K. Review of Recent Advances in the Biodegradability of Polyhydroxyalkanoate (PHA) Bioplastics and their Composites. *Green Chem.* **2020,** *22* (17), 5519–5558.

Mellinas, C.; Ramos, M.; Jiménez, A.;Garrigós, M. C. Recent Trends in the Use of Pectin From Agro-Waste Residues as a Natural-Based Biopolymer for Food Packaging Applications. *Materials* **2020,** *13* (3), 673.

Mihalca, V.; Kerezsi, A. D.; Weber, A.; Gruber-Traub, C.; Schmucker, J.; Vodnar, D. C.; ... others. Protein-Based Films and Coatings for Food Industry Applications. *Polymers* **2021,** *13* (5), 769.

Nair, M. S.; Tomar, M.; Punia, S.; Kukula-Koch, W.;Kumar, M. Enhancing the Functionality of Chitosan-and Alginate-Based Active Edible Coatings/Films for the Preservation of Fruits and Vegetables: A Review. *Int. J. Biol. Macromol.* **2020,** *164,* 304–320.

Nilsen-Nygaard, J.; Fernández, E. N.; Radusin, T.; Rotabakk, B. T.; Sarfraz, J.; Sharmin, N.; ... Pettersen, M. K. Current Status of Biobased and Biodegradable Food Packaging Materials: Impact on Food Quality and Effect of Innovative Processing Technologies. *Compr. Rev. Food Sci. Food Safety* **2021,** *20* (2), 1333–1380.

Oduro, K. O.-A. Edible Coating. 2021.

Osorio, F.; Valdés, G.; Skurtys, O.; Andrade, R.; Villalobos-Carvajal, R.; Silva-Weiss, A.; ... Lopez, J. Surface Free Energy Utilization to Evaluate Wettability of Hydrocolloid Suspension on Different Vegetable Epicarps. *Coatings* **2017,** *8* (1), 16.

Oyom, W.; Xu, H.; Liu, Z.; Long, H.; Li, Y.; Zhang, Z.; ... Prusky, D. Effects of Modified Sweet Potato Starch Edible Coating Incorporated with Cumin Essential Oil on Storage Quality of 'Early Crisp.' *LWT* **2022,** *153,* 112475.

Pandey, S.; Sharma, K.;Gundabala, V. Antimicrobial Bio-Inspired Active Packaging Materials for Shelf Life and Safety Development: A Review. *Food Biosci.* **2022,** 101730. DOI: 10.1016/j.fbio.2022.101730

Picchio, M. L.; Linck, Y. G.; Monti, G. A.; Gugliotta, L. M.; Minari, R. J.;Igarzabal, C. I. A. Casein Films Crosslinked by Tannic Acid for Food Packaging Applications. *Food Hydrocoll.* **2018,** *84,* 424–434.

Prabha, K.; Ghosh, P.; Abdullah, S.; Joseph, R. M.; Krishnan, R.; Rana, S. S.;Pradhan, R. C. Recent Development, Challenges, and Prospects of Extrusion Technology. *Future Foods*. **2021,** *3,* 100019.

Pranata, M. P.; González-Buesa, J.; Chopra, S.; Kim, K.; Pietri, Y.; Ng, P. K. W.; ... Almenar, E. Egg White Protein Film Production Through Extrusion and Calendering Processes and Its Suitability for Food Packaging Applications. *Food Bioproc. Technol.* **2019,** *12* (4), 714–727.

Riahi, Z.; Priyadarshi, R.; Rhim, J.-W.; Bagheri, R. Gelatin-Based Functional Films Integrated with Grapefruit Seed Extract and TiO_2 for Active Food Packaging Applications. *Food Hydrocoll.* **2021,** *112,* 106314.

Ritchie, H.;Roser, M. Plastic Pollution. Our World in Data. 2020.

Said, N. S.; Howell, N. K.;Sarbon, N. M. A Review on Potential Use of Gelatin-Based Film as Active and Smart Biodegradable Films for Food Packaging Application. *Food Rev. Int.* **2021,** 1–23.

Sameen, D. E.; Ahmed, S.; Lu, R.; Li, R.; Dai, J.; Qin, W.; ... Liu, Y. Electrospun Nanofibers Food Packaging: Trends and Applications in Food Systems. *Critic. Rev. Food Sci. Nutr.* **2021,** 1–14.

Samui, A. B.; Kanai, T. Polyhydroxyalkanoates Based Copolymers. *Int. J. Biol. Macromol.* **2019,** *140,* 522–537.

Schmid, M.;Müller, K. Whey Protein-Based Packaging Films and Coatings. In *Whey Proteins*. Elsevier, 2019; pp 407–437.

Settier-Ramírez, L.; López-Carballo, G.; Gavara, R.; Hernández-Muñoz, P. Effect of Casein Hydrolysates on the Survival of Protective Cultures of Lactococcus Lactis and Lactobacillus Sakei in PVOH Films. *Food Hydrocoll.* **2021,** *121,* 107012.

Shaikh, S.; Yaqoob, M.;Aggarwal, P. An Overview of Biodegradable Packaging in Food Industry. *Curr. Res. Food Sci.* **2021,** *4,* 503–520.

Sharma, H. P.; Chaudhary, V.; Kumar, M. Importance of Edible Coating on Fruits and Vegetables: A Review. *J. Pharmacogn. Phytochem.* **2019,** *8,* 4104–4110.

Soo, P. Y.; Sarbon, N. M. Preparation and Characterization of Edible Chicken Skin Gelatin Film Incorporated with Rice Flour. *Food Packag. Shelf Life* **2018,** *15,* 1–8.

Suderman, N.; Isa, M. I. N.;Sarbon, N. M. The Effect of Plasticizers on the Functional Properties of Biodegradable Gelatin-Based Film: A Review. *Food Biosci.* **2018,** *24,* 111–119.

Suhag, R.; Kumar, N.; Petkoska, A. T.; Upadhyay, A. Film Formation and Deposition Methods of Edible Coating on Food Products: A Review. *Food Res. Int.* **2020,** *136* (March), 109582.

Tahir, H. E.; Xiaobo, Z.; Mahunu, G. K.; Arslan, M.; Abdalhai, M.; Zhihua, L. Recent Developments in Gum Edible Coating Applications for Fruits and Vegetables Preservation: A Review. *Carbohydr. Polym.* **2019,** *224,* 115141.

Tan, S. X.; Andriyana, A.; Ong, H. C.; Lim, S.; Pang, Y. L.; Ngoh, G. C. A Comprehensive Review on the Emerging Roles of Nanofillers and Plasticizers Towards Sustainable Starch-Based Bioplastic Fabrication. *Polymers* **2022,** *14* (4), 664.

Tang, Y.; Zhou, Y.; Lan, X.; Huang, D.; Luo, T.; Ji, J.; ... Wang, W. Electrospun Gelatin Nanofibers Encapsulated with Peppermint and Chamomile Essential Oils as Potential Edible Packaging. *J. Agric. Food Chem.* **2019,** *67* (8), 2227–2234.

Tsai, M.-J.; Weng, Y.-M. Novel Edible Composite Films Fabricated with Whey Protein Isolate and Zein: Preparation and Physicochemical Property Evaluation. *LWT* **2019,** *101*, 567–574.

Ulusoy, B. H.; Yildirim, F. K.; Hecer, C. Edible Films and Coatings: A Good Idea from Past to Future Technology. *J. Food Technolo. Res.* **2018,** *5* (1), 28–33.

Velaga, S. P.; Nikjoo, D.; Vuddanda, P. R. Experimental Studies and Modeling of the Drying Kinetics of Multicomponent Polymer Films. *AAPS PharmSciTech.* **2018,** *19* (1), 425–435.

Vidal, C. P.; Muñoz-Shugulí, C.; Vidal, M. P.; Galotto, M. J.; de Dicastillo, C. L. Active Electrospun Mats: A Promising Material for Active Food Packaging. **2022**.

Wang, Q.; Chen, W.; Zhu, W.; McClements, D. J.; Liu, X.; Liu, F. A Review of Multilayer and Composite Films and Coatings for Active Biodegradable Packaging. *npj Sci. Food.* **2022,** *6* (1), 1–16.

Wu, X.; Liu, Q.; Luo, Y.; Murad, M. S.; Zhu, L.; Mu, G. Improved Packing Performance and Structure-Stability of Casein Edible Films by Dielectric Barrier Discharges (DBD) Cold Plasma. *Food Packag. Shelf Life* **2020,** *24*, 100471.

Yue, T.-T.; Li, X.; Wang, X.-X.; Yan, X.; Yu, M.; Ma, J.-W.; … Long, Y.-Z. Electrospinning of Carboxymethyl Chitosan/Polyoxyethylene Oxide Nanofibers for Fruit Fresh-Keeping. *Nanoscale Res. Lett.* **2018,** *13* (1), 1–8.

Zibaei, R.; Hasanvand, S.; Hashami, Z.; Roshandel, Z.; Rouhi, M.; de Toledo Guimarães, J.; … Mohammadi, R. Applications of Emerging Botanical Hydrocolloids for Edible Films: A Review. *Carbohydr. Polym.* **2021,** *256,* 117554.

CHAPTER 2

PHYSICAL AND MECHANICAL PROPERTIES OF BIOPOLYMERS USED FOR PACKAGING AGRICULTURAL PRODUCE

AJIT KUMAR SINGH and YOUN SUK LEE

Department of Packaging, Yonsei University, Wonju, Gangwon-do, South Korea

ABSTRACT

Every year, nearly 1300 million tons of foods are wasted worldwide, owing in part to a lack of appropriate packaging. Packaging is not only an effective supply chain tool for maintaining the quality, shelf life, nutritional characteristics, and authenticity of fresh agricultural products urged by end consumers, but it is also a suitable strategy for enhancing the product's value and profits for agricultural producers and processors. While there are numerous advantages for petroleum-based plastics being used for agricultural produces, challenges such as rising environmental issues, sustainability concerns, inadequate end-of-life (EoL) approaches, and raw material inflation is prime concerns among packaging stakeholders. To address this, biopolymers as an alternative packaging material have been gaining significant momentum in recent years, driven by sustainability and circular economy, and fit into the paradigm of achieving an eco-friendly environment. However, packaging materials based on biopolymers have lower physical, barrier, and mechanical properties compared to conventional packaging materials. This chapter discusses the key characteristics such as physical and mechanical properties, which are

Novel Packaging Systems for Fruits and Vegetables. Kirtiraj K. Gaikwad & Suman Singh (Eds.)

crucial for the implementation of appropriate biopolymers for agricultural produce packaging to maintain the quality and shelf life. Knowledge of the physical and mechanical characteristics of biopolymers that govern their functionality and manufacturability is critical to identifying the suitable material for agricultural produce packaging. Furthermore, the factors and measures that influence biopolymer's physical and mechanical properties have been deliberated for enhancing the characteristics and strength from the standpoints of food safety and sustainability. In addition, this chapter briefly discusses recent innovative trends in enhancing the physical and mechanical properties of biopolymers in to investigate the advanced approach to be followed in the development of competent biopolymer packaging materials for agricultural produce.

2.1 INTRODUCTION

The majority of agricultural products, especially fruits and vegetables, are perishable and have limited shelf-life. Fresh agricultural produce contains around 90% moisture or even more, with an equilibrate humidity of up to 95%, causing them to dry quickly owing to transpiration under typical atmospheric conditions. In general, around 20–30% of the fresh produces are wasted in terms of qualitative and quantitative losses during various storage and distribution chains, making high instability and cost competitiveness of the final product in the market. In this context, packaging imparts a vital role in extending shelf life by meeting requirements at various storage and distribution stages, such as moisture and weight loss protection, reduced rate of respiration in delaying early ripening, proper gas exchange, control of ethylene concentrations in the package, and protection from microbial contamination and deterioration (Elik et al., 2019). Currently, non-biodegradable plastics have been the preferred packaging options for agricultural produce due to their ease of processing, relatively inexpensive, and significant resistance to physical and mechanical stress. Plastics derived from petroleum-based products, such as polystyrene, polyethylene, polypropylene, and polyvinyl chloride, offer excellent physical, mechanical, and barrier properties, including density, tensile and tear strength, elongation at break, and water vapor, and gas permeability. Nevertheless, novelty is deemed to be accomplished with the adoption of materials suitable for economic, environmental, and sociological perspectives.

To address the growing problem of littering, recycling chaos, and safety concerns associated with accumulation on land and water, packaging scientists and manufacturers have been encouraged to gradually replace petrochemical-based polymers with polymers from renewable sources. From the economic viewpoint, the implementation of packaging materials through biopolymers would enhance the renewable selection of resources and thereby reducing the dependency on petroleum-based polymers and the cost of recycling. From an environmental perspective, biopolymers have the potential to promote eco-friendly packaging materials as a solution to central concerns such as climate-related challenges as well as land and marine pollution, which pose a threat to human and aquatic life. With 300 million tons of plastic waste generated each year and approximately 80% of marine litter coming from plastic waste, one of the biggest environmental challenges for governments and policymakers are currently facing, is uncontrolled plastic pollution (Di Bartolo et al., 2021). From a sociological viewpoint, the adoption of biopolymers in packaging practices reduces carbon dioxide emissions, municipal plastic waste, and dependence on fossil-based resources, resulting in a more sustainable society. With increasing societal preference toward recyclable, biodegradable, and "eco-friendly" materials, it is expected that the use of biopolymers in agricultural produce packaging would expand in near future (Mendes and Pedersen, 2021). Overall, tailoring packaging materials from renewable sources by adopting biopolymers in practice is essential to address the challenges of reducing food wastage and sustainable food consumption.

Biopolymers have numerous benefits over petroleum-based plastics, including abundant access to raw materials, renewability, processability, enhanced functionalities, and recycling options. As a result, the global market for biopolymers is expected to grow by several folds in the near future. Figure 2.1 depicts the current and projected worldwide scenario of biopolymer application by material type and market segment. In comparison to the plastic industry, biopolymers currently have a limited production volume. According to European Bioplastics, the global production of biopolymers in 2021 was around 2.42 million tons (Mt), while the global production for plastics was nearly 370 Mt. However, it is predicted that the global market for biopolymers is expected to register more than 200% growth rate within the next 5 years, increasing in volume significantly from 2.42 Mt in 2021 to 7.59 Mt in 2026. Biopolymers, which include starch blends, PLA, PHA, and others, account for over 64% of global biopolymers production capacity. In general, total production is further expected to expand to over 5.2 Mt in 2026, owing to considerable development of polymers such as PBAT (polybutylene adipate terephthalate) and PBS (polybutylene succinate), as well as constant growth of PLA. Further, biopolymers continued to be used in numerous applications including but not limited to, packaging, agriculture and horticulture, coating and adhesives, and consumer goods. With almost 1.2 Mt (more than 48%), packaging remains the leading sector of application for biopolymers in the total biopolymers market in 2021.

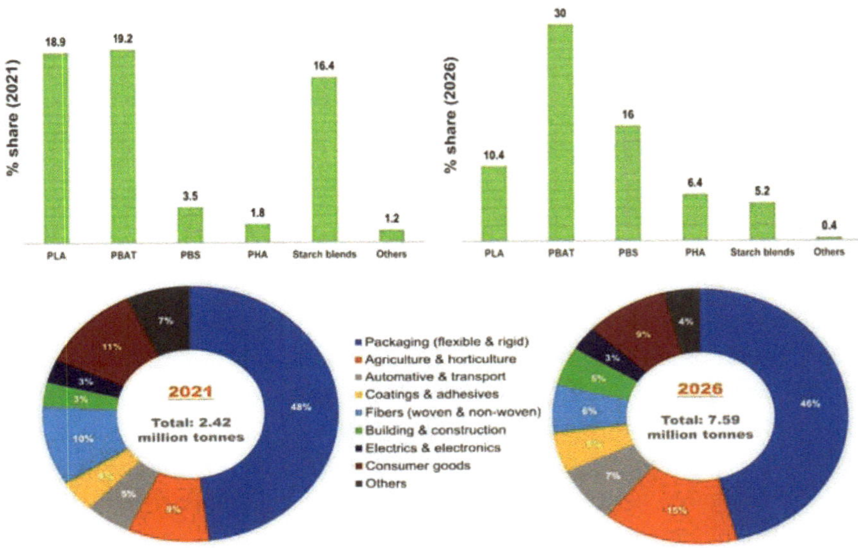

FIGURE 2.1 Global production share (%) of biopolymers by material type and market segment in 2021 and anticipated in 2026.

Source: Adapted from Bioplastics, 2021

It is evident that biopolymer manufacturing is increasing, owing to a greater emphasis on sustainable alternatives as a response to public demands and government legislation. However, developing materials with comparable or even better physical, mechanical, and barrier properties than petroleum-based plastics is a major challenge. Plasticizers (natural-based and biodegradable plasticizers), blending with other biopolymers, and the use of nanoparticles as reinforcing fillers have been investigated as innovative strategies to improve the structure, barrier properties, strength, and flexibility of biopolymer packaging materials in recent years (Fernandes et al., 2020). Furthermore, crosslinking techniques can be used to enhance natural biopolymer films, leading to an improved three-dimensional network and overcoming biopolymer's inherent physical, mechanical, and barrier properties (Garavand et al., 2017). This chapter explores the major properties, such as physical and mechanical properties, that are critical for the use of effective biopolymers for agricultural produce packaging in order to retain quality and shelf life.

2.2 BIOPOLYMERS: DEFINITION AND THEIR CLASSIFICATIONS

Biopolymers are polymers developed from derived biomass (plants, trees, algae, marine organisms, and microorganisms) rather than fossil-based sources, which is the conventional source of polymers. Because of their distinct and valuable characteristics such as inexhaustible accessibility, renewability, and ecological approachability, biopolymers have received considerable attention in recent years in the context of packaging applications with low environmental effects (Christian, 2020; Sadasivuni et al., 2020). Moreover, the term "biopolymer" represents bio-based polymers that may or may not be biodegradable but effective in replacing non-renewable resources with renewable ones, resulting in a more sustainable packaging sector. Biopolymers, unlike synthetic polymers, have a unique molecular assembly that results in a well-defined three-dimensional structure and can mimic synthetic polymer features. They are often extracted and separated from biomass; however, they can also be produced from bio-derived monomers and even microbes (Jayan et al., 2021). As illustrated in Figure 2.2, the production of packaging materials using biopolymers from renewable resources can be done in a variety of methods. Biopolymers can be characterized as natural or synthetic depending on the materials from which they are produced. Because they are widely available, renewable, economical, ecofriendly, and biocompatible, biopolymers derived from renewable sources such as starch, cellulose, chitosan, and various proteins from living organisms' origins have been viewed as potential alternatives to non-biodegradable petroleum-based plastic packaging materials (Ponnamma et al., 2017). Besides this, the other properties of biopolymers include neutral, nontoxicity, noncarcinogenic, and nonimmunogenic. Biopolymer films can be designed to preserve the value and shelf life of fresh agricultural produce by acting as gas and solute barriers and integrating other packaging methods. Furthermore, biopolymer-based packaging materials may offer some additional benefits as packaging materials in terms of retaining freshness and keeping quality by delaying microbiological activity by employing antimicrobial components.

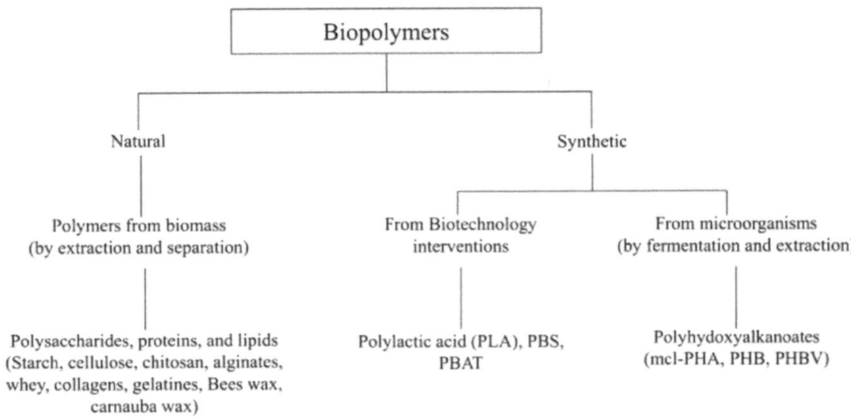

FIGURE 2.2 Schematic classification of biopolymers based on the source of production.

Biopolymers are commonly employed as coatings and films and can be used in a variety of applications. The term "coating" is referred to the process of forming a layer directly on the surface of agricultural produce, whereas "films" are materials that are formed independently and then used simultaneously (Basavegowda and Baek, 2021). Films are independent layers made by casting and drying into specific forms as isolated materials that can be used to wrap or cover agricultural products between several layers. Whereas, coatings are applied to products by immersing them in a coating solution or spraying them directly on their surfaces (Amin et al., 2021). Polysaccharides, proteins, and lipids are the most abundant natural biopolymers derived from plant and animal sources, and they offer enormous potential as a feedstock for biodegradable packaging. Nontoxicity, mechanical stability, good gas, aroma, and lipid barrier qualities are all coupled with polysaccharides (starch, cellulose, chitosan, and alginates, etc.) films and coatings. Synthetic biopolymers are further produced from biotechnological interventions including Polylactic acid (PLA), Polybutylene Succinate (PBS), Poly(butylene adipate-co-terephthalate) (PBAT) and biopolymers derived from microbial production include Poly hydroxy alkanoates (PHAs), Poly(3-hydroxybutyrate) (PHB), and Poly(3-hydroxylbutyrate-co-3-hydroxyvalerate) (PHBV). These biopolymers are being explored as sustainable packaging materials for fresh and processed agricultural products as a substitute for synthetic polymers derived from non-renewable petroleum resources.

2.3 BIOPOLYMER PROPERTIES

The primary objective of biopolymers as sustainable packaging materials is to reduce and replace many of the current petroleum-derived plastics. To be used in a wide range of applications, they must have attributes that are comparable to, if not superior to, the materials they are replacing. Because of parameters including density, degree of polymerization, nature, the concentration of additives, and inclusion of reinforcing agents, many of the property measurements of biopolymers differ. Although there is limited information about the properties of biopolymers than conventional polymers, there is still a lot of research into their physical and mechanical properties. In order to maintain the shelf life and quality, packaging materials for agricultural products are required to possess low water permeability, as well as suitable mechanical strength and flexibility. Natural biopolymer-based packages, on the other hand, have lower physical, mechanical, and barrier properties than synthetic polymers, limiting their commercial applications. Flexibility, strength, hydrophobicity, gas barrier, mechanical characteristics, and water resistance are among the most desirable characteristics of biopolymer-based packaging materials, while the applied properties of the material are determined by its intended use (Agarwal, 2021). For instance, in the case of a carry bag, flexibility and strength are more crucial than barrier properties, whereas, in the context of agricultural products packaging film, water and gas barrier properties are more essential than strength, which could negatively influence the fruits and vegetables contained within it. Since physical and mechanical properties are so fundamental, the key properties of biopolymers are discussed in detail in the following subsections. Table 2.1 summarizes the properties and applications of biopolymers derived from various sources as packaging materials, with the idea of studying sustainable packaging for agricultural products.

TABLE 2.1 Characteristics and Application of Biopolymers from Different Sources as Packaging Materials.

Biopolymers	Sources	Characteristics	Applications	References
Starch	Corn, potato, and tapioca	Excellent capacity to form films; Improved physical properties and mechanical strength with large starch granules and high amylose content	Bio-based packaging materials and edible films	(Chen et al., 2019; Shih and Zhao, 2021)
Cellulose	Fruits and vegetables processing waste, softwood cellulose fibers	Enhanced mechanical properties using plasticizers and microfluidization under high pressure	Green packaging films, edible and biodegradable bioplastics	(Otoni et al., 2018; Sirviö et al., 2018)
Chitosan	Crustaceans (crabs and shrimps)	Antimicrobial qualities inherent in the material, intermediate mechanical strength, and low barrier properties	Biodegradable and edible films, coatings for fruits and vegetables	(Huang et al., 2019; Pavinatto et al., 2020)
Alginates	Seaweed	Good film-forming and gelling characteristics; Low water vapor and oxygen permeability, flexibility, and moderate mechanical strength	Bio-based wrapping packaging materials	(Dai et al., 2022)
Whey	Milk, cheese, and yogurt	Enhance mechanical strength, high moisture absorption, and effective barriers to oxygen and aromas	Wraps or packaging material for food products and edible films	(Erdem et al., 2019)
Collagen	Waste fisheries	Improve functional properties like antioxidant capacity, antibacterial activity, and bioavailability; high-water absorption	Bio-based active packaging	(Valencia et al., 2019)
Gelatin	Skin, bones of land and marine animals	Reduce lipid oxidation and dehydration of food products, increase mechanical strength, and improve light and atmospheric oxygen barrier qualities	Biocomposite films for food packaging applications	(Yadav et al., 2020)

TABLE 2.1 *(Continued)*

Biopolymers	Sources	Characteristics	Applications	References
Beeswax	Honeycomb of the honeybee and other bees	Improve moisture migration barriers and enable densely packed crystalline structure	Edible films, and food packaging applications	(Bahrami et al., 2019)
Carnauba wax	Fronds of the carnauba palm	Enhance the water vapor permeability, and tensile strength, and smoothen the microstructure	Bio-based food packaging materials	(de Oliveira Filho et al., 2021)
Polylactic acid (PLA)	Corn starch and sugar cane	Tensile strength nearly equal to traditional plastics like PS and PET; Intermediate water vapor and moderate oxygen and carbon dioxide permeability; brittleness and poor thermal stability	Bio-based packaging materials	(Lu et al., 2021; Thiyagu et al., 2021)
Polybutylene Succinate (PBS)	Renewable feedstocks such as glucose and sucrose through fermentation	Good processability and biodegradability, as well as desirable physical and mechanical properties	Biodegradable packaging material for food products	(Ayu et al., 2018; Nazrin et al., 2020)
Poly(butylene adipate-co-terephthalate) (PBAT)	A synthetic polymer based on fossil resources, made from 1,4–butanediol, adipic acid, and terephthalic acid	LDPE-like mechanical characteristics in a flexible material; high elongation at break	Biocomposite film for agri-food packaging applications	(Sciancalepore et al., 2022; Thiyagu et al., 2021)
Medium-chain-length PHA (mcl-PHA)	Monomer units with 6–14 carbon atoms; biosynthesized by various types of microorganisms	Good processability, effective barriers to oxygen, water, and carbon dioxide	Promising biodegradable food packaging material for a wide range of food products	(Rekhi et al., 2021; Rigouin et al., 2019)

TABLE 2.1 *(Continued)*

Biopolymers	Sources	Characteristics	Applications	References
Poly (3-hydroxybutyrate) (PHB)	Food waste, and produced through different microbial fermentation processes	Excellent mechanical properties (stiffness and strength) equivalent to or superior to some materials (like PP) and good barrier properties (comparable to PET); limiting processability; high cost; high degree of crystallinity	Applications for sustainable food packaging as an alternative to synthetic polymers like PP and PE	(McAdam et al., 2020; Sánchez-Safont et al., 2018)
Poly(3-hydroxyl-butyrate-co-3-hydroxyvalerate) (PHBV)	Different types of carbon sources using microbial fermentation processes.	biocompatible and good gas barrier properties; decreased brittleness and better processability in comparison to PHB; brittle, low-thermal stability, high cost.	Sustainable packaging material for agri-food products.	(Angellier-Coussy et al., 2020; Li et al., 2020)

2.3.1 PHYSICAL PROPERTIES

Packaging of fresh agricultural produce is a dynamic system that includes the range of specific products to be packed, selection of appropriate packaging material, and optimizing physical and structural designs for efficient handling, transporting, and marketing. A multitude of biopolymer packaging materials is available to offer required functions for a variety of fresh and processed agricultural products within the packaging system. For this, biopolymer characteristics have been extensively investigated in recent decades in order to provide a sustainable alternative to non-biodegradable plastic-based packaging materials. Primary characteristics such as density, molecular weight, molecular structure (MW), and biopolymer type must be considered while understanding the physical properties and specifications of biopolymers. Also, the color and visual properties of packaging materials have a direct impact on consumer preferences and can also benefit from UV protection. In addition, consumers prefer packaging that is more transparent and has minimal color changes because it provides a more authentic image of the product. Some materials, however, are light-sensitive and need less transparent packaging to visible light and UV to avoid oxidative degradation (Martins et al., 2019). Water solubility is another significant factor that affects the packaging material's integrity, as lower solubility results in higher stability and less humidity interaction. This characteristic is entirely dependent on the material structure and hydrophobicity of the material's surface (Khodaei et al., 2021).

Generally, because of the hydrophilic properties of several biopolymers, they are permeable to water and can gain or lose moisture from or to the atmosphere. Contact angles are another significant physical property to determine the hydrophilicity, or wettability, of biopolymer-based packaging materials (Agarwal, 2021). As shown in Figure 2.3, the angle formed by the intersection of the liquid–solid interface and the liquid–vapor interface is referred to as the contact angle. Young's equation expresses the contact angle of a liquid drop on a solid surface when equilibrium is established, which was originally presented by Thomas Young in 1805 with the relationships between θ and three interfacial tensions:

$$\gamma_{SV} = \gamma_{SL} + \gamma_{LV} {}^{\cdot \cos\theta} \tag{2.1}$$

where γ_{sv}, γ_{sl}, and γl_v are the interfacial tensions between solid–vapor, solid–liquid, and liquid–vapor, respectively, and θ is the contact angle. The contact angle determines the extent of surface hydrophilicity/wettability. For instance, a contact angle of less than 90° suggests hydrophilicity, which means the liquid will extend over a large area on the surface, whereas a contact angle of greater than 90° indicates hydrophobicity, which implies the liquid will restrict its contact with the surface and form a compact liquid droplet.

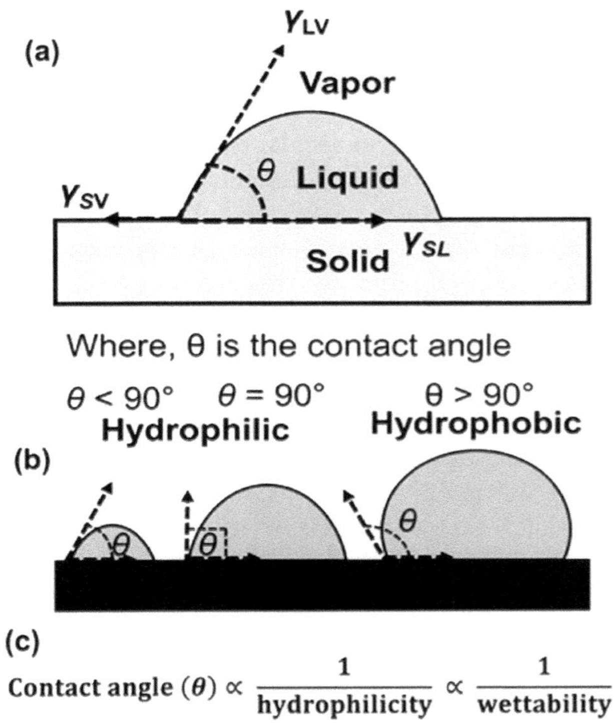

FIGURE 2.3 Schematic representation of (a) contact angle formed by the liquid spreads over the material's surface; (b) hydrophilicity and hydrophobicity as a function of contact angle; (c) correlation between contact angle, hydrophilicity, and wettability.

Furthermore, barrier properties such as water vapor permeability, gas permeability, lipid, aroma, and light barrier are factors to take into account when evaluating packaging materials since they limit interactions with the environment. Water vapor permeability (WVP) is an important property for packaging materials used in fresh agricultural products, such as fruits and vegetables to avoid dehydration. Understanding the possible mass transfer mechanisms of biopolymer materials can aid in the development of appropriate packaging materials for the agricultural product to ensure their quality and shelf life. WVP is measured using the ASTM E96/E96M gravimetric method, commonly called the "cup method," and calculated using the following equation:

$$WVP = \frac{WVTR \cdot x}{\Delta p} \tag{2.2}$$

where WVTR is water vapor transmission rate (kg/m^2·s), x is the thickness of the material, and Δp is the differential water vapor partial pressure across the material. Similarly, the prime emphasis in gas barrier analyses is oxygen permeability, which is generally assessed using the ASTM D1434-82 standard and may be computed as:

$$OP = \frac{OTR \cdot x}{\Delta p} \tag{2.3}$$

where OTR is water vapor transmission rate (cm^3/m^2·s), x is the thickness of the material, and Δp is the differential oxygen partial pressure across the material. The molecular structure, free volume, moisture affinity, degree of crystallinity, and cross-linking of biopolymers are all reflected by these barrier parameters. Moreover, the microstructure properties of biopolymers are directly correlated to their physical and barrier properties. Table 2.2 highlights the barrier properties of some common biopolymers. Generally, it has been observed that the barrier properties of most the biopolymers are poor or average when compared to petroleum-based non-biodegradable polymers (PE, PP, and PVC), which have either a high-water vapor barrier or an excellent oxygen barrier or both (Farah et al., 2016). As a result, biopolymer barrier properties must be addressed to meet the criteria for agricultural produce packaging and other food packaging applications that require a good water vapor/gas barrier. Multi-layer films made from different polymers derived from traditional petroleum-based plastics like PET and PP are utilized in industry to cater to the requirements for packaging fresh and processed fruits and vegetables. However, because biodegradable polymers' barrier performance is inadequate to achieve the required water vapor/gas permeability, simply blending different biodegradable polymers to produce a multi-layer film will not be enough for the intended applications (Wu et al., 2021). Therefore, in recent years, functional modifications or innovative structural designs have been investigated to improve the barrier properties of biopolymer packaging materials. Functional materials and innovative techniques, as well as their potential for packaging materials, are discussed in Section 3.1 to improve the physical, barrier, and mechanical properties of biopolymers.

2.3.2 MECHANICAL PROPERTIES

Selection of suitable biopolymer to be used for packaging applications primarily depends on numerous aspects such as availability, compatibility, cost, physical, mechanical, photodegradation stability, and thermal properties, etc. Exploring the mechanical properties of packaging materials is crucial since identifying them enables structural integrity, which is important for ensuring the physical protection of the product enclosed inside (Gaikwad et al., 2017). Mechanical properties such as tensile strength, elongation at break, and Young's modulus (Figure 2.4) are used to evaluate biopolymer as a packaging material for the intended applications of packaging fresh agricultural produce. Among all mechanical properties, tensile strength (TS), and elongation at break (EAB) of packaging materials are the most studied properties. TS characterizes the maximum stress force that a material can resist when it is pulled or stretched before breaking, whereas EAB is the flexibility to stretch under applied load. Dumbbell and rectangular bar-shaped specimens are commonly used for tensile testing, as shown in Figure 2.4b. The extent of stretching and bending a material can sustain before breaking is measured by its elongation at break. The estimated elongation at break values is a measure of a biopolymer's ductility. Further, Young's modulus indicates the stiffness (the ratio between stress and strain) of material at the elastic region of the tensile test and is a very important material parameter for its relevance to material deformation under load.

Figure 2.4c demonstrates an idealized stress-strain curve for biopolymer tensile testing. The tensile test's stress-strain curve can be used to characterize biopolymer mechanical properties. The following equations are used to determine these mechanical properties:

$$Tensile\ strength = \frac{F_{max}}{t \times w} \quad (2.4)$$

where F_{max} is the maximum load at failure (the force at which the material breaks), t is the thickness of the material, and w is the width.

$$Strain\ at\ break = \frac{L' - L_0}{L_0} \tag{2.5}$$

where L' is the material's final length at failure and L_0 is the material's initial length between grips. Further, using eq 2.2, elongation at break (%) can be calculated as:

$$Elongation\ at\ break\ (\%) = \frac{L' - L_0}{L_0} \times 100 \tag{2.6}$$

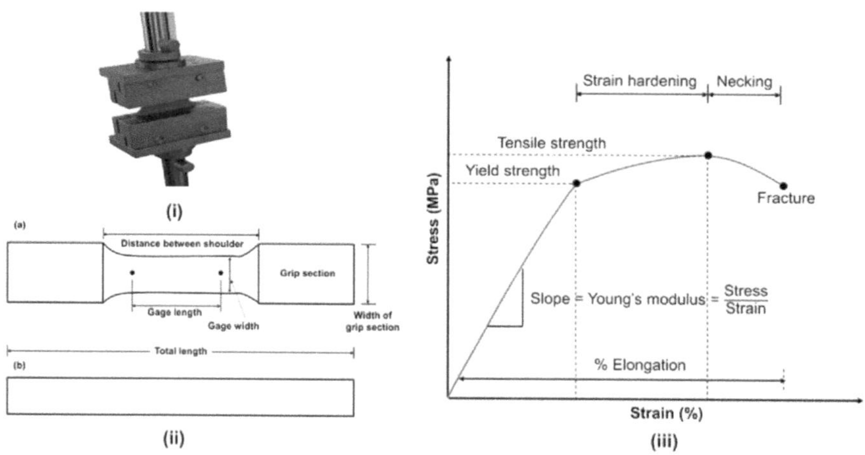

FIGURE 2.4 (i) Testing apparatus for tensile testing; (ii) Specimen shape for tensile testing (a) Dumbbell shaped and (b) Rectangular bar; (iii) A typical stress-strain curve for biopolymer film undergoing tensile strength testing.

Furthermore, the slope of the initial linear region on the stress-strain curve where the material begins elastic deformation can be used to compute Young's modulus (elastic modulus). The trapezoid rule can be used to measure the material's toughness from the area under the stress–strain curve. The mechanical properties of biopolymer packaging materials must have good tensile strength, flexibility, suitable elongation, and penetrating resistance in order to create a physical barrier and extend the shelf life of the food (Sadasivuni et al., 2020). Films or coatings prepared from biopolymers can be used in a wide variety of applications such as packaging materials, heat shrink wrap, bags, trays, and fruit or vegetable containers. The end-use of the material may require either high stretching (packaging wrap) or low stretching (containers or protective coatings). Because tensile strength is an indicator of a film's resistance to stretching, it is an important property to consider when evaluating EoL performance. The mechanical properties of biopolymers can range from elastic to stiff, high-strength materials, depending on various criteria such as biopolymer structure, molecular weight, morphology, molecular orientation, crystallinity, product formulation (blends, plasticizers, and additives, etc.), the thickness of material, and relative humidity of the atmosphere. Table 2.2 exemplifies the physical, mechanical, and barrier properties of some common biopolymers used in packaging to indicate their influence in developing packaging material for fresh and processed agricultural products.

2.4 PHYSICAL AND MECHANICAL PROPERTIES OF BIOPOLYMERS

2.4.1 STARCH

Starch is one of the most prevalent, affordable, and extensively employed natural polysaccharides, and its application as a film and coating material has increased significantly over the years due to its edibility, strength, low oxygen permeability, and processability. Biopolymers are produced from a variety of starches, including corn, rice, potato, tapioca, and cassava. Starch is commonly used as a thermoplastic, and it is extruded after being restructured in the presence of certain amounts of water or plasticizers and heat. Amylose and amylopectin polymers are usually observed in starch. Common starches, such as wheat, corn, and potato starches, include generally 25% amylose and 75% amylopectin. Because of the high-ordered hydrogen bonding network supported by the alternating layers of amylose and amylopectin, pure starch films have high-tensile strength. Generally, the presence of higher content of amylose in starch improves the tensile strength and gas barrier qualities of the biopolymeric film. However, the brittleness of starch film limits its extensive use in the application of agricultural produce (Othman et al., 2021).

2.4.2 CELLULOSE

Cellulose is a natural polymer that consists of long chains of unbranched β-(1$\rightarrow$4)-glycosidic bond linked D-anhydroglucopyranose, and it is one

TABLE 2.2 Physical, Mechanical, and Barrier Properties of Biopolymers in Packaging Material.

Biopolymers	Physical properties	Mechanical properties	Barrier properties	References
Starch	Transparent, high-water solubility, and strong film-forming ability	Tensile strength = 8.2–8.3 MPa; Elongation at break = 100–105%; Young's modulus = 140–142 MPa	High-water vapor permeability of the order of $6.75–6.95 \times 10^{-8}$ g/Pa·h·m; Oxygen permeability of the order of 0.02 cm^3/m·day.Pa	Othman et al. (2021)
Cellulose	Semicrystalline structure, but limited processability	Tensile strength = 32.8–77.0 MPa; Elongation at break = 3.1–4%;	cellulose film having a high-water vapor transfer rate in the range of 1.766–1.969 kg/m^2·day and gas transmission rate of 9325.8–10015.4 ml/m^2·day	Ai et al. (2021)
Chitosan	Average molecular weights (50–2000 kDa) and degrees of deacetylation (40–98%); Films with a slight yellow appearance; Solubility = 16–17%; density = 1.12–1.13 g/cm^3 of 0.070 mm thick film	1% chitosan in the film-forming solution having TS = 109.9–129.2 MPa; EAB = 4.3–5.3%	WVP values range between 0.53–0.65 $g \cdot m^{-1} \cdot h^{-1} \cdot Pa^{-1}$; OP values of chitosan films range from $5.98–6.62 \times 10^{-5}$ cc/m·24 h·atm	Liu et al. (2019)
Alginates	Ionic crosslinking, film-forming ability, pH responsiveness, gelling, hydrophilicity, non-toxic, and processability	2.5% Alginate extracted from *Sargassum fulvellum* in the film-forming solution having TS = 17.70–19.78 MPa; EAB = 20.34–22.42%	1.5% Alginate film-forming solution exhibited high water vapor permeability of the order of $6.24–6.52 \times 10^5$ g·mm/m^2·h·Pa at 97% RH and 25°C	de Oliveira Filho et al. (2019); Kim et al. (2018)
Whey	Characterized by transparency and lack of aroma and flavor; Solubility = 95–96%; Films with a slightly yellow appearance having a thickness of 0.12–0.14 mm	Whey film having TS = 3.72–3.88 MPa; EAB = 141.5–146.75%	5% whey film-forming solution exhibited WVP ranging between 1.09–1.27 g·mm/m^2·day·kPa	Socaciu et al. (2020); Fernandes et al. (2020)

TABLE 2.2 *(Continued)*

Biopolymers	Physical properties	Mechanical properties	Barrier properties	References
Polylactic acid (PLA)	3% (w/w) PLA film-forming solution exhibited 0.20–0.26 mm having density of 1.107–1.871 g/mL and 2.057–2.583% moisture	PLA film having TS = 19.21–20.91 MPa and EAB = 21.08–22.92%	WVP of PLA film 0.0131–0.0143 $g \cdot mm/m^2 \cdot day \cdot kPa$	Sucinda et al. (2021); Lu et al. (2021)
Polybutylene Succinate (PBS)	PBS has a semi-crystalline structure with density:1.24 g/cm^3; and a melting point: of 110–120°C	PBS film having TS = 23.25 MPa; EAB = 14.70%	WVTR of PBS film 83.8 $g/m^2 \cdot day$ and OTR is 737.7 $cc/m^2 \cdot day$	(Xu et al., 2019)
Poly(butylene adipate-co-terephthalate) (PBAT)	PBAT has a density in the range of 1.24–1.26 g/cm^3 and a low degree of crystallinity has excellent film-forming properties with 2.8% solution PBAT, the thickness of film observed around 0.090–0.106 mm	TS = 9.08–10.06 MPa; EAB = 509.05–526.01%	WVP of PBAT having 7.01–7.17 × 10^{12} $mol/m \cdot s \cdot Pa$ and OP = 4.40–4.42 × 10^{16} $mol/m \cdot s \cdot Pa$	de Andrade et al. (2020); de Matos Costa et al. (2020)
Polyhydroxyalkanoate (PHA)	Easy processing, UV rays' resistance, stiffness, high degree of polymerization; insolubility in water	PHA film having TS = 2.5–2.9 MPa and EAB = 21.08–22.92%	WVP of PHA = 1.19–3.99 × 10^{-13} $kg \cdot m/s^{-1} \cdot Pa^{-1} \cdot m^{-2}$	Pérez-Arauz et al. (2019)

of the most important components of lignocellulosic materials, accounting for 30–60% of the total weight. The hydroxyl groups on the cellulose chain tend to stick out laterally along the stretched molecule due to their alignment and are readily available for hydrogen bonding. By hydrogen bonding, the chains are linked together in a highly organized form. Because cellulose is hydrophilic and crystalline in its natural form, it is required to alter its structure, which involves replacing hydroxyl groups with acetate or methyl functional groups, to make it suitable to be used as a packaging material (Asgher et al., 2020; Yuvaraj et al., 2021). It is commonly used in packaging as paper or cardboard due to the presence of hydroxyl structures and a regular structure that leads to the formation of crystalline microfibrils and fibers with strong hydrogen bonds. Cellophane, a versatile thin transparent film that has long been used in commercial food packaging for products that require a long shelf life, is the most commonly used cellulose-based food packaging film. In one of the studies, cellulose film synthesized from delignified banana stem fibers for mangos preservation was reported as promising packaging material for extending the shelf life of tropical fruits (Ai et al., 2021). The moisture vapor transmission rate of the synthesized cellulose film was reported in the range of 1.766–1.969 kg/m^2day, with a gas transmission rate of 9325.8–10,015.4 mL/m^2day, which was favorable to the release of ethylene for extending the ripening period of the fruit. Cellulose is a biodegradable and abundant polysaccharide that has been used to produce several types of biopolymers, making it a potential alternative to petroleum-based plastics for packaging applications of agricultural products. In the near future, cellulose and its derivatives are expected to gain widespread acceptance due to its unique characteristics, biodegradability, and good physical and mechanical properties.

2.4.3 CHITOSAN

Chitosan is a naturally occurring cationic polymer produced by alkaline deacetylation of chitin. Chitin is a crucial constituent of the exoskeletons of arthropods such as crustaceans, shrimps, crabs, and insect skeletons, and it is the second most abundant biopolymer after cellulose in nature. Because of its biodegradability, non-toxicity, low cost, and antibacterial activity, chitosan is a promising polymer for packaging material, particularly in the form of biopolymer films and coatings (Amin et al., 2021). Because of its excellent film-forming ability, high transparency, and antibacterial action, chitosan is one of the most investigated biopolymers. Chitosan films and coatings, like other polysaccharides, have poor water barrier characteristics; nonetheless, they have strong mechanical strength and barrier properties to the gas and aroma (Table 2.2). Despite chitosan's remarkable film-forming ability and mechanical strength, the usage of chitosan films on a commercial scale is impractical due to the use of highly corrosive treatments during the processing method. Furthermore, the non-thermoplastic nature of chitosan films makes the biopolymer difficult to extrude, shape, or stretch (since it degrades at temperatures below those needed for prior melting), raising manufacturing costs and confining application (Khodaei et al., 2021). Because of its distinctive glucosamine structure and cationic tendency, chitosan can be blended with a variety of polysaccharides and other biopolymers to improve interfacial interaction, hydrophobicity, and contact angle properties of formed films. Several phytoconstituents and essential oils can also be incorporated into chitosan films to enhance their functioning for active food packaging applications (Basavegowda and Baek, 2021; Khodaei et al., 2021).

2.4.4 POLYLACTIC ACID (PLA)

Polylactic acid (PLA) is one of the most well-known biodegradable polymers derived from renewable resources among synthetic biopolymers. It provides a wide range of packaging applications due to its renewability, recyclability, compostability, and biodegradability. PLA is typically made from lactic acid via ring-opening polymerization or polycondensation. Lactide monomer is synthesized from the fermentation of natural resources such as wheat, rice, corn, and sugarcane (Sucinda et al., 2021). PLA's high molecular weight, transparency, water solubility resistance, processability (easily thermoformed), and biodegradability are all features that make it a desirable food packaging material. As shown in Table 2.2, PLA is a hydrophobic polymer due to the presence of $-CH_3$ side groups with an elongation at break of about 21–23% and tensile strength of 19–21 MPa. PLA is not water-soluble; however, it can be degraded into water and carbon dioxide by microbes in marine environments. PLA has become more popular as an eco-friendly food packaging material since it has been found that it exceeds synthetic plastics in a variety of circumstances. Whereas PLA possesses the tensile strength, flavor and odor barrier, and processability of polyethylene (PET) or flexible PVC, the properties of printability and grease resistance are similar to polyethylene. However, it has limitations, including brittleness, hydrophobicity, poor gas permeability, temperature sensitivity, lower thermal stability, and increased cost.

2.5 IMPROVING BIOPOLYMERS PROPERTIES: TRENDS AND INNOVATIONS

As mentioned earlier, reducing and substituting non-biodegradable petroleum-based plastics with biopolymer-based packaging materials (films and coatings) has become a major concern in recent years. Accordingly, scientists and manufacturers have been striving to improve the physical, barrier, and mechanical properties of biopolymers. Further, novel approaches like blending, plasticizers, crosslinking as well as nanofillers have a substantial impact on physical and mechanical properties like thickness, density, transparency, contact angle, crystallization, permeability, tensile strength, elongation at break, elastic modulus, and toughness. In the following subsections, these innovative techniques have been discussed briefly to understand the function and possible impact on the basic and functional properties of biopolymers-based packaging materials.

2.5.1 BIOPOLYMER BLENDING

In recent years, biopolymers have been acclaimed as sustainable material, but most of the packaging materials produced from biopolymers still fall short of petroleum-based plastics in terms of water resistance, poor gas/water vapor barrier, brittleness, and mechanical properties. Blending is one of the significant strategies for modifying biopolymer characteristics at a reasonable cost to tailor them to the intended use (Lisitsyn et al., 2021). More specifically, blending can be referred to as the process of integrating polymeric components without causing a chemical reaction in order to produce novel materials with the desired properties at a low cost and in a short amount of time. Researchers are focusing on incorporating various biopolymer blends to overcome the limits of individual biopolymers and achieve the desired barrier with better physical and mechanical properties. The hydrophobic–hydrophilic characteristics of certain biopolymers can be used to alter the properties of packaging films and coatings. Hydrophobic biopolymers, for example, have a beneficial effect on moisture barrier properties. Hydrophilic biopolymers, on the other hand, aid in the formation of materials with higher strength and lower gas permeability.

Farajpour et al. (2020) investigated the structural and physico-mechanical properties of potato starch films when olive oil (lipid) was used. It was reported that with the incorporation of olive oil into composite films, the wettability of the composite films decreased, and the surface of the composite films became more hydrophobic than that of neat starch films. The findings point to an optimal configuration for biopolymer-based packaging materials, as well as judicious material selection for blending. Another study examined at how PLA could be used to overcome the issues that PHB biopolymers endure because of their brittleness and poor mechanical characteristics (Aydemir and Gardner, 2020). The mechanical properties of neat PHB improved substantially as a result of the good interaction between the biopolymers established by blending PLA. Blending appropriate biopolymers based on their nature, structure, and characteristics, together with advanced processing techniques, is likely to provide films and coatings with excellent properties in the near future.

2.5.2 PLASTICIZERS (NATURAL-BASED AND BIODEGRADABLE PLASTICIZERS)

Plasticizers are a class of non-volatile low-molecular-weight compounds that are extensively applied in polymer modification and are the most typical additive ingredients in packaging material formulation. The major function of such additives is to increase the flexibility and processability of polymer blends (Decroix et al., 2020). A plasticizer is defined as "a substance or additive incorporated in a material (commonly a plastic or elastomer) to enhance its flexibility, processability, or elastic properties," according to the IUPAC (International Union of Pure and Applied Chemistry). Plasticizers are available in many different forms, each with its own set of characteristics that can be tailored to fit specific material requirements. Traditional petroleum phthalate, as well as other synthetic plasticizers, are in high demand and recognized worldwide. However, the use of plasticizers has been criticized because of the migration of residues from packing materials into food products and the environment, posing a serious threat of toxicity and causing a risk to human health and the environment.

Considering the increasing acceptance of packaging materials composed of natural and bio-based materials, it seems more reasonable to incorporate plasticizers derived from the same sources. Natural and biodegradable plasticizers with good compatibility, nontoxicity, and minimal migration must be explored as an alternative to conventional plasticizers such as phthalates and other synthetic plasticizers. Some of the most commonly used natural plasticizers for improving the fundamental properties of biopolymers are epoxidized triglyceride vegetable oils from sunflower oil, soybean oil, castor oil, sunflower oil, linseed oil, and fatty acid esters (Jia et al., 2018). Tarique et al. (2021) investigated the physical, barrier, and mechanical properties of arrowroot (*Maranta arundinacea*) starch biopolymers when they were loaded with glycerol plasticizer. In this study, the researcher found that increasing the plasticizer concentration increased film thickness, moisture content, and water solubility, as well as water vapor permeability while lowering density and water absorption. Furthermore, biopolymer films' tensile strength and modulus of elasticity were dramatically reduced, while elongation at break was substantially increased, as expected. Since biopolymers account for only 5–10% of the present plastic market, natural-based plasticizers are still in their initial phases of development, but they are expected to become more popular in near future as a way to enhance the properties of biopolymers as a substitute for existing plasticizers like phthalates.

2.5.3 CROSSLINKING

Although various techniques to enhance biopolymer film-forming properties have been explored and identified as significant advancements, current biopolymer modification approaches are insufficient for industrial scale-up and application. In recent years, using a crosslinking process to improve the physical, barrier, and mechanical properties of biopolymer films and coatings has become a novel strategy. Crosslinking is the process of the formation of chemical bonds by binding the molecular chains with covalent or non-covalent bonds to create a stronger three-dimensional network (Garavand et al., 2017; Lisitsyn et al., 2021). Heating or chemical crosslinking agents such as glutaraldehyde, transglutaminase, formaldehyde, and polyphenol can be used to crosslink biopolymers. Although crosslinking enhances the characteristics of materials, the majority of crosslinking agents cause undesirable changes in biopolymer functioning or cause cytotoxicity. The molecular weight of biopolymers, chemical arrangement, presence of the active group, and compatibility to interact with each other are all factors to be considered while integrating cross-linking agents into the structure of biopolymers.

Qin et al. (2020) prepared electrospun chitosan/pullulan composite nanofiber films by crosslinking two green methods (heating and cinnamaldehyde). Increased contact angle and water resistance, as well as improved barrier and mechanical properties, were found in crosslinked films. For the preparation of the improved film, this study used a green crosslinking method in the form of heat treatment and Cinnamaldehyde, the primary bioactive component of cinnamon essential oil, instead of the toxic glutaraldehyde. Another study used sodium carboxymethyl cellulose (CMC) and hydroxyethyl cellulose (HEC) with citric acid to develop cellulose-based films (Singh et al., 2019). The prepared film's swelling, mechanical, and thermal characteristics were improved significantly. This study also adopted citric acid as a natural cross-linker that can be safely used for food packaging applications. The improvement in biopolymer properties by crosslinking method as reinforcing strategies is considerably simpler than integrating nanomaterials. However, because of the toxicity of synthetic crosslinking agents, natural crosslinkers are becoming more acceptable due to environmental and health benefits.

2.5.4 NANOFILLERS

For the past few years, there is a surge of interest in the combination of biopolymer and nanofiller-based materials to improve the properties of the resulting packaging materials. Because of their increased surface area, aspect ratio, and enhanced functionalities, innovations and approaches in nanotechnology have paved the way for nanofillers to be used as both active constituent and reinforcement. "Nanofiller" is referred to as a form of material, usually particulate, that is incorporated in a polymeric matrix to produce composite materials and has a geometric size of less than 100 nm in at least one dimension (Shankar and Rhim, 2018). Depending on the nature of biopolymers and nanofiller properties, high-quality films and coatings with improved physical and barrier properties, greater mechanical strength, and functionality such as antibacterial and antioxidant characteristics can be produced. Figure 2.5 shows a graphical representation of nanofillers in the biopolymer matrix and their influence on the physical, barrier, mechanical, and functional properties. The nanofillers can be either organic or inorganic such as clay (montmorillonite, kaolinite, and halloysite, natural biopolymers [cellulose, starch, collagen, and chitosan], natural antimicrobial agents [nisin and grape seed extract], metal [silver, copper, and gold], and metal oxides [ZnO, CuO, and TiO_2]).

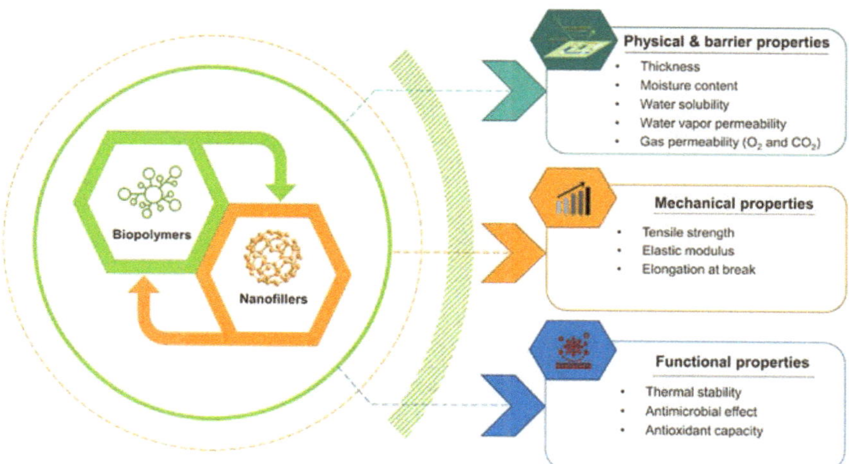

FIGURE 2.5 Potential impact of nanofillers incorporation on the essential properties of biopolymers.

Since nanofillers have been recommended as a way to improve biopolymer properties, several studies have found that incorporating nanofillers into the matrix of biopolymers improved their physical, barrier, and mechanical properties. In one of the studies related to the creation of biodegradable fenugreek seed gum-based active nanocomposite film reinforced with montmorillonite nanoclay, Memiş et al. (2017) reported that increasing the amount of nanoclay improved oxygen barrier, mechanical characteristics, and a uniform and smooth film structure. Deshmukh et al. (2022) carried out a study to develop antioxidant films using guar gum/carboxymethyl cellulose, halloysite nanotubes (HNT), and litchi shell extract (LSE). The benefits of combining HNT and LSE improved film compactness with higher elongation at break, enhanced UV–light barrier properties, and increased antioxidant activity, according to the findings of this study. In similar studies, silver and chitosan nanoparticles were applied to enhance the basic and functional properties of developed materials (Cui et al., 2020; Vishnuvarthanan and Rajeswari, 2019).

In summary, the aforementioned techniques, such as blending, natural/bio-based plasticizers, crosslinking, and nanofillers, offer a significant advantage in terms of improving physical, barrier, and mechanical properties. Researchers are expected to delve deeply into developing environmentally friendly packaging materials obtained from nature to replace petroleum-based non-renewable plastics in the near future to promote sustainable packaging in the circular economy.

2.6 BIOPOLYMER APPLICATION FOR PACKAGING AGRICULTURAL PRODUCES

Agricultural products are living things and go through a variety of biological processes that continue even after they are harvested. This involves rapid product deterioration and decomposition, as well as undesired changes over time. To retain the quality and preserve optimal shelf-life, it is also essential to avoid physical damage to agricultural products during handling and transportation. Thus, two critical components of the food supply chain that ensure the quality and shelf life of agricultural products are processing and packaging. Since packaging plays a crucial role in maintaining the physicochemical and sensory properties of agricultural products such as fruits and vegetables, the biopolymers employed as sustainable packaging materials should have comparable or better properties than petroleum-based plastics. Processability, quality, strength, barrier property, and safety are all key attributes in ensuring the product's quality and shelf life. Biopolymer-based films and/or coatings are sustainable alternatives to traditional food packaging. However, when compared to conventional plastic packaging, these biopolymers originating from plant, animal, and microbial sources lack essential fundamental physical, barrier, and mechanical properties. As mentioned in the previous section, novel interventions like blending, plasticizers, crosslinking, and nanofillers, etc., for enhancing water resistance, strength, toughness as well as barrier property have been explored for effective packaging materials.

Li et al. (2019) employed soybean protein isolate (SPI), cinnamaldehyde (CIN), and zinc oxide nanoparticles (ZnONP) to create a bionanocomposite coating that was applied to bananas. They observed that coating had the optimum fresh-keeping properties on bananas and that further biochemical changes and the shelf-life of banana samples were improved greatly by his research. The results showed that SPI nanocomposite film could significantly delay banana ripening and weight loss. In comparison to other SPI-based coating treatments, SPI/CIN/ZnONP nanocomposite coating could also prevent adverse changes in fruit firmness, total soluble sugar, titratable acidity, and sensory quality throughout storage. Further, Kanikireddy et al. (2019) reported that sodium alginate-gum acacia-silver nanocomposite films exhibited good mechanical strength, thermally more stable, with antimicrobial and antioxidant activity that maintained the quality and freshness of packed grape even after storage for 18 days. Biopolymers are also being studied in the context of active and intelligent packaging techniques, which include releasing systems, absorbers, sensors, and other components that help to maintain and improve the quality and safety of fruits and vegetables. Innovative films and coatings are expected to meet the growing demand for fresh and cut fruits and vegetables using biopolymers while also providing appropriate shelf life during the supply chain.

2.7 CONCLUSIONS AND FUTURE PROSPECTS

Biopolymers have garnered considerable attention as a replacement for petroleum-based synthetic plastics, and demand for bio-based products in the packaging industry is substantially growing due to increasing awareness of health and environmental concerns. However, biopolymers confront numerous constraints in terms of practical packaging applications, the most significant of which is a poor water vapor/gas barrier, which considerably limits the use of such sustainable packaging material options. Other features, such as hydrophilicity and lower strength, also influence the acceptability of biopolymers for widespread use in agricultural produce packaging. This chapter outlined the physical and mechanical properties of biopolymers that must be understood to implement packaging material from diverse renewable sources that include natural biopolymers (starch, cellulose, chitosan, alginates, whey, and collagen, etc.), and synthetic biopolymers (PLA, PBS, PBAT, PHA, PHB, and PHBV, etc.). As discussed, advanced techniques such as biopolymer blending, natural/bio-based plasticizers, crosslinking, and nanofillers are expected to make commercially viable high-quality films and coatings. Although novel materials, additives, and advanced processes have been reported to improve the physical and mechanical properties of biopolymers, implementing strategies for retaining the physicochemical and sensory aspects of agricultural products remains a challenge. The technological advancement outlined here to enhance the biopolymer properties and their applications are only the tip of the iceberg. Biopolymer's prospects are contingent on a continuous focus on enhancing performance, integrating safety, and regulation, and reducing the cost to prolong the shelf life of fresh and processed agricultural products providing a green alternative to fulfill the global demand.

KEYWORDS

- **mechanical properties**
- **water vapor permeability**
- **fresh produce**
- **biopolymer**
- **food packaging**

REFERENCES

Agarwal, S. Major Factors Affecting the Characteristics of Starch Based Biopolymer Films. *Eur. Polym. J.* **2021,** *160,* 110788.

Ai, B.; Zheng, L.; Li, W.; Zheng, X.; Yang, Y.; Xiao, D.; Shi, J.; Sheng, Z. Biodegradable Cellulose Film Prepared from Banana Pseudo-Stem Using an Ionic Liquid for Mango Preservation. *Front. Plant Sci.* **2021,** *12,* 234.

Amin, U.; Khan, M. U.; Majeed, Y.; Rebezov, M.; Khayrullin, M.; Bobkova, E.; Shariati, M. A.; Chung, I. M.; Thiruvengadam, M. Potentials of Polysaccharides, Lipids and Proteins in Biodegradable Food Packaging Applications. *Int. J. Biol. Macromol.* **2021,** *183,* 2184–2198.

Angellier-Coussy, H.; Kemmer, D.; Gontard, N.; Peyron, S. Physical–Chemical and Structural Stability of PHBV/Wheat Straw Fibers Based Biocomposites Under Food Contact Conditions. *J. Appl. Polym. Sci.* **2020,** *137* (40), 49231.

Asgher, M.; Qamar, S. A.; Bilal, M.; Iqbal, H. M. Bio-Based Active Food Packaging Materials: Sustainable Alternative to Conventional Petrochemical-Based Packaging Materials. *Food Res. Int.* **2020,** *137,* 109625.

Aydemir, D.; Gardner, D. J. Biopolymer Blends of Polyhydroxybutyrate and Polylactic Acid Reinforced with Cellulose Nanofibrils. *Carbohydr. Polym.* **2020,** *250,* 116867.
Ayu, R. S.; Khalina, A.; Harmaen, A. S.; Zaman, K.; Jawaid, M.; Lee, C. H. Effect of Modified Tapioca Starch on Mechanical, Thermal, and Morphological Properties of PBS Blends for Food Packaging. *Polymers* **2018,** *10* (11), 1187.
Bahrami, A.; Mokarram, R. R.; Khiabani, M. S.; Ghanbarzadeh, B.; Salehi, R. Physico-Mechanical and Antimicrobial Properties of Tragacanth/Hydroxypropyl Methylcellulose/Beeswax Edible Films Reinforced with Silver Nanoparticles. *Int. J. Biol. Macromol.* **2019,** *129,* 1103–1112.
Basavegowda, N.; Baek, K.-H. Advances in Functional Biopolymer-Based Nanocomposites for Active Food Packaging Applications. *Polymers.* **2021,** *13* (23), 4198.
Bioplastics, E. *Bioplastics Market Data.* 2021. https://www.european-bioplastics.org/market/ (accessed Apr 29, 2022).
Chen, Q.; Liu, Y.; Chen, G. A Comparative Study on the Starch-Based Biocomposite Films Reinforced by Nanocellulose Prepared from Different Non-Wood Fibers. *Cellulose* **2019,** *26* (4), 2425–2435.
Christian, S. J. Natural Fibre-Reinforced Noncementitious Composites (Biocomposites). In *Nonconventional and Vernacular Construction Materials.* Elsevier, 2020; pp 169–187.
Cui, H.; Surendhiran, D.; Li, C.; Lin, L. Biodegradable Zein Active Film Containing Chitosan Nanoparticle Encapsulated with Pomegranate Peel Extract for Food Packaging. *Food Packag. Shelf Life* **2020,** *24,* 100511.
Dai, Q.; Huang, X.; Jia, R.; Fang, Y.; Qin, Z. Development of Antibacterial Film Based on Alginate Fiber, and Peanut Red Skin Extract for Food Packaging. *J. Food Eng.* **2022,** 111106.
de Andrade, M. F.; de Lima Silva, I. D.; da Silva, G. A.; Cavalcante, P. V. D.; da Silva, F. T.; de Almeida, Y. M. B.; Vinhas, G. M.; de Carvalho, L. H. A Study of Poly (Butylene Adipate-co-Terephthalate)/Orange Essential Oil Films for Application in Active Antimicrobial Packaging. *LWT* **2020,** *125,* 109148.
de Matos Costa, A. R.; Crocitti, A.; Hecker de Carvalho, L.; Carroccio, S. C.; Cerruti, P.; Santagata, G. Properties of Biodegradable Films Based on Poly (Butylene Succinate) (PBS) and Poly (Butylene Adipate-co-Terephthalate)(PBAT) Blends. *Polymers* **2020,** *12* (10), 2317.
de Oliveira Filho, J. G.; Albiero, B. R.; Cipriano, L.; de Oliveira Nobre Bezerra, C. C.; Oldoni, F. C. A.; Egea, M. B.; de Azeredo, H. M. C.; Ferreira, M. D. Arrowroot Starch-Based Films Incorporated with a Carnauba Wax Nanoemulsion, Cellulose Nanocrystals, and Essential Oils: A New Functional Material for Food Packaging Applications. *Cellulose* **2021,** *28* (10), 6499–6511.
de Oliveira Filho, J. G.; Rodrigues, J. M.; Valadares, A. C. F.; de Almeida, A. B.; de Lima, T. M.; Takeuchi, K. P.; Alves, C. C. F.; de Figueiredo Sousa, H. A.; da Silva, E. R.; Dyszy, F. H. Active Food Packaging: Alginate Films with Cottonseed Protein Hydrolysates. *Food Hydrocoll.* **2019,** *92,* 267–275.
Decroix, C.; Chalamet, Y.; Sudre, G.; Caroll, V. Thermo-Mechanical Properties and Blend Behaviour of Cellulose Acetate/Lactates and Acid Systems: Natural-Based Plasticizers. *Carbohydr. Polym.* **2020,** *237,* 116072.
Deshmukh, R. K.; Akhila, K.; Ramakanth, D.; Gaikwad, K. K. Guar Gum/Carboxymethyl Cellulose Based Antioxidant Film Incorporated with Halloysite Nanotubes and Litchi Shell Waste Extract for Active Packaging. *Int. J. Biol. Macromol.* **2022.**

Di Bartolo, A.; Infurna, G.; Dintcheva, N. T. A Review of Bioplastics and their Adoption in the Circular Economy. *Polymers* **2021,** *13* (8), 1229.

Elik, A.; Yanik, D. K.; Istanbullu, Y.; Guzelsoy, N. A.; Yavuz, A.; Gogus, F. Strategies to Reduce Post-Harvest Losses for Fruits and Vegetables. *Strategies* **2019,** *5* (3), 29–39.

Erdem, B. G.; Dıblan, S.; Kaya, S. Development and Structural Assessment of Whey Protein Isolate/Sunflower Seed Oil Biocomposite Film. *Food Bioprod. Process.* **2019,** *118,* 270–280.

Farah, S.; Anderson, D. G.; Langer, R. Physical and Mechanical Properties of PLA, and their Functions in Widespread Applications—A Comprehensive Review. *Adv. Drug Deliv. Rev.* **2016,** *107,* 367–392.

Farajpour, R.; Djomeh, Z. E.; Moeini, S.; Tavakolipour, H.; Safayan, S. Structural and Physico-Mechanical Properties of Potato Starch-Olive Oil Edible Films Reinforced with Zein Nanoparticles. *Int. J. Biol. Macromol.* **2020,** *149,* 941–950.

Fernandes, L. M.; Guimarães, J. T.; Silva, R.; Rocha, R. S.; Coutinho, N. M.; Balthazar, C. F.; Calvalcanti, R. N.; Piler, C. W.; Pimentel, T. C.; Neto, R. P. Whey Protein Films Added with Galactooligosaccharide and Xylooligosaccharide. *Food Hydrocoll.* **2020,** *104,* 105755.

Gaikwad, K. K.; Lee, S. M.; Lee, J. S.; Lee, Y. S. Development of Antimicrobial Polyolefin Films Containing Lauroyl Arginate and their Use in the Packaging of Strawberries. *J. Food Meas. Charact.* **2017,** *11* (4), 1706–1716.

Garavand, F.; Rouhi, M.; Razavi, S. H.; Cacciotti, I.; Mohammadi, R. Improving the Integrity Of Natural Biopolymer Films Used in Food Packaging by Crosslinking Approach: A Review. *Int. J. Biol. Macromol.* **2017,** *104,* 687–707.

Huang, Q.; Qian, X.; Jiang, T.; Zheng, X. Effect of Chitosan and Guar Gum Based Composite Edible Coating on Quality of Mushroom (Lentinus Edodes) During Postharvest Storage. *Sci. Hortic.* **2019,** *253*, 382–389.

Jayan, J. S.; Deeraj, B.; Saritha, A.; Joseph, K. Biopolymer-Derived Carbonaceous Composites and their Potential Applications. In *Hybrid Natural Fiber Composites.* Elsevier, 2021; pp 253–280.

Jia, P.; Xia, H.; Tang, K.; Zhou, Y. Plasticizers Derived from Biomass Resources: A Short Review. *Polymers* **2018,** *10* (12), 1303.

Kanikireddy, V.; Kanny, K.; Padma, Y.; Velchuri, R.; Ravi, G.; Jagan Mohan Reddy, B.; Vithal, M. Development of Alginate-Gum Acacia-Ag0 Nanocomposites via Green Process for Inactivation of Foodborne Bacteria and Impact on Shelf Life of Black Grapes (Vitis vinifera). *J. Appl. Polym. Sci.* **2019,** *136* (15), 47331.

Khodaei, D.; Álvarez, C.; Mullen, A. M. Biodegradable Packaging Materials from Animal Processing Co-products and Wastes: An Overview. *Polymers* **2021,** *13* (15), 2561.

Kim, S.; Baek, S.-K.; Song, K. B. Physical and Antioxidant Properties of Alginate Films Prepared from Sargassum Fulvellum with Black Chokeberry Extract. *Food Packag. Shelf Life* **2018,** *18*, 157–163.

Li, J.; Sun, Q.; Sun, Y.; Chen, B.; Wu, X.; Le, T. Improvement of Banana Postharvest Quality Using a Novel Soybean Protein Isolate/Cinnamaldehyde/Zinc Oxide Bionanocomposite Coating Strategy. *Sci. Hortic.* **2019,** *258,* 108786.

Li, Z.; Reimer, C.; Wang, T.; Mohanty, A. K.; Misra, M. Thermal and Mechanical Properties of the Biocomposites of Miscanthus Biocarbon and Poly(3-hydroxybutyrate-co-3-hydroxyvalerate)(PHBV). *Polymers* **2020,** *12* (6), 1300.

Lisitsyn, A.; Semenova, A.; Nasonova, V.; Polishchuk, E.; Revutskaya, N.; Kozyrev, I.; Kotenkova, E. Approaches in Animal Proteins and Natural Polysaccharides Application for Food Packaging: Edible Film Production and Quality Estimation. *Polymers* **2021,** *13* (10), 1592.
Liu, F.; Chang, W.; Chen, M.; Xu, F.; Ma, J.; Zhong, F. Tailoring Physicochemical Properties of Chitosan Films and their Protective Effects on Meat by Varying Drying Temperature. *Carbohydr. Polym.* **2019,** *212*, 150–159.
Lu, W.; Cui, R.; Zhu, B.; Qin, Y.; Cheng, G.; Li, L.; Yuan, M. Influence of Clove Essential Oil Immobilized in Mesoporous Silica Nanoparticles on the Functional Properties of Poly (Lactic Acid) Biocomposite Food Packaging Film. *J. Mater. Res. Technol.* **2021,** *11*, 1152–1161.
Martins, V. G.; Romani, V. P.; Martins, P. C.; da Silva Filipini, G. Innovative Packaging that Saves Food. In *Saving Food.* Elsevier, 2019, pp 171–202.
McAdam, B.; Brennan Fournet, M.; McDonald, P.; Mojicevic, M. Production of Polyhydroxybutyrate (PHB) and Factors Impacting Its Chemical and Mechanical Characteristics. *Polymers* **2020,** *12* (12), 2908.
Memiş, S.; Tornuk, F.; Bozkurt, F.; Durak, M. Z. Production and Characterization of a New Biodegradable Fenugreek Seed Gum Based Active Nanocomposite Film Reinforced with Nanoclays. *Int. J. Biol. Macromol.* **2017,** *103,* 669–675.
Mendes, A. C.; Pedersen, G. A. Perspectives on Sustainable Food Packaging:–is Bio-Based Plastics a Solution? *Trends Food Sci. Technol.* **2021,** *112,* 839–846.
Nazrin, A.; Sapuan, S.; Zuhri, M.; Ilyas, R.; Syafiq, R.; Sherwani, S. Nanocellulose Reinforced Thermoplastic Starch (TPS), Polylactic Acid (PLA), and Polybutylene Succinate (PBS) for Food Packaging Applications. *Front. Chem.* **2020,** *8,* 213.
Othman, S. H.; Nordin, N.; Azman, N. A. A.; Tawakkal, I. S. M. A.; Basha, R. K. Effects of Nanocellulose Fiber and Thymol on Mechanical, Thermal, and Barrier Properties of Corn Starch Films. *Int. J. Biol. Macromol.* **2021,** *183,* 1352–1361.
Otoni, C. G.; Lodi, B. D.; Lorevice, M. V.; Leitão, R. C.; Ferreira, M. D.; de Moura, M. R.; Mattoso, L. H. Optimized and Scaled-Up Production of Cellulose-Reinforced Biodegradable Composite Films Made Up of Carrot Processing Waste. *Indus. Crops Produc.* **2018,** *121,* 66–72.
Pavinatto, A.; de Almeida Mattos, A. V.; Malpass, A. C. G.; Okura, M. H.; Balogh, D. T.; Sanfelice, R. C. Coating with Chitosan-Based Edible Films for Mechanical/Biological Protection of Strawberries. *Int. J. Biol. Macromol.* **2020,** *151,* 1004–1011.
Pérez-Arauz, A.; Aguilar-Rabiela, A.; Vargas-Torres, A.; Rodríguez-Hernández, A.-I.; Chavarría-Hernández, N.; Vergara-Porras, B.; López-Cuellar, M. Production and Characterization of Biodegradable Films of a Novel Polyhydroxyalkanoate (PHA) Synthesized from Peanut Oil. *Food Packag. Shelf Life* **2019,** *20,* 100297.
Ponnamma, D.; Sadasivuni, K. K.; AlMaadeed, M. Introduction of Biopolymer Composites: What to do in Electronics? In *Biopolymer Composites in Electronics.* Elsevier, 2017; pp 1–12.
Qin, Z.; Jia, X.; Liu, Q.; Kong, B.; Wang, H. Enhancing Physical Properties of Chitosan/ Pullulan Electrospinning Nanofibers via Green Crosslinking Strategies. *Carbohydr. Polym.* **2020,** *247,* 116734.
Rekhi, P.; Goswami, M.; Ramakrishna, S.; Debnath, M. Polyhydroxyalkanoates Biopolymers Toward Decarbonizing Economy and Sustainable Future. *Crit. Rev. Biotechnol.* **2021,** 1–25.

Rigouin, C.; Lajus, S.; Ocando, C.; Borsenberger, V.; Nicaud, J. M.; Marty, A.; Avérous, L.; Bordes, F. Production and Characterization of Two Medium-Chain-Length Polydroxyalkanoates by Engineered Strains of Yarrowia Lipolytica. *Microb. Cell Factories* **2019,** *18* (1), 1–9.

Sadasivuni, K. K.; Saha, P.; Adhikari, J.; Deshmukh, K.; Ahamed, M. B.; Cabibihan, J. J. Recent Advances in Mechanical Properties of Biopolymer Composites: A Review. *Polym. Compos.* **2020,** *41* (1), 32–59.

Sánchez-Safont, E. L.; Aldureid, A.; Lagarón, J. M.; Gámez-Pérez, J.; Cabedo, L. Biocomposites of Different Lignocellulosic Wastes for Sustainable Food Packaging Applications. *Compos. B. Eng.* **2018,** *145,* 215–225.

Sciancalepore, C.; Togliatti, E.; Giubilini, A.; Pugliese, D.; Moroni, F.; Messori, M.; Milanese, D. Preparation and Characterization of Innovative Poly(Butylene Adipate Terephthalate)-Based Biocomposites for Agri-Food Packaging Application. *J. Appl. Polym. Sci.* **2022,** 52370.

Shankar, S.; Rhim, J.-W. Bionanocomposite Films for Food Packaging Applications. *Ref. Module Food Sci.* **2018,** *1,* 1–10.

Shih, Y.-T.; Zhao, Y. Development, Characterization and Validation of Starch Based Biocomposite Films Reinforced by Cellulose Nanofiber as Edible Muffin Liner. *Food Packag. Shelf Life* **2021,** *28,* 100655.

Singh, P.; Magalhães, S.; Alves, L.; Antunes, F.; Miguel, M.; Lindman, B.; Medronho, B. Cellulose-Based Edible Films for Probiotic Entrapment. *Food Hydrocoll.* **2019,** *88,* 68–74.

Sirviö, J. A.; Visanko, M.; Ukkola, J.; Liimatainen, H. Effect of Plasticizers on the Mechanical and Thermomechanical Properties of Cellulose-Based Biocomposite Films. *Ind. Crops Prod.* **2018,** *122,* 513–521.

Socaciu, M.-I.; Fogarasi, M.; Semeniuc, C. A.; Socaci, S. A.; Rotar, M. A.; Mureşan, V.; Pop, O. L.; Vodnar, D. C. Formulation and Characterization of Antimicrobial Edible Films Based on Whey Protein Isolate and Tarragon Essential Oil. *Polymers* **2020,** *12* (8), 1748.

Sucinda, E.; Majid, M. A.; Ridzuan, M.; Cheng, E.; Alshahrani, H.; Mamat, N. Development and Characterisation of Packaging Film from Napier Cellulose Nanowhisker Reinforced Polylactic Acid (PLA) Bionanocomposites. *Int. J. Biol. Macromol.* **2021,** *187,* 43–53.

Tarique, J.; Sapuan, S.; Khalina, A. Effect of Glycerol Plasticizer Loading on the Physical, Mechanical, Thermal, and Barrier Properties of Arrowroot (Maranta Arundinacea) Starch Biopolymers. *Sci. Rep.* **2021,** *11* (1), 1–17.

Thiyagu, T. T.; Sai Prasanna Kumar., J. V.; Sathiyamoorthy, V.; Arun Prakash, V. R. Effect of Cashew Shell Biomass Synthesized Cardanol Oil Green Compatibilizer on Flexibility, Barrier, Thermal, and Wettability of PLA/PBAT Biocomposite Films. *Biomass Convers. Biorefin.* **2021,** 1–11.

Valencia, G. A.; Luciano, C. G.; Lourenço, R. V.; Bittante, A. M. Q. B.; do Amaral Sobral, P. J. Morphological and Physical Properties of Nano-Biocomposite Films Based on Collagen Loaded With Laponite®. *Food Packag. Shelf Life* **2019,** *19,* 24–30.

Vishnuvarthanan, M.; Rajeswari, N. Food Packaging: Pectin–Laponite–Ag Nanoparticle Bionanocomposite Coated on Polypropylene Shows Low O_2 Transmission, Low Ag Migration and High Antimicrobial Activity. *Environ. Chem. Lett.* **2019,** *17* (1), 439–445.

Wu, F.; Misra, M.; Mohanty, A. K. Challenges and New Opportunities on Barrier Performance of Biodegradable Polymers for Sustainable Packaging. *Prog. Polym. Sci.* **2021,** *117,* 101395.

Xu, J.; Manepalli, P. H.; Zhu, L.; Narayan-Sarathy, S.; Alavi, S. Morphological, Barrier and Mechanical Properties of Films From Poly (Butylene Succinate) Reinforced with Nanocrystalline Cellulose and Chitin Whiskers Using Melt Extrusion. *J. Polym. Res.* **2019,** *26* (8), 1–10.

Yadav, S.; Mehrotra, G.; Bhartiya, P.; Singh, A.; Dutta, P. Preparation, Physicochemical and Biological Evaluation of Quercetin Based Chitosan-Gelatin Film for Food Packaging. *Carbohydr. Polym.* **2020,** *227,* 115348.

Yuvaraj, D.; Iyyappan, J.; Gnanasekaran, R.; Ishwarya, G.; Harshini, R.; Dhithya, V.; Chandran, M.; Kanishka, V.; Gomathi, K. Advances in Bio Food Packaging—An Overview. *Heliyon* **2021,** *7* (9), e07998.

CHAPTER 3

RECENT TRENDS IN FILMS AND GASES FOR MODIFIED ATMOSPHERE PACKAGING OF FRESH PRODUCE

SHEFALI TRIPATHI, LOKMAN HAKIM, DEEPIKA GUPTA, RAM KUMAR DESHMUKH, and KIRTIRAJ K. GAIKWAD

Department of Paper Technology, Indian Institute of Technology Roorkee Saharanpur Campus, Saharanpur, India

ABSTRACT

In this chapter, the new development in modified atmospheric packaging (MAP) has been discussed. In recent developments, MAP has been used with several other technologies. This technique has been used with active and intelligent packaging, which helps the customers to know about the food safety of the food product. Silicon membrane has also been lately used as a packaging material with MAP. This technique—when used with several other methods such as irradiation, UV light, and biodegradable films—has enhanced the quality and shelf life of fruits and vegetables. Microperforated films with a pore size of <200 μm are one of the latest technologies in MAP. MAP increased the shelf life by marinating the permeability of the gases. Several types of technologies have been introduced for the formation of microperforated films, but the most used is laser technology in fresh food products. Recent developments have come up with several other gases such as ozone, carbon-mono-dioxide, sulfur oxides, and noble gases such as xenon, argon, and helium. These gases have increased the quality and shelf life of freshly used fruits

Novel Packaging Systems for Fruits and Vegetables. Kirtiraj K. Gaikwad & Suman Singh (Eds.)

and vegetables. The use of MAP with several other techniques such as dehydration, freeze drying, non-thermal methods such as high-pressure processing, cold plasma technology, and electric pulse field could also be explored in the coming future.

3.1 INTRODUCTION

Due to harvesting respiration, fresh, and minimally processed products are susceptible to diseases and physiological disorders. Because these products continue their metabolic process, which facilitates their shape, size, color, flavor, and commercial value, but after peak maturity, it leads to quality deterioration due to over-ripening or senescence. The shelf life of fresh fruits and vegetables under ambient or room conditions is limited due to their physiological metabolism behavior. The quality deterioration results from transpiration, wound-initiated, ripening process, and senescence after post-harvest (Sandhya, 2010). Each product has specific characteristics such as respiration rate, ethylene production capacity, physical damage endurance, and most importantly, shelf life, distinguishing from one product to another. The shelf life of one product is highly dependent on the individual metabolism and its external factors. The low temperature and hygienic processing or handling of fresh produce can significantly contribute to product preservation.

The respiration rate of fresh produces greatly depends upon external factors such as temperature, humidity, moisture content, microbial proliferation, and the nature of produce. According to the respiration behavior, fresh agricultural produce can be classified into two categories such as climacteric and non-climacteric produce (Kou and Wu, 2018). Peaks in respiration rate and autocatalytic ethylene production occur in climacteric fruits even after harvesting; in contrast, non-climacteric produce involves cyanide-insensitive respiration, so after the harvesting, they produce no or too little ethylene hormone (Liu et al., 2015). Climacteric produces such as tomatoes, banana, apricot, peach, mango, apple, and pears, whereas non-climacteric produce is citrus, grapes, cherry, litchi, watermelon, and strawberry (Pech et al., 2012). Climacteric fruits and vegetables are triggered in the respiration rate during ripening; this co-occurs with the rise in ethylene generation rate or follows soon afterward. The respiration of fruits and vegetables can be controlled or reduced by many preservation methods such as low-temperature storage, canning, dehydration, freeze drying, controlled atmosphere packaging (CAP), hypobaric storage, and modified atmosphere packaging (MAP) (Sandhya, 2010).

MAP is well investigated and accepted post-harvest technology for preserving the natural quality and integrity of fresh and minimally processed produce and extending the shelf life under well-equipped MAP design and storage conditions. MAP technology is applied mainly for fresh and minimally processed fruits and vegetables, termed ready-to-use products. MAP is generally involved in removing and replacing the gases in the surrounding atmosphere to interplay in the respiration of packed products in gas barrier or moisture vapor barrier materials to inhibit the proliferation of microorganisms and control metabolic activity (Zhang et al., 2015a). Modifying gas compositions for individual products is referred to as MAP, and for bulk storage is MA. MAP and MA techniques are designed to lower the product's respiration rate and control the metabolic activity or ripening; in other words, they can stay fresh for longer.

The first application of MAP was for transporting lamb meat from Australia to England in the early 12th century. After that, this technology was applied to apple fruit in a controlled atmosphere around the product with reduced O_2 and elevated CO_2 concentration in 1927 (Zhang et al., 2015a). In 1930 the MAP was used for bulk shipping of the fruits and meat carcass with the increased CO_2 concentration for long-distance transportation, and it was observed that the shelf life of the products found significantly improved (Davies, 1995). However, the MAP technique was not commercially available until 1970 due to its primary limitation of MAP as a lack of gas concentration control, gas and vapor permeability of the packaging material, and a more comprehensive range of physical and mechanical properties of the package. After 1979 there have been several new packaging materials were developed to maintain the gaseous components in the MAP storage. In the United Kingdom (UK), Mark & Spenser commercially MAP techniques for meat products; this successful venture led to many MAP application interventions for bacon, fish, sliced meat, and cooked shellfish. Owing to this success, several food manufacturers and supply chains started following the same, resulting in the availability of MAP sharply increased food storage. In recent years, MAP has been widely explored in the packaging and storage of raw and cooked meat, fresh and chilled foods, fresh and cut fruits-vegetables, coffee, tea, and bakery products (Church and Parsons, 1995).

MAP of fresh and minimally processed fruits and vegetables relies on the internal atmosphere of the package achieved by natural interplay between two factors one is the respiration activity of the produce, and the second is the gas barrier property of the packaging materials. This atmospheric condition inside the package significantly reduces respiration, ethylene sensitivity, production, ripening, softening compositional changes, decay, and physiological changes such as oxidation and browning (Kargwal et al., 2020). MAP involves exposing fresh or minimally processed products to the atmosphere generated by the interaction of the produce, package, and external atmosphere through the packaging material. Atmospheric modification inside the package highly depends on the film permeability, product respiration rate, gas diffusion mechanism, product weight, surface area, initial free volume, and initial atmospheric composition inside the package. The effectiveness of the MAP system for fresh or processed food storage highly depends on the packaging material applied for MAP storage. Generally, MAP involves the application of polymer film such as low-density polyethylene (LDPE), high-density polyethylene (HDPE), polyethylene terephthalate (PET), and polypropylene (PP), etc. The gas permeability of the packaging material is the main factor in selecting any packaging material for MAP storage. However, many factors affect the permeability of the packaging material, such as polymer type and thickness, are the most important. The different food respiration rates must decide the packaging material type for MAP storage to get special equilibrium in gaseous compositions. In recent years, extensive research will develop customized packaging materials such as antioxidant packaging film, biodegradable films, and higher and O_2 lower, varying according to the production flowchart as shown in Figure 3.1. microperforated films to achieve the optimum MAP storage condition inside the package (Zhang et al., 2016).

This book chapter encompasses the MAP storage for fresh and minimally processed foods. The gaseous composition has been advanced in the MAP for fruits and vegetable packaging in recent years. The packaging materials for the MAP application have mainly focused on finding inclusive previous research and development for the MAP application in fresh and minimally processed fruits and vegetables.

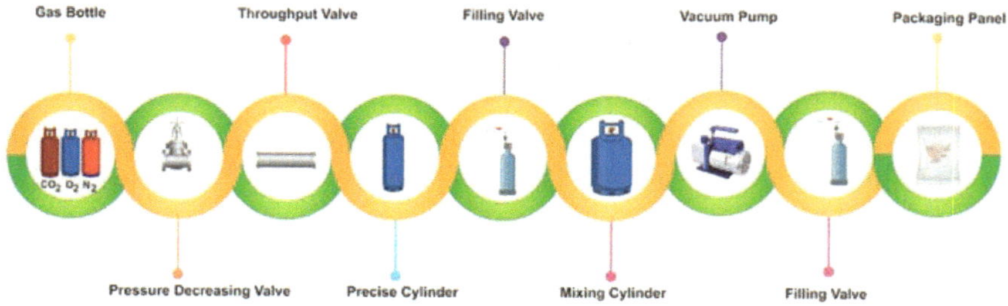

FIGURE 3.1 The production line flow chart of modified atmospheric (MA) packaging (MAP) in fruits and vegetables processing.

3.2 RECENT DEVELOPMENTS IN MAP FILMS

In the past decades, MAP has gained a lot of attention in preserving and increasing the shelf life of fruits and vegetables. Active and passive packaging are the most common types of MAP, but recently there have been advances in the types of MAP (Zhang et al., 2015b). There have been studies in which a higher concentration of oxygen, also known as high oxygen MAP >70% O_2, is considered effective in inhibiting the anaerobic fermentations reactions, discoloration of the fresh produce by the enzymes helps in the prevention of odor loss and moisture loss (Van Der Steen et al., 2002). This type of packaging was mostly seen in the ready-to-eat vegetable. For example, in vegetables that are minimally processed such as grated celeriac, slices of mushrooms, and chicory, it was observed that high oxygen concentration reduces enzymatic loss and microbial growth. This type of packaging, when combined with the different concentrations of nitrogen and carbon dioxide, becomes more effective. It was developed to overcome the losses in low Equilibrium atmospheric packaging (Jacxsens et al., 2001).T. Jiang, in his study, observed the impact of alginate coating on physiochemical sensory qualities of button (*Agaricus biporous*) mushrooms kept under a high oxygen atmosphere, in his results, it was seen that the coating of alginate under high oxygen-MA maintained the integrity and increased the shelf life of the mushrooms as shown in Figure 3.2 (Jiang, 2013). There have also been studies on the impact of high oxygen-MA types of packaging on the nutritional content of fresh fruits and vegetables. It was observed that there is a significant reduction in the vitamin C content and elevation in the phenolic compounds (Ghidelli and Pérez-Gago, 2018). High-modified atmospheric packaging (MAP) has also been combined with ultraviolet light C (UV-C) irradiation. Li et al., in their study, observed the synergistic effect of UV-C irradiation with high oxygen MAP on carrots. The results indicated that UV-C and MAP together help in reducing microbial growth, maintaining physiological quality, slowing down the senescence process, and less whitening the surface. It maintained the overall quality of the carrots and increased their shelf life (Li et al., 2021).This type of MAP is still being researched due to biochemical changes in fruits and vegetables.

FIGURE 3.2 High oxygen–MAP is used in combination with biodegradable packaging material.

Another type of recently developed MAP is the controlled modified atmospheric packaging (CMAP) which increases the efficiency of MAP. In this type of packaging, the packaging films are formed to regulate the gaseous concentration. It is based on the selective diffusion of gases through the membrane. Here, the oxygen is taken inside through the membrane, and carbon dioxide is released out simultaneously (Irtiza et al., 2019). The degree of diffusion of gases is subjected to the membrane permeability to gases and concentration gradient. The resulting concentration is dependent on the produce, that is, the quantity present and the respiratory rate of the produce, temperature, concentration gradient, and permeability. This type of packaging controls the atmospheric conditions inside the package. Silicon membrane is most commonly used, which has a high permeability to gases, as shown in Figure 3.3. There has been research on different fruits and vegetables with silicon membranes. Srivastava and Said worked on MAP using silicon membrane on pomegranate. The pomegranate was stored at different pressure in HDPE package combined with the silicon-MAP. The results showed that the shelf life of pomegranate extended to 96 days at 10°C, while in the ambient conditions, it was 60 days. The quality parameters and sensory characteristics were also better than the normal conditions (Srivastava and Said, 2019). Silicon gum-based films with MAP have also been in use. The quality of mushrooms was observed by Li et al. under cold storage. The results showed that the shelf life increased by more than 8 days in comparison to the control sample. The gas concentration of 10% CO_2 and 5% O_2 were considered optimal for maintaining the quality of mushrooms. Different types of silicon gum-based windows have been introduced. Tiehua Li and Min Zang in their research study on asparagus (*Asparagus officianlis*). They observed the effect of MAP with a silicon gum window. The optimum atmospheric conditions observed were CO_2 below 157 mL L^{-1}, O_2 above 21 mL L^{-1}, and ethylene 15.84 μL L^{-1}.

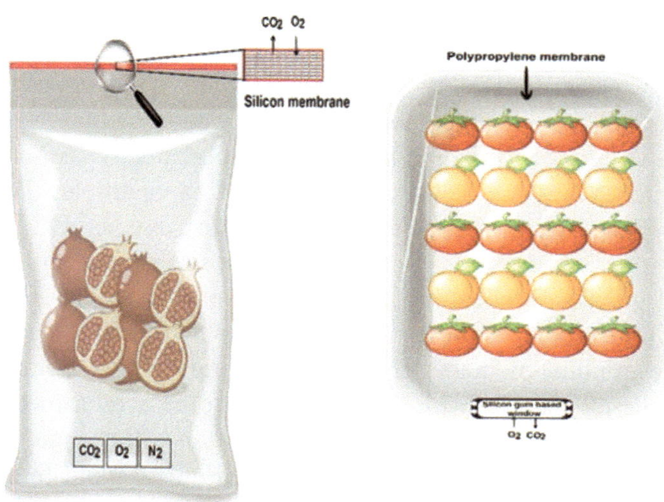

(a) (b)

FIGURE 3.3 Schematic diagrams of controlled MAP. (a) Packaging in a silicon-based membrane-based pouch. (b) Packaging tray with a silicon gum-based window and polypropylene membrane.

It was concluded that the package prohibited the anaerobic respiration of asparagus, decreased the loss of ascorbic acid, and held on to a relatively low respiration rate. The low ethylene concentration inhibited chlorophyll degradation which enhanced the appearance of asparagus.

Intelligent MAP is another approach to MAP which has been seen lately. This type of packaging has been developed due to the growing concerns of consumers regarding food safety and food quality. Intelligent packaging is a type of packaging in which the package consists of real-time detectors such as indicators, sensors, RFID tags which will determine the quality of the products (Zhang et al., 2015b). The various types of indicators include time-temperature indicators, pH indicators, CO_2 indicators, leakage indicators, freshness indicators, etc. This is a novel technique that reduces food spoilage. It monitors the condition of the product inside the package and tracks down the information of the product as well as helps in accessing the data. The function of MAP is to maintain the concentration level of gases inside the packages. However, if there are some leaks it will lead to deterioration of the food product. Therefore, leak indicators can be helpful in determining the leaks and maintaining the integrity of the packages. Gas detecting devices such as carbon-dioxide sensors and oxygen sensors can be used in MAP. The changes in the concentration level of these two gases can degrade the shelf life and quality of the fruits and vegetables and can cause microbial spoilage (Kalpana et al., 2019). Research studies have shown that ultraviolet (UV)-based oxygen indicators with alginate polymers can provide a good water-resistant coating which is observable from the decrease in thionine leakage. A similar study was done by Mill et al. with an oxygen indicator made up of nanoparticles TiO_2 with a coating of methylene blue which acts as a redox dye and DL-Threitol. A pressure-operated oxygen indicator has also been designed with a different blend of glucose, NaOH, and methylene blue at atmospheric conditions for 12-h time period, as shown in Figure 3.4. When NaOH was dissociated from the mixture of methylene blue and gases with a delicate barrier showed higher stability which demonstrated to be a calorimetric oxygen indicator. The change in concentration of oxygen will change the color of the proportion. Similarly, carbon dioxide calorimetric indicators have also been introduced with a mixture of anthocyanin, ε-polylysine, and lysine to determine changes in the pH levels. The working of the indicator is based on the formation of the carbonic acid derivatives by an irreversible reaction (Kalpana et al., 2019). Intelligent packaging is a newer technology and has still been researched with MAP to enhance the quality and safety of food products.

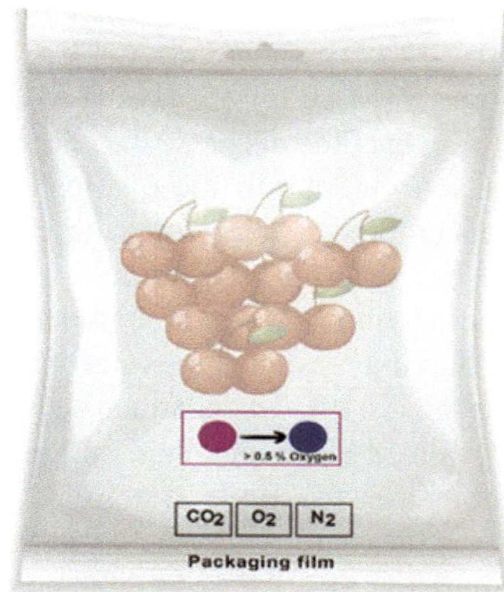

FIGURE 3.4 Intelligent packaging used in combination with MAP–oxygen sensor showing the color change from pink to blue as the concentration of oxygen increases.

MAP has been used with different types of polymers such as polypropylene ethylene-vinyl acetate, LDPE, and HDPE, etc., Sustainable packaging is high in demand these days, and therefore the development of biodegradable films with MAP is being used. Studies have shown that the use of edible coating with MAP has increased the shelf life of freshly-cut fruits (Wilson et al., 2019). The fruits and vegetables are coated with edible film and kept under MAP. Through this process, the degradation of fruits and vegetables can be minimized physically, chemically, and microbiologically. Karagöz and Demirdöven (2019) studied fresh-cut apples which were coated with stevia and chitosan combination and kept under MAP. The results were analyzed based on the polyphenol oxidase (PPO), microbiological, antioxidant activity, total phenolic content, and pH. It was observed that the combinations of stevia and chitosan gave better results than any other samples, it was also concluded that the quality of the apples was increased by using MAP and edible coatings in combinations (Karagöz and Demirdöven, 2019). A similar study was done on fresh-cut papaya by using alginate coating under a MA at low temperatures. Essential oils such as thyme and oregano were added to the alginate coating to improve the quality. The results showed that an increased concentration of essential oils enhances the quality, thus increasing the shelf life of the papaya (Tabassum and Khan, 2020). Yasaman Esmaeili et al. worked on the packaging determining the interdependent effects of MAP and aloe vera gel on strawberry fruit. The aloe vera gel was combined with other essential oils and stored in MAP for 15 days at 4°C. The results showed that aloe vera gel packaging could decrease the deterioration by decreasing the microbes, and a MA could stop the quality loss. Another research study was done on cucumber by coating it with chitosan and packaging it in a nitrogen argon-based MA. The cucumber was stored at 5°C for 12 days. The results concluded that the argon-based MA was better than the nitrogen-based MA, thus increasing the overall quality and shelf life of cucumber (Olawuyi et al., 2019). The coatings such as soy protein isolate have also been used with MAP on eggplant and persimmon. The results depicted that a combination of edible starch coating with MMT-MAP helped in preserving the total antioxidant activity and organic and volatile acids. The new combination of MAP and edible coating will increase the quality of the food product by decreasing the microbial load. The fruits and vegetables which are more prone to enzymatic browning could be explored by this technique.

Nanometals, including nanosilver, zinc oxide, and titanium oxide, could be incorporated into the packaging films. The antimicrobial films loaded with nanomaterial work on the principle of inducing disruption of the microbial membrane and by changing the permeability of the cell membrane through catalysis of the enzyme, which will kill the microorganisms. Guimaraes et al. (2016) worked on the coating of fresh-cut carrots with natural montmorillonite (MMT) and starch under passive MAP. The dispersion of MMT nanoparticles in the starch changed the barrier properties of the film. Sobhani et al. (2021) worked on the analysis of the physiological and microbiological qualities of cherry tomatoes. Nanohydroxyapetite (3%) was incorporated with polypropylene in combination with a MA during a cold storage temperature of about 4 ± 1°C. The results showed nanohydroxyapatite incorporation with MAP was able to slow down the consumption of antioxidants, total phenolic compounds, and vitamin C content, it was because of the slow consumption of oxygen uptake and carbon-dioxide emissions (Sobhani et al., 2022). Some of the research studies have included two nanomaterial particles to form an active film packaging material.

MAP is being used in combination with other techniques such as UV irradiation and ozone. It has been observed that UV-C treatment with the MAP has lowered the post-harvest aging of fruits and vegetables. A research study was conducted on red cherry tomatoes; it was concluded that UV-C treatment increased the antioxidant activity. When UV-C was combined with MAP, it enhanced the overall quality of tomatoes in terms of color changes, physiological changes, respiration rate, and total soluble solids contents (Olveira-Bouzas et al., 2021). The optimum UV-C treatment is found to be 0.2–6.0 kJ m^{-2} which is capable of passing UV light through the polymers (Wilson et al., 2019).

Irradiation techniques with MAP have also been discovered and researched for fresh fruits and vegetables. For the fruits such as cherry and blueberry, a low dose of irradiation (0.6 kGy) has been recommended, and a high dose of irradiation (1 kGy) for fruits which are refrigerated such as watermelon. A research study has been conducted on carrots to know the effects of the antimicrobial coating of the fruits and vegetables + high level of oxygen + gamma irradiation at a concentrated dose of 0.25 or 0.5 kGy with the inoculation of *Listeria inocua.* The results showed that the use of high oxygen inhibited *L. inocua* as shown in Figure 3.5a. The antimicrobial coating reduced the microbial growth in minimally processed peeled carrots when loaded under air.

The use of ozone gas is the most recent technology used with MAP. It is a non-thermal technology. Ozone (O_3) has the property of disinfecting and sterilizing the microbe. It has been included as generally regarded as a safe (GRAS) category. It is common in citrus fruits to prevent the deterioration of fruits from *Penicillium italicum* or *Penicillium digitatum.* Higher doses of ozone were seen to be more effective against spore-forming microorganisms. Researchers have shown the treatment of ozone in combination with MAP on strawberries, blueberries, and raspberries was able to reduce yeasts and fungi (Fang and Wakisaka, 2021). Research conducted on minimally processed cauliflower showed that the pretreatment of ozonated water and antimicrobials helped to improve the shelf life of the cauliflower. Ozone has also been combined with microperforated MAP (Fasake et al., 2021). A research study combining MAP with ozone on chilies has shown an increase in shelf life. Ozone is a non-poisonous gas, easy-to-use gas, and shows no major detrimental effects on food; thus can be used effectively to enhance the quality of fruits and vegetables, as shown in Figure 3.5b.

(a) (b)

FIGURE 3.5 (a) Irradiation technique used in combination with MAP. (b) Ozone gas used in MAP.

Microperforated MAP is another type of technology that has been developed in recent years. Some of the polymeric films have less permeability to gas. Microporous-based MAP is the emerging technology that overcomes the hindrance of polymeric films. This type of film modifies the permeability of the packaging.

3.3 MICROPERFORATED FILMS

Microperforated films are the types of films that consist of microholes, which allow them to maintain the permeability of the gases. It helps to maintain the respiration rate, thus increasing the shelf life of the fruits or vegetables. The exchange of the gases is dependent on the thickness of the packaging material. The transfer of gases increases if there is less thickness, nevertheless, the strength of the material may decrease, which could be a problem during the transportation of the product (Qu et al., 2022). These types of films had a major application in MAP. The microholes present in the films maintain the equilibrium levels of oxygen, carbon dioxide, and relative humidity. The diameter of the microholes should not be more than 200 μm. It prevents the anaerobic microbial growth and the condensation of water vapor inside the film. The rate of the respiration of fruits and vegetables is dependent on the following factors, that is, Size of the microhole, number of microholes, respiration rate, size of the packaging, and storage conditions (Boonthanakorn et al., 2020a). Microperforated MAP is a clean type of preservation technique. It is safe for consumers also.

Microperforated film holes can be obtained in two ways. Firstly, it can be done by perforation on the ordinary film where the aperture is <200 μm, as shown in Figure 3.6; secondly, by directly obtaining the microporous film with 0.01–10 μm. There are three types of perforation techniques, that is, physical perforation technique, chemical foaming technology, inorganic filler polymer drawing technology, and semi-crystalline polymer drawing technology. Physical perforation techniques include three types of techniques; cold or hot needle perforation, electric spark perforation, and laser perforation. This type of method is mainly used for the perforation of the film. Cold or hot needle type of physical perforation is a manual or semi-automatic method. It is a slow technique in which the holes formed are large, that is, greater than 1 mm. In hot needles, the plastic melts to form the holes, and it accumulates as a large brim around the edges, while in cold needle, the needles are punched into the film, and the material removed from the film remain attached and can be used as a flap. The holes formed are also not regular in shape. It is flexible and easy to use. The formation of large and irregular holes can be a disadvantage to the film. The electric spark perforation technique uses electrode discharge. It creates holes with a large depth-to-diameter ratio, but the disadvantage of this type of perforation method is that it is difficult to procure the desirable number of holes. A new technique has been developed in this method which is known as the electro microelectric discharge method (Micro-EDM).In this type, a spark is created between two electrodes which are separated by oil or water for the creation of a small hole. The particles formed are washed and constantly flooded with water or oil. The laser type of perforation method is most commonly used for food. In laser perforation, smaller sizes of micropores are obtained, which does not result in thermal deformation or mechanical damage to the film. This method is a very clean method that produces clean small holes (Allan-wojtas et al., 2008; Fan et al., 2021; Qu et al., 2022).

3.3.1 MATHEMATICAL MODELS FOR MICROPERFORATED FILMS

There have been few research studies where mathematical models have been developed. Mathematical models help to determine the input parameters.

The input parameters may include respiratory concentration gradient, transpiration rate, the thickness of the film, the permeability of gases, time, storage temperature, and weight of the product. Many mathematical models have been proposed to know the kinetics of the products and gases exchanged through microperforated packaging. It has also been researched that the diffusion model is more effective in predicting the permeability and estimation of the surrounding atmosphere. On the basis of the diffusion model of Fick's law, there are three steps, that is, adsorption, and at high concentration, the molecules get absorbed by the film at one side of the film, after the absorption, it gets diffused in the film and lastly desorbed from the other side of the film as illustrated in Figure 3.7. This law has been used recently in the utilization of microporous films. Stephan Maxwell's equation depicts the relation between flux and concentration gradient (Qu et al., 2022). Various other models are depicted in Table 3.1 below.

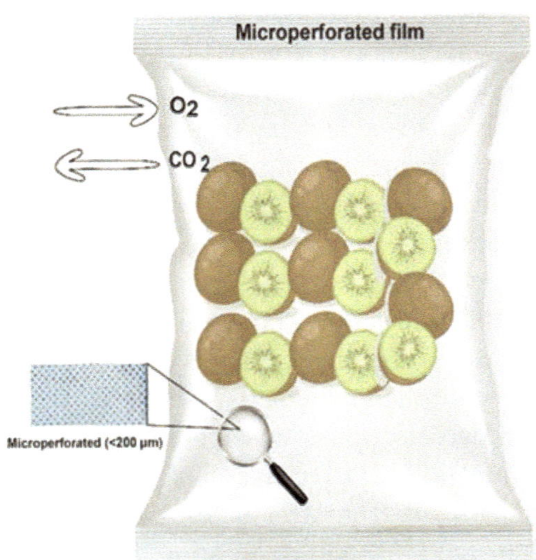

FIGURE 3.6 Microperforated packaging films in MAP depict the permeability of oxygen (O_2) and carbon dioxide (CO_2).

There have been several research studies on microperforated films. A recent research study has been done on mushrooms with high oxygen and microperforated films. The effects on the quality of the button mushrooms were observed while stored at room temperature. The results concluded

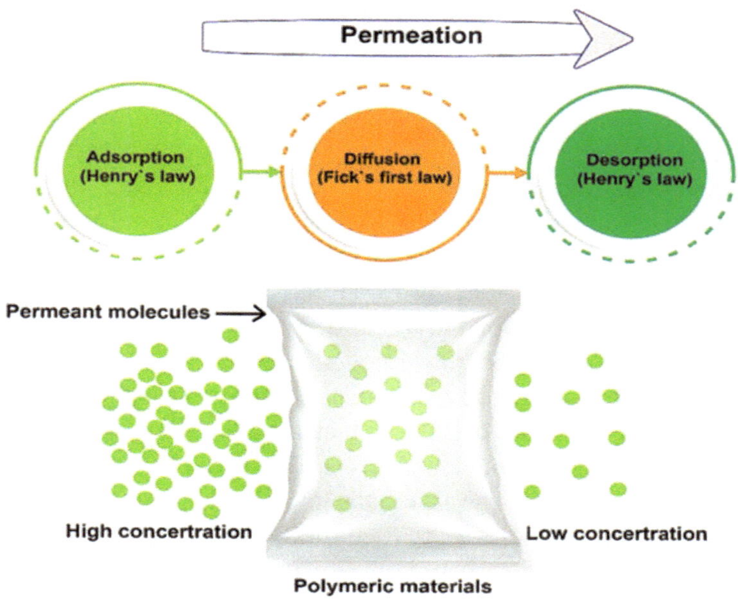

FIGURE 3.7 The transfer mechanism of permeation of gases from high concentration to low concentration in polymeric materials.

TABLE 3.1 Various Models of Microperforated MAP.

Model names	Description	Reference
Comprehensive Mathematical model	Depicts the relation between the microporous films and storage cost. It depicted that the microporous films were able to reduce storage costs.	Qu et al., 2022
Spatial–temporal dependent mathematical model	It is based on the fundamental laws and their effect on the solubility of carbon dioxide and on heat transfer.	
The mathematical model of Del-Valle et al.	This model considered the gas mass transfer rate and respiration rate through plastics. The change in the environmental factors was also observed.	

that the respiration rate was inhibited by the high oxygen presence. The integrated technique of high oxygen atmosphere and perforated film maintained the white color of the mushrooms; it did not allow the condensation of the mushrooms inside the package; also, it did not lead to any volatile compounds formation; however, there was a decrease in the pH levels. The mushroom samples consisting of only perforated films were also analyzed, and it was observed that the tested factors such as vitamin C, total phenolic content, antioxidant capacity, MDA content, and weight loss did not affect the textural properties of the mushrooms (Pogorzelska-Nowicka et al., 2020). A similar study was done on mushrooms with different microperforated films, that is, 0, 2, 4, 8, and 16. The quality and the volatile profile were analyzed, and it was observed that microperforated film decreases the browning action. The eight perforations were found to be effective aroma varieties (Yujiao Cheng et al., 2022).

Another study was done by Lu et al. (2018), on the red globe table grapes, and the effect on fatty acid-derived volatile metabolism under storage in 50 days. The physicochemical properties such as texture profile, and decay incidence aroma related, were also analyzed. The use of microperforated film packaging enhanced the shelf life. A laser-produced microperforated MAP with chitosan carbon dot coating has been done on the fresh-cut cucumber. The size of the microporous film was 100 μm, and different numbers of pores were used (0, 2, 4, 6, 8). The concentration of oxygen was 9.8%, and carbon dioxide was 10.3%. The results concluded that the porous-modified packaging consisting of four micropores used with the chitosan coating in number were able to maintain the firmness, weight, and aerobic plate count. It was also observed that there was no degradation of the flavor and maintained the distribution of water on the 15th day. Winotapun et al. (2021), in their research study, used laser perforation technology on polyethylene terephthalate/polyethylene laminated film for fresh mixed salad. It was concluded that the mixed salad was fresh after 9 days (Winotapun et al., 2021).

An ethylene-treated banana was packed in microporous compostable packaging by Rodov et al. (2022). It has been observed that the compostable packaging film without the pores caused off flavor, impaired ripening, low-level oxygen fermentation, and softening of the banana but the microporous-based packaging film overcame these problems and also controlled the aging of the fruit. The results also showed that this type of packaging was able to control water condensation and low crown rot incidence (Rodov et al., 2022). There have been few pieces of research where the effect of laser microperforated packaging on ethylene propagation rate on different food packages such as propylene which is used for fresh fruits and vegetables, has been observed. The pore diameter size was taken to be 50.00–148.75 μm and OPP thickness 20–30 μm. It was seen that the increase in the size of the pores increased the ethylene transmission rate exponentially, whereas there was no such impact of the thickness of the film on the ethylene transmission (Sonawane et al., 2022).

A study on pak choi (*Brassica rapa* subsp. *chinensis*) was done by Yuan et al. (2021), observing the impact of microperforated packaging on the quality attributes of pak choi. The number of microperforation was taken to be 12 with a diameter of 100 µm packaged in a BOPP film. The microperforations delayed the yellowing of leaves, increased antioxidant activity, and increased shelf life when combined with cold storage (Yuan et al., 2021).

Boonthanakorn et al. (2020) worked on freshly cut durian cv. Monthong and preserved it using microperforated PET/PE films in combination. In this research study, the film was perforated with carbon dioxide of the microsize of 113.5 µm. Three samples were taken. The ripe durian in a pulp tray, the microperforated film, and a non-microperforated film. The samples were observed, and the microperforated films showed a proper MAP condition at a steady rate. However, there was no change in the color and total soluble solids, but the textural properties of the non-perforated package were of low quality in comparison to the perforated and off-flavor observed in the PET/PE film. Some of the gases were released from all three packages at different time periods (Boonthanakorn et al., 2020b).

Microperforated films have replaced the conventional packaging of using polymeric films. From the recent studies, it is evident that it provides better quality fruits and vegetables. However, with various advantages, there is a certain limitation that can also arise with this technique. Sometimes, the exchange of unwanted gases may take place; if the fruit or vegetable is kept for a long duration of time, there are chances that it may develop odd flavors. The exchange of the gases depends on the perforation, the perforation size should be exact. If the size of the microholes varies, then it will lead to the exchange of a high amount of gases which will eventually lead to the deterioration of the fruits and vegetables.

3.4 PERMEABILITY MECHANISM FOR GASES

MAP storage is highly influenced by the gas permeability of the packaging film or sachet. The gas permeability of the packaging material depends upon the film structure, permeability to specific gas, thickness, area, the concentration gradient of the gases across the packaging film, temperature, pressure differences, and the relative humidity of the atmosphere condition. Gaseous molecules can penetrate a polymeric film through microscopic holes. However, in the absence of the crakes and pinholes, the mechanism of gas permeability at a molecular level is known as activated diffusion. The activated diffusion has four steps such as the absorption of permeant gas into the film, penetrant dissolution at a higher concentration gradient of the film matrix, molecular diffusion through film driven by the concentration gradient, and desorption of the gaseous molecule from the other side of the film (Mangaraj and Goswami, 2009).

The permeability of the gas of the polymeric films is due to the solution diffusion coefficient ratio of the equilibrium concentration of the permeant and its partial vapor pressure. The equilibrium concentration depends upon the polymer's polarity, cohesive energy to the penetrant gases, and the free volume for pore filling. The solubility difference of the gas in a particular packaging material may determine the gas diffusion or permeation rate through the packaging film. It requires the activation energy to open ample space to allow the permeant to penetrate from one sorption site to another (Yasuda et al., 1968).

3.5 RECENT DEVELOPMENT IN MAP FOR FOOD PACKAGING GAS RESEARCH

There are several gases designed for the MAP to enhance the storage or shelf-life of the foods and to retain the original taste, texture, and appearance of the product being packaged. Oxygen (O_2), nitrogen (N_2), and carbon dioxide (CO_2) are major gases used in MAP for foods. Different mixtures of two or three gases have been employed for the food products. The gases are used separately or in combined form for each kind of product. Oxygen is highly reactive and supports combustion. In the presence of oxygen rise growth of fungi, and spoilage bacteria decreases the storage life of fresh vegetables and fruits. CO_2 is particularly effective at preventing the growth of microorganisms (such as mold and other common aerobic bacteria) and penetrates biological membranes without any change in permeability and functionality. Sometimes, carbon dioxide displaces oxygen and, at high concentrations, will cause asphyxiation. Nitrogen is a kind of unreacted with less density than air, is non-flammable and maintains pressure inside the packs. The addition of N_2 in the gas composition is used to prohibit pack and decrease the volume of carbon dioxide. Nitrogen is largely utilized in packaging as an inert gas to replace oxygen and stop oxidation. Nitrogen also contributes to preventing package collapse by retaining internal volume because of its low solubility in water. When figuring out a food's gaseous composition, it is important to remember that low O_2 concentrations in some foods can cause quality and safety issues (such as unfavorable color, senescence in fruits and vegetables, and growth of food poisoning bacteria). Hydrogen is the first gas in the periodic table and is also used to retain the freshness of food. Hydrogen plays a good role in food packaging for the preservation of fresh food by removing excess oxygen by more than 2%. In the chemical reaction of O_2 and H_2, the concentration of hydrogen should be more than 5% for the removal of O_2. To quickly release any remaining oxygen from the headspace of the packaging, H_2 is converted to water. In the mixture of CO_2 and N_2 with 4% hydrogen for the fresh food, packaging has been used. Some other gases, such as oxides and noble gases (the most unreactive gases on the earth), have diverse impacts on the freshness and storage life of fresh food used in the packaging. Most of the gases are argon, neon, xenon, ethylene, chlorine, ozone, helium, and gas oxides such as carbon monoxide, propylene oxide, nitrous oxide, sulfur dioxide, and nitric oxide recently used nowadays to extend the shelf life of foods.

3.5.1 OXIDES

Nitrous oxide (N_2O) is an inflammable, colorless gas used for fresh food packaging, including fresh-cut kiwifruit and pineapple (Rocculi et al., 2009). N_2O prohibits ethylene production in post-harvest climacteric foods storage under controlled conditions and extends the shelf life. Experiments using N_2O with a mixture of single showed that low oxygen levels on green banana fruit after harvest indicated the prolonged shelf life of the fruit without affecting physicochemical attributes (Palomer et al., 2005). Nitrous oxide-enriched surroundings reduced both respiration and ethylene formation in fresh-cut pineapple. Nitrous oxide is also used for maintaining pears, carnations, strawberries, and onion rings.

Sulfur dioxide is an effective gas for the MAP to increase the characteristics of grapes and some other foods. Sodium pyro sulfate ($Na_2S_2O_5$) particles emit SO_2 gas. Sodium pyro sulfate is present in the form of white powder, and the odor is the same as rotten eggs. Sulfur dioxide is mostly preserved in dried fruits, vinegar, and vegetable juices. Some other sources of sodium dioxide, such as salt and pad formulations, can be used to release SO_2. The quality or freshness of grapes is retained by the covering of a microperforated sulfur dioxide emission pad, and that pad helps as an obstacle to water and moisture. The mixtures of carbon dioxide and SO_2 gases help increase the storage periods of the foods at two-degree temperatures. Under two degrees Celsius, the level of glucose, sucrose, and fructose remained constant during storage under all the tests, while citric and tartaric acids increased dramatically. Fungal infection, odor, composition, and weight were slowed by the action of sulfur dioxide on table grapes (Pretel et al., 2006). Seiya Saito, in his research, found that sulfur dioxide liner box was a standard method and had various applications in food packaging such as sheets and pads. The new implementation (gas permeability) of the sulfur dioxide/modified atmosphere (SO_2/MA) liners appears to be appropriate in that sulfur dioxide prevents the high concentration of carbon dioxide for increasing the shelf-life of blueberry and enhances blueberry quality for storage.

Nitric oxide (NO), a reactive free radical gas, was first recognized as an industrial pollutant but is now understood to play a role in horticultural products' ability to withstand vegetative stress and senescence. Low concentrations of nitrous oxide gas or its donor molecules can enhance the post-harvest life of certain intact vegetables and fresh fruits. Simple diatomic gas molecule carbon monoxide (CO) has limited solubility in water. It readily combines hemoglobin, slowing the delivery of oxygen and damaging organisms. Therefore, it used to be thought of as a hazardous gas to nature and life. However recent studies have discovered that CO, such as NO and H_2S, may function as a gaseous signal molecule involved in stomata that are tightly controlled by plant guard cells and the development of lateral roots. Shaoying Zhang in his research, found that an optimum concentration of carbon monoxide improved the quality, nutrition, and antioxidant properties of post-harvest peaches during the storage time (Zhang et al., 2014).

3.5.2 *NOBEL GASES*

A class of chemical elements in group 18 known as noble or inert gases are colorless, odorless, and have very little chemical reactivity. Nobel gases do not combine chemically with other elements to form compounds, and they do not usually interact with anything. For food packaging, inert noble gases such as helium and argon are frequently utilized. These gases will retain your food fresh and, according to studies, would not chemically react with foods or create any dangerous consequences. When dissolved in water, noble gases can create clathrate hydrates and prevent water molecules from moving around in food, extending the shelf life of that item (Floros and Matsos, 2005).

Argon is colorless, odorless, tasteless, and–third most abundant gas on the earth. When foods are packaged, carbon dioxide is frequently added to kill microorganisms, but it also degrades the flavor and freshness of the product. Argon helps to prevent microbial oxidase and also inhibits enzymes that increase the rate of oxidation. When pressurized argon action is applied to freshly cut fruit and vegetables, firstly, the argon gas passes into H_2O and forms hydrates of clathrates, which impede the mobility of the water. Meng et al. (2012), in their research on fresh green peppers that were treated with 2–6 megapascals pressure argon and deterioration of quality were also postponed by the argon treatment, which minimized water, and ascorbic acid, chlorophyll, and color loss. Argon pressurization method was used to preserve a good quality of fresh peppers by inhibiting the activities of ant oxidative enzymes and membrane damage (membrane permeability) as well as enzymes that damage membranes (Meng et al., 2012). Similar outcomes were seen when fresh-cut green pineapples were stored (Lu et al., 2018). Helium is inert and not soluble in H_2O. He used a tracer gas in MAP to find out the leaks in foods. Helium is implemented as a non-reactive gas in the MA packaging and is also used in fresh foods for antioxidants. Different combinations of gas with helium help to extend the shelf or storage life of the foods.

Helium and oxygen-enriched MAP are helpful gases for continuing the quality of freshly red leaves and expanding the storage life of the food by decreasing the life of microflora throughout 7 days. Helium and oxygen-MA are both effective methods for maintaining the value of freshly cut baby leaves and show less vitamin C loss (Tomás-Callejas et al., 2011). Treatments, which included Hydrogen peroxide and helium, and argon-enriched atmospheres, reduced respiratory activity, successfully slowed down microbial growth, preserved color features, increased the concentration of bioactive compounds, and increased protection arugula for 7 days. Asparagus spears were preserved in MAP (5% oxygen and 5% carbon dioxide), and frigid storage with the composition of compressed pressure one mega Pascal's argon and xenon helped to extend storage life (Zhang et al., 2008).

TABLE 3.2 Application of MAP (MAP) in Fruits and Vegetables.

Gas type	Product	Shelf-life (days)	Reference
Helium	Broccoli	7	(Jamie and Saltveit, 2002)
Nitrous oxide	Banana	10	(Palomer et al., 2005)
Nitrous oxide (N_2O)	Minimally processed kiwifruit	12	(Rocculi et al., 2009)
Sulfur dioxide	Grapes	120	(Zutahy et al., 2008)
Argon and Xenon (Ar, Xe)	Slim green spears	12	(Zhang et al., 2008)
Nitrous oxide(N_2O)	Pineapple	4	(Rocculi et al., 2009)
Helium (He)	Fresh red baby leaves	7	(Tomás-Callejas et al., 2011)
Argon	Green papers	10	(Zhang et al., 2015b)
Carbon monoxide (CO)	Peach	4	(Zhang et al., 2014)
Sulfur dioxide (SO_2)	Blueberry	7	(Abeli et al., 2021)

3.6 FUTURE TRENDS OF MAP FOR FRESH FOODS

Fresh foods, such as fruits, vegetables, meat, and seafood, are in high demand and have large market potential. The usage of MAP technology is widespread, which maintains the food's freshness until it reaches to consumer. In this context, it is necessary to improve the MAP technique to get optimum results. Researchers are trying their best to get new ideas and modify the MAP technology, that is, ethylene scavenger on the ripening and shelf life of fresh fruits; increased O_2 concentrations may encourage the oxidation of Mb into MbO, which keeps the product fresh, Increasing the levels of CO_2 can prevent or delay the growth of a variety of aerobic microbes that degrade fresh foods, Protein denaturation is encouraged by carbonic acid, which reduces the water-holding capacity of proteins (Olatunde and Benjakul, 2018). A practical alternative for storing is at super chilled temperatures (between freezing and chilling, Micropores on the packaging film with a specific diameter, thickness, and density can lengthen shelf life, improving flavor quality by mixing mono terpenoids (linalool and geraniol) (Tyagi et al., 2020).

These studies have indicated several novelties and primarily positive effects of MAP and have indicated the need for additional research. Freshly prepared produce could benefit from novel MAP by having its quality and microbiological safety maintained. The commercialization of these novel technologies and their success could promote greater consumption of neatly packaged, freshly prepared vegetables and fruits and contribute to customers' improved health and wellbeing. The impact of micropores on food flavor, quality, and microbiological safety has to be further researched. More research needs to be done to fasten the process, maintain a reasonable price, and use appropriate films to laminate or coextruded to get the proper barriers such as gas, moisture vapor, and seal ability. MAP with irradiation and MAP with Antimicrobial films, which are a combination of unique methods of food preparation and packaging system, needs to be developed. The antibacterial effect of super atmospheric O_2 on the safety of fresh-cut food, storage temperature, and atmosphere on post-cutting shelf life needs to be studied. It is necessary to investigate how enteric infections survive and how foodborne viruses and protozoan parasites behave on MAP produce and antimicrobial substances that can be added to the coating. A new phase in the development of food science and technology has begun with the synergistic integration of several disciplines to produce the best outcome in terms of food quality and safety (Kerry, 2012).

3.7 CONCLUSION

This chapter concludes that the MAP has been of great importance in the preservation of fruits and vegetables. There have been developments that have been used in combination with this technology such as biodegradable films, intelligent packaging, UV –light, irradiation, the use of active films by adding antioxidants, antimicrobials, use of nanomaterial, etc. Traditionally it was observed that only oxygen, carbon dioxides, and nitrogen were used as gases in the MAP, but there have been recent advances in gases such as helium, argon, xenon sulfur oxides, carbon monoxide, ozone gas have been used in increasing the shelf life of the fruits and vegetables. The use of microperforated films has been of great significance these days. These types of films have microholes of less than 200 μm, which control the permeability of the gases. This technology is helpful in increasing the shelf life from days to weeks in comparison with the traditional preservation methods. The future scope could be the use of MAP with other non-thermal techniques and thermal methods such as high-pressure processing, cold plasma technology pulse electric field.

KEYWORDS

- **modified atmospheric packaging**
- **micro-perforated**
- **non-thermal treatment**
- **agriculture produce**
- **minimal processing**

REFERENCES

Abeli, P. J.; Fanning, P. D.; Isaacs, R.; Beaudry, R. M. Blueberry Fruit Quality and Control of Blueberry Maggot (Rhagoletis mendax Curran) Larvae after Fumigation with Sulfur Dioxide. *Postharvest Biol. Technol.* **2021,** *179*, 111568.

Allan-wojtas, B. P.; Forney, C. F.; Moyls, L.; Moreau, D. L. Structure and Gas Transmission Characteristics of Microperforations in Plastic Films and Science. *Packag. Technol. Sci.* **2008,** *21* (4), 217–229.

Boonthanakorn, J.; Daud, W.; Aontee, A.; Wongs-Aree, C. Quality preservation of fresh-cut durian cv. 'Monthong' using micro-perforated PET/PE films. *Food Packag. Shelf Life* **2020a,** *23*, 100452. https://doi.org/10.1016/j.fpsl.2019.100452

Boonthanakorn, J.; Daud, W.; Aontee, A.; Wongs-Aree, C. Quality preservation of fresh-cut durian cv. 'Monthong' using micro-perforated PET/PE films. *Food Packag. Shelf Life* **2020b,** *23*, 100452. https://doi.org/10.1016/j.fpsl.2019.100452

Church, I. J.; Parsons, A. L. Modified Atmosphere Packaging Technology: A Review. *J. Sci. Food Agric.* **1995,** *67* (2), 143–152. https://doi.org/10.1002/JSFA.2740670202

Davies, A. R. Advances in Modified-Atmosphere Packaging, In *New Methods of Food Preservation*, pp 304–320. https://doi.org/10.1007/978-1-4615-2105-1_14

Fan, K.; Zhang, M.; Guo, C.; Dan, W.; Devahastin, S. Laser-Induced Microporous Modified Atmosphere Packaging and Chitosan Carbon-Dot Coating as a Novel Combined Preservation Method for Fresh-Cut Cucumber. *Food Bioproc. Technol.* **2021,** *14*(5), 968–983. https://doi.org/10.1007/s11947-021-02617-y

Fang, Y.; Wakisaka, M. A Review on the Modified Atmosphere Preservation of Fruits and Vegetables with Cutting-Edge Technologies. *Agriculture* **2021,** *11* (10), 1–16. https://doi.org/10.3390/agriculture11100992

Fasake, V.; Dash, S. K.; Dhalsamant, K.; Sahoo, N. R.; Pal, U. S. Effect of Ozone and Antimicrobial Treatments on the Shelf Life of Cauliflower Under Modified Atmosphere

Packaging. *J. FoodSci. Technol.* **2021,** *59* (8), 2951–2961. https://doi.org/10.1007/s13197-021-05326-8

Floros, J. D.; Matsos, K. I. Introduction to Modified Atmosphere Packaging. In *Innovations in Food Packaging*; Elsevier, 2005; pp 159–172.

Ghidelli, C.; Pérez-Gago, M. B. Recent Advances in Modified Atmosphere Packaging and Edible Coatings to Maintain Quality of Fresh-Cut Fruits and Vegetables. *Crit. Rev. Food Sci. Nutr.* **2018,** *58* (4), 662–679. https://doi.org/10.1080/10408398.2016.1211087

Irtiza, S.; Bhat, A.; Wani, A. B.; Khan, F.; Murtaza, I.; Younus-Wani, Y. Physiological and Biochemical Interactions for Extending the Shelf Life of Fruits and Vegetables : A Review. *Int. J. Chem. Stud.* **2019,** *7* (1), 2153–2166. https://www.researchgate.net/publication/332028749

Jacxsens, L.; Devlieghere, F.; Van der Steen, C.; Debevere, J. Effect of High Oxygen Modified Atmosphere Packaging on Microbial Growth and Sensorial Qualities of Fresh-Cut Produce. *Int. J. Food Microbiol.* **2001,** *71* (2–3), 197–210. https://doi.org/10.1016/S0168-1605(01)00616-X

Jamie, P.; Saltveit, M. E. Post-Harvest Changes in Broccoli and Lettuce During Storage in Argon, Helium, and Nitrogen Atmospheres Containing 2% Oxygen. *Postharvest Biol. Technol.* **2002,** *26* (1), 113–116.

Jiang, T. Effect of Alginate Coating on Physicochemical and Sensory Qualities of Button Mushrooms (Agaricus bisporus) Under a High Oxygen Modified Atmosphere. *Postharvest Biol. Technol.* **2013,** *76*, 91–97. https://doi.org/10.1016/j.postharvbio.2012.09.005

Kalpana, S.; Priyadarshini, S. R.; Maria Leena, M.; Moses, J. A.; Anandharamakrishnan, C. Intelligent Packaging: Trends and Applications in Food Systems. *Trends Food Sci. Technol.* **2019,** *93*, 145–157. https://doi.org/10.1016/j.tifs.2019.09.008

Karagöz, Ş.; Demirdöven, A. Effect of Chitosan Coatings with and Without Stevia Rebaudiana and Modified Atmosphere Packaging on Quality of Cold Stored Fresh-Cut Apples. *LWT* **2019,** *108*, 332–337. https://doi.org/10.1016/j.lwt.2019.03.040

Kargwal, R.; Garg, M.; Singh, V.; Garg, R.; Kumar, N. Principles of Modified Atmosphere Packaging for Shelf Life Extension of Fruits and Vegetables: An Overview of Storage Conditions. *Int. J. Chem. Stud.* **2020,** *8* (3), 2245–2252. https://doi.org/10.22271/chemi.2020.v8.i3af.9545

Kerry, J. P. In *Advances in Meat, Poultry and Seafood Packaging*, 2012.

Kou, X.; Wu, M. Characterization of Climacteric and Non-climacteric Fruit Ripening. *Methods Mol. Biol.* **2018,** *1744*, 89–102. https://doi.org/10.1007/978-1-4939-7672-0_7/COVER

Li, L.; Li, C.; Sun, J.; Xin, M.; Yi, P.; He, X.; Sheng, J.; Zhou, Z.; Ling, D.; Zheng, F.; Li, J.; Liu, G.; Li, Z.; Tang, Y.; Yang, Y.; Tang, J. Synergistic Effects of Ultraviolet Light Irradiation and High-Oxygen Modified Atmosphere Packaging on Physiological Quality, Microbial Growth and Lignification Metabolism of Fresh-Cut Carrots. *Postharvest Biol. Technol.* **2021,** *173*, 111365. https://doi.org/10.1016/j.postharvbio.2020.111365

Liu, M.; Pirrello, J.; Chervin, C.; Roustan, J. P.; Bouzayen, M. Ethylene Control of Fruit Ripening: Revisiting the Complex Network of Transcriptional Regulation. *Plant Physiol.* **2015,** *169* (4), 2380–2390. https://doi.org/10.1104/PP.15.01361

Lu, H.; Wu, W.; Limwachiranon, J.; Yang, D.; Xiao, G.; Luo, Z.; Li, L. Effect of Micro-Perforated Film Packing on Fatty Acid-Derived Volatile Metabolism of "Red Globe" Table Grapes. *Food Bioproc. Technol.* **2018,** *11*(10), 1807–1817. https://doi.org/10.1007/s11947-018-2142-1

Mangaraj, S.; Goswami, T. K. Modified Atmosphere Packaging of Fruits and Vegetables for Extending Shelf-Life-A Review. *Fresh Produce* **2009,** *3* (1), 1–31.
Meng, X.; Zhang, M.; Adhikari, B. Extending Shelf-Life of Fresh-Cut Green Peppers Using Pressurized Argon Treatment. *Postharvest Biol. Technol.* **2012,** *71*, 13–20.
Olatunde, O. O.; Benjakul, S. Non-thermal Processes for Shelf-Life Extension of Seafoods: A Revisit. *Compr. Rev. Food Sci. Food Saf.* **2018,** *17* (4), 892–904.
Olawuyi, I. F.; Park, J. J.; Lee, J. J.; Lee, W. Y. Combined Effect of Chitosan Coating and Modified Atmosphere Packaging on Fresh-Cut Cucumber. *Food Sci. Nutr.* **2019,** *7* (3), 1043–1052. https://doi.org/10.1002/fsn3.937
Olveira-Bouzas, V.; Pita-Calvo, C.; Lourdes Vázquez-Odériz, M.; Ángeles Romero-Rodríguez, M. Evaluation of a Modified Atmosphere Packaging System in Pallets to Extend the Shelf-Life of the Stored Tomato at Cooling Temperature. *Food Chem.* **2021,** *364,* 130309. https://doi.org/10.1016/j.foodchem.2021.130309
Palomer, X.; Roig-Villanova, I.; Grima-Calvo, D.; Vendrell, M. Effects of Nitrous Oxide (N2O) Treatment on the Post-Harvest Ripening of Banana Fruit. *Postharvest Biol. Technol.* **2005,** *36* (2), 167–175.
Pech, J. C.; Purgatto, E.; Bouzayen, M.; Latché, A. Ethylene and Fruit Ripening. *Annu. Plant Rev. Plant Horm. Ethyl.* **2012,** *44*, 275–304.
Pogorzelska-Nowicka, E.; Hanula, M.; Wojtasik-Kalinowska, I.; Stelmasiak, A.; Zalewska, M.; Półtorak, A.; Wierzbicka, A. Packaging in a High O2 or Air Atmospheres and in Microperforated Films Effects on Quality of Button Mushrooms Stored at Room Temperature. *Agriculture* **2020,** *10* (10), 1–19. https://doi.org/10.3390/agriculture 10100479.
Pretel, M. T.; Martinez-Madrid, M. C.; Martinez, J. R.; Carreno, J. C.; Romojaro, F. Prolonged Storage of 'Aledo'Table Grapes in a Slightly CO2 Enriched Atmosphere in Combination with Generators of SO2. *LWT Food Sci. Technol.* **2006,** *39* (10), 1109–1116.
Qu, P.; Zhang, M.; Fan, K.; Guo, Z. Microporous Modified Atmosphere Packaging to Extend Shelf Life of Fresh Foods: A Review. *Crit. Rev. Food Sci. Nutr.* **2022,** *62* (1), 51–65. https://doi.org/10.1080/10408398.2020.1811635
Rocculi, P.; Cocci, E.; Romani, S.; Sacchetti, G.; Dalla Rosa, M. Effect of 1-MCP Treatment and N2O MAP on Physiological and Quality Changes of Fresh-Cut Pineapple. *Postharvest Biol. Technol.* **2009,** *51* (3), 371–377.
Rodov, V.; Porat, R.; Sabag, A.; Kochanek, B.; Friedman, H. Microperforated Compostable Packaging Extends Shelf Life of Ethylene-Treated Banana Fruit. *Foods* **2022,** *11* (8). https://doi.org/10.3390/foods11081086.
Sandhya. Modified Atmosphere Packaging of Fresh Produce: Current Status and Future Needs. *LWT Food Sci. Technol.* **2010,** *43* (3), 381–392. https://doi.org/10.1016/J.LWT. 2009.05.018.
Sobhani, N.; Zamindar, N.; Aarabi Najvani, F. Effect of Polypropylene Packaging Containing Nano-hydroxyapatite and Modified Atmosphere on the Physicochemical and Microbial Properties of Cherry Tomatoes. *J. Food Meas. Charact.* **2022,** *16* (1), 307–323. https://doi.org/10.1007/s11694-021-01160-w
Sonawane, A.; Pathak, N.; Weltzien, C.; Mahajan, P. Role of Laser Micro-Perforations on Ethylene Transmission Rate in Packaging Materials used for Fresh Produce. *Packag. Technol. Sci.* **2022,** 621–627. https://doi.org/10.1002/pts.2674.
Srivastava, S.; Said, P. Application of Modified Atmosphere Packaging Using Silicone Membrane System for Shelf Life Extension of Pomegranate (Punica granatum L.) and

its Effect on Physico-Chemical Properties. *Food Qual. Saf.* **2019,** *3* (3), 145–155. https://doi.org/10.1093/fqsafe/fyz014.

Tabassum, N.; Khan, M. A. Modified Atmosphere Packaging of Fresh-Cut Papaya using Alginate Based Edible Coating: Quality Evaluation and Shelf Life Study. *Sci. Hortic.* **2020,** *259*, 108853. https://doi.org/10.1016/j.scienta.2019.108853.

Tomás-Callejas, A.; Boluda, M.; Robles, P. A.; Artés, F.; Artés-Hernández, F. Innovative Active Modified Atmosphere Packaging Improves Overall Quality of Fresh-Cut Red Chard Baby Leaves. *LWT Food Sci. Technol.* **2011,** *44* (6), 1422–1428.

Tyagi, K.; Maoz, I.; Vinokur, Y.; Rodov, V.; Lewinsohn, E.; Lichter, A. Enhancement of Table Grape Flavor by Post-harvest Application of Monoterpenes in Modified Atmosphere. *Postharvest Biol. Technol.* **2020,** *159*, 111018.

Van Der Steen, C.; Jacxsens, L.; Devlieghere, F.; Debevere, J. Combining High Oxygen Atmospheres with Low Oxygen Modified Atmosphere Packaging to Improve the Keeping Quality of Strawberries and Raspberries. *Postharvest Biol. Technol.* **2002,** *26* (1), 49–58. https://doi.org/10.1016/S0925-5214(02)00005-4

Wilson, M. D.; Stanley, R. A.; Eyles, A.; Ross, T. Innovative Processes and Technologies for Modified Atmosphere Packaging of Fresh and Fresh-Cut Fruits and Vegetables. *Crit. Rev. Food Sci. Nutr.* **2019,** *59* (3), 411–422. https://doi.org/10.1080/10408398.2017.1375892

Winotapun, C.; Aontee, A.; Inyai, J.; Pinsuwan, B.; Daud, W. Laser Perforation of Polyethylene Terephthalate/Polyethylene Laminated Film for Fresh Produce Packaging Application. *Food Packag. Shelf Life* **2021,** *28*, 100677. https://doi.org/10.1016/j.fpsl. 2021.100677

Yasuda, H.; Clark, H. G.; Stannett, V. Permeability. *Encyl. Polym. Sci. Technol.* **1968,** *9*, 794–807.

Yuan, S.; Zuo, J.; Li, X.; Fan, X.; Li, X.; Wang, Q.; Zheng, S. Micro-Perforated Packaging Delays Leaf Yellowing and Maintains Flavor of Post-Harvest Pak Choi (Brassica rapa subsp. chinensis) Following Low-Temperature Storage. *Food Packag. Shelf Life* **2021,** *29*, 100681. https://doi.org/10.1016/j.fpsl.2021.100681

Zhang, S.; Li, Y.; Pei, F. Carbon Monoxide Fumigation Improved the Quality, Nutrients, and Antioxidant Activities of Post-Harvest Peach. *Int. J. Food Sci.* **2014,** *2014*.

Zutahy, Y.; Lichter, A.; Kaplunov, T.; Lurie, S. Extended Storage of 'Red Globe' Grapes in Modified SO2 Generating Pads. *Postharvest Biol. Technol.* **2008,** *50* (1), 12–17.

Zhang, M.; Meng, X.; Bhandari, B.; Fang, Z. Recent Developments in Film and Gas Research in Modified Atmosphere Packaging of Fresh Foods. *Crit. Rev. Food Sci. Nutr.* **2016,** *56* (13), 2174–2182. https://doi.org/10.1080/10408398.2013.819794

Zhang, M.; Meng, X.; Bhandari, B.; Fang, Z.; Chen, H. Recent Application of Modified Atmosphere Packaging (MAP) in Fresh and Fresh-Cut Foods. *Food Rev. Int.* **2015a,** *31* (2), 172–193. https://doi.org/10.1080/87559129.2014.981826

Zhang, M.; Meng, X.; Bhandari, B.; Fang, Z.; Chen, H. Recent Application of Modified Atmosphere Packaging (MAP) in Fresh and Fresh-Cut Foods. **2015b,** *31* (2), 172–193. https://doi.org/10.1080/87559129.2014.981826

Zhang, M.; Zhan, Z. G.; Wang, S. J.; Tang, J. M. Extending the Shelf-Life of Asparagus Spears with a Compressed Mix of Argon and Xenon Gases. *LWT Food Sci. Technol.* **2008,** *41* (4), 686–691.

CHAPTER 4

NATURAL ANTIMICROBIAL AGENTS FOR ACTIVE PACKAGING OF FRUITS AND VEGETABLES

SAURABH KUMAR KARDAM and DHARM DUTT

Department of Paper Technology, Indian Institute of Technology Roorkee, Saharanpur Campus, Saharanpur, India

ABSTRACT

Consumers and the food industry have been paying close interest in using natural antimicrobial agents in food. Moreover, as consumers become more conscious of synthetic antimicrobials' potentially harmful health effects against the benefits of natural antimicrobials, researchers have developed more interest in creating and using natural antimicrobials in foods. This article discusses the antimicrobial activity of natural chemicals from plants (extract/essential oils), animals, and microorganisms and their possible application in food packaging. Because they are comparatively harmless and do not pose harmful effects to consumers. Their use as substitutes for artificial antimicrobials can open up various new food safety and quality preservation methods. Antimicrobial agents are derived from natural sources, their main bioactive components, and antimicrobial mechanisms. Natural antimicrobials can be incorporated into the active packaging system. Low quantities of antimicrobials can be utilized in the fruits and vegetables during their postharvest life, alleviating several added concerns and ensuring customer acceptance. Humans have been connected to various health benefits from eating fruits and vegetables. Nonetheless, pathogenic or spoilage bacteria and fungi degrade the

Novel Packaging Systems for Fruits and Vegetables. Kirtiraj K. Gaikwad & Suman Singh (Eds.)

nutritional composition, affecting food preservation and commercial value. Harmful bacteria cause thousands of fatalities and many infections due to foodborne diseases every year. Pathogenic microorganisms isolated from fruits and vegetables were included in the study. Despite the multiple types of antimicrobial packaging available for foods, several issues have arisen throughout research and development, like techno-economic feasibility and indispensable *in vivo* studies.

4.1 INTRODUCTION

"Active packaging" is a solution where the food product, headspace, and packaging system interact in the desired way. These systems actively alter the environments of the packed food by microbial, chemical, and physical activity, extending shelf-life and significantly improving microbial safety and organoleptic properties while maintaining its quality. Various antibacterial and antifungal packaging are also used in active packaging, known as "antimicrobial packaging." Antimicrobial agents increased the lag phase, the microbial growth rate was slowed, and microbial spoilage decreased. By incorporating an antimicrobial agent or employing a polymer with such inherent antimicrobial activity, packaging can potentially hinder the development of microbes or directly destroy them to limit the progress of microflora. It allows a product's shelf life to be extended and ensures its acceptable microbiological state is maintained for a more extended period (Wyrwa and Barska, 2017).

Fruit and vegetable consumption gives various health benefits to humans. The estimated losses for fruit and vegetables are about 55% (FAO, 2019). Pathogenic bacteria and fungi are the primary causes of food deterioration in nutritional quality, impacting food preservation, shelf-life, and commercial value (Aguilar-Veloz et al., 2020). Minimally, processed foods are becoming more popular in both consumption and manufacturing. The world's top 10 fruits and vegetable-producing countries are shown in Figure 4.1. These countries can be found on four of the world's major continents (Balali et al., 2020).

Because of the economic and social consequences of foodborne disease outbursts, there is excellent awareness of developing innovative methods for eliminating pathogens from fruits and vegetables. Concerns about the toxicity of chemically manufactured preservatives and developing microorganism resistance to these chemicals have made the need to find new natural antimicrobials more vital (Guo et al., 2021).

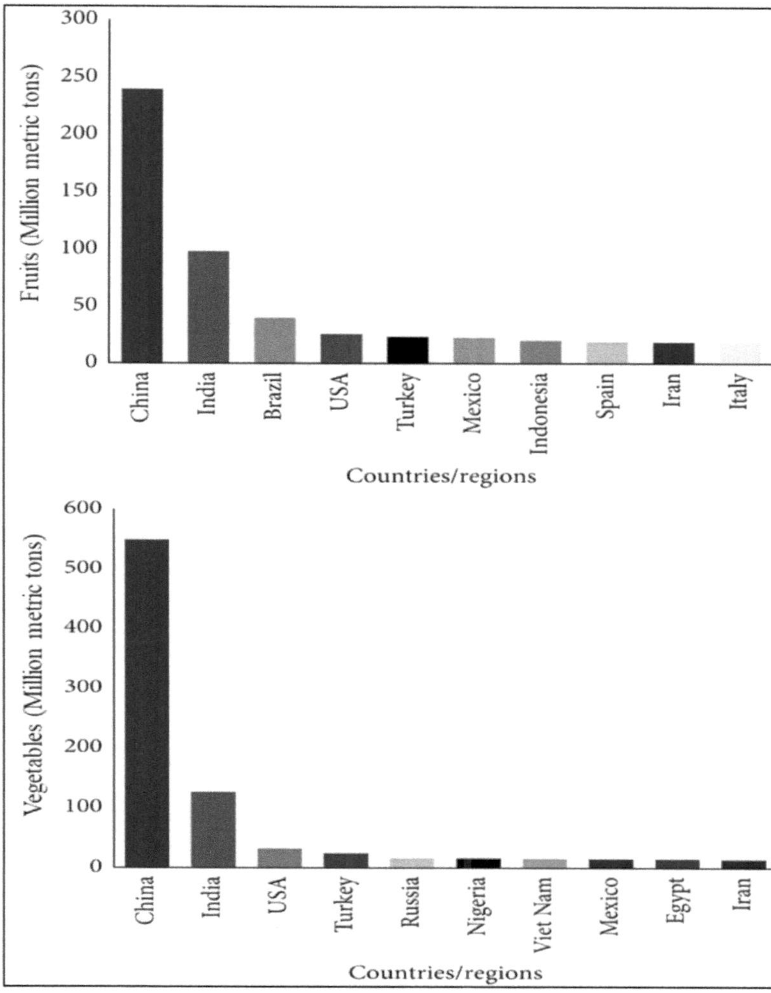

FIGURE 4.1 Top 10 fruits and vegetables growers in the world.

Source: Reprinted with permission from Balali et al. (2020). Copyright © 2020 Gadafi Iddrisu Balali et al. https://creativecommons.org/licenses/by/4.0/

For food preservation, by suppressing microorganisms, a variety of synthetic preservatives are used. Specific microorganisms include *E. coli* O157:H7, *Campylobacter spp., Salmonella spp., Clostridium perfringens, B. cereus, S. aureus,* and *L. monocytogenes* not spoilage the food but also pose a health risk by causing foodborne illness. Foodborne illness is becoming a rising public health concern around the world. In the United States alone, 31 pathogenic species are thought to be accountable for 9.4 million instances of foodborne sickness every year (Gutiérrez-del-Río et al., 2018). Customers increasingly reject artificial additives and need higher-quality food devoid of artificial preservatives while keeping a lengthy shelf life. Scientists have been looking for natural options for standard treatments. A good natural antimicrobial preservation must meet specific criteria, including (1) sufficiently effective at low concentration; (2) being affordable; (3) not causing sensory alterations in the food; (4) broad antimicrobial spectrum; and (5) being non-toxic.

The food industry is motivated to explore natural-based alternatives with high antimicrobial properties to satisfy customer demands, ensure food safety, and regain public confidence. Essential oils, plant extracts, enzymes, chitosan, and peptides nisin can all be employed as possible antimicrobial agents to increase the food's shelf life and safety (Aziz and Karboune, 2018). Natural antimicrobials from plants (extracts or essential oils), animals (enzymes, peptides, chitosan), and microorganisms (Glucose oxidase and nisin); their major bioactive compounds, characteristics, and antimicrobial mechanism are summarized in Table 4.1 (Aziz and Karboune, 2018; Pisoschi et al., 2018).

4.2 NATURAL ANTIMICROBIAL AGENTS' UTILIZATION

Natural antimicrobials can be combined into packaging systems or added to food directly to extend its shelf life. There are many benefits to the latter. Several natural antimicrobials for microbial safety and extended storage periods without significant harm to fruit and vegetables are summarized in Table 4.2 (Al-Tayyar et al., 2020; Guo et al., 2021; Salehi, 2020).

4.2.1 DIRECT ADDITION OF NATURAL ANTIMICROBIALS TO FOODS

Chemical antimicrobials are often employed in food items to inhibit microbial decay and escalate the product's shelf life. However, the food industry is looking for natural alternatives because of the potential health dangers of synthetic antimicrobials and consumers' demand for natural antimicrobials. The natural antimicrobial can be added directly to food. However, the flavor constituent of using plant-based antimicrobials

TABLE 4.1 Natural Antimicrobials their Sources, Major Bioactive Compounds, and Antimicrobial Mechanisms.

Source		Example	Major constituents/ bioactive molecule	Main characteristics/ antimicrobial action
Plants	Extracts	Grape seed	Monomeric phenolic compounds (Epicatechin, catechin epicatechin-3-O-gallate)	These extracts have antibacterial properties and a high antioxidant capacity, which helps prevent oxidative deterioration.
		Green tea	Epicatechin, epicatechin gallate, teaflavin, epigallocatechin, teaflavin digallate, teaflavin gallate, monogallate A and B,	
		Cranberry	Phenolic Compounds (tannins, phenolic acids, proanthocyanidins,) Flavonoids (anthocyanins)	
	Essential Oils	Lemongrass, Clove, Garlic, Cinnamon, Coriander, Rosemary, Parsley, and Oregano	Carvacrol, Thymol, Eugenol, Citral	• Complex natural blends including 20–60 components in varying quantities • Chemical structure and bioactive concentrations are both crucial factors in antimicrobial effectiveness • Attacks the cell membrane's phospholipid bilayer, altering enzyme systems, damaging bacteria's genetic material, and oxidizing unsaturated fatty acids, culminating in fatty acid hydroperoxide production • More active toward Gram-positive bacteria
Animal	Enzymes	Lysozyme	Single peptide enzyme produced by animals and humans	• The hydrolyzed N-acetylmuramic acid and N-beta acetylglucosamin's 1,4-glucosidic links result in bacterial cell lysis • Effective against Gram-negative bacteria

TABLE 4.1 *(Continued)*

Source	Example	Major constituents/ bioactive molecule	Main characteristics/ antimicrobial action
			• Exhibit good stability over a broad value of pH and temperature
Antimicrobial Peptides	Lactoperoxidase system	Hypothiocynate and hypothiocyanous acid produced by oxidation of thiocyanate with H_2O_2 in the presence of lactoperoxidase catalyst	• (Lactoperoxidase + thiocyanate + H_2O_2) secreted in mammary glands (saliva, milk, and tears) • Inhibit microbes by the oxidation of sulfhydryl function group and proteins • Exhibit antimicrobial action against bacteria (Gram-negative/Gram-positive), fungi, and parasites
	Lactoferrin	Iron binding globular glycoprotein	• Inhibit microbes by (1) sequestering iron from bacteria and (2) large, cationic electrostatic patches on the lactoferrin surface interact with anionic Lipid A
Polysaccharides	Chitosan	(1,4)-linked 2-aminodeoxy-b-D-glucan	• It can be extracted from the crustacean, yeast, fungi, protozoa, and green microalgae • Inhibits microbes when (1) positively charged amino group reacts with the negatively charged microbial cell membrane, (2) Binding trace metals inhibit the generation of toxins and microbial growth, and (3) Chitosan reaches the nuclei of bacteria, interfering with protein and mRNA production

TABLE 4.1 *(Continued)*

Source	Example	Major constituents/ bioactive molecule	Main characteristics/ antimicrobial action
			• When the molecular weight of chitosan was reduced, antimicrobial activity against Gram-negative *E. coli* increased, whereas Gram-positive bacteria *S. aureus* had the reverse effect.
Microbial -	-	Glucose oxidase	• Produced by molds *Aspergillus niger* and *Penicillium spp* • The cytotoxicity of H_2O_2 produced by glucose oxidase is primarily responsible for its antimicrobial activity; however, the lowering of pH caused by the synthesis of D-gluconic acid may also impact microbial growth
-	-	Nisin	• Antimicrobial peptides made by fermenting a modified milk medium with particular strains of *Lactococcus lactis*, a lactic acid bacteria • Interaction with phospholipids in bacteria's cytoplasmic membrane, disrupting membrane function and interrupting the germination process, impeding spore outgrowth.

Source: Adapted from Aziz and Karboune (2018) and Pisoschi et al. (2018).

TABLE 4.2 Natural Antimicrobials for Microbial Control and Shelf Life Increment of Fruits and Vegetables.

Natural antimicrobial-based ingredient	Fruit and vegetables	Main benefits and microorganisms targeted
Chitosan + Alginate	Fresh-cut melon	• Reduced yeast, fungi, and bacteria • CO_2 generation reduced
Honey + Soy Protein Isolate	Fresh-cut Kajari melon	• Microbial spoilage delayed • Honey maintains the color and taste
Arabic gum + (Lemongrass, Cinnamon EO)	Papaya	• Microbial spoilage reduced
Alginate + (Thyme, Oregano) EO	Fresh-cut papaya	• Antimicrobial activity enhanced • Less weight loss, pH changes slower, delayed senescence
Cinnamon EO	Strawberries	Effective antifungal action against *R. stolonifera* and *B. cinerea*
Chitosan + Beeswax	Strawberries	• Effective for fungal infections • Less weight loss, Good retention of color, firmness, and pH
Aloe vera + Ascorbic acid	Strawberries	• Microbial load reduced
(Quinoa protein and Chitosan) + Thymol	Strawberries	• Microbial load of fungi and yeast was lowered
Pullulan + Cinnamon EO	Strawberries	• Strong antimicrobial effects on molds and bacteria • Less strawberry weight reduction, reduced firmness, and total soluble solids
Sodium alginate + Lemongrass EO	Fresh-cut Fuji apples	• Psychrophilic bacteria growth was delayed
Pectin + Cinnamon EO	Peach	• *L. monocytogenes, S. Choleraesuis, E. coli O157:H7* and *S. aureus*, and bacterial growth reduced • Increased antioxidant and insignificant odor change

TABLE 4.2 *(Continued)*

Natural antimicrobial-based ingredient	Fruit and vegetables	Main benefits and microorganisms targeted
Alginate + Rhubarb extract	Peach	• Excellent antifungal activity
Starch + Cinnamon EO	Fresh-cut apple	• Microbial growth reduced
		• Decreased production of CO_2, C_2H_4, and respiration rate
Alginate + Thyme EO	Fresh-cut apple	• Excellent bactericidal effect against *S. Typhimurium, S. aureus, L. monocytogenes, E. coli O157:H7,*
Sodium alginate + Sweet orange EO	Tomato	• Strong antibacterial results against mesophilic bacteria, *S. typhi, L. monocytogenes*
		• Treatment shows less weight loss and higher firmness
(Methylcellulose + Beeswax)	Cherry tomato	• Gray mold development reduced
Arabic gum + (Sodium caseinate and Cinnamon, Lemongrass EO)	Guava	• Microbial load reduced
Chitosan + (Bergamot thyme, Tea Tree EO)	Orange	• Fungal growth delayed
Alginate + Basil Oil	Okra	• Delayed fungi growth
		• Maintains color, texture, and overall acceptability
Tragacanth gum + Zataria multiflora Boiss EO	Button mushrooms	• Molds and yeasts microbial count decreased
Carnauba wax + Lemongrass EO	Grape	• *Salmonella* and *E. coli* O157:H7 contamination retarded
Chitosan + Mandarin EO	Green beans	• *L. innocua r*educed throughout the storage of green beans
Carnauba Wax + Oregano EO	Cucumber	• Inhibited pathogenic bacteria *E. coli O157:H7, Alternaria alternate, S. typhi, mesophilic bacteria, and fungi*

(extracts or essential oils) as preservatives in food products can be restricted, mainly when used in high quantities.

4.2.2 INCORPORATION OF NATURAL ANTIMICROBIALS INTO PACKAGING SYSTEMS

Incorporating natural antimicrobials into packaging material can give additional advantages apart from protecting against microbial spoilage and meeting growing consumer desires for safe and synthetic chemical-free food. Active antimicrobial packaging is an innovative method to protect foods by using migrating or non-migrating antimicrobials. The release rate of these natural antimicrobials can be controlled over a prolonged period to maintain food's nutritional quality and storage period without the direct addition of these antimicrobials into the food. Additionally, incorporating natural antimicrobials into packaging material prevents them from interacting with other food molecules such as proteins and lipids, potentially reducing their effectiveness. Sustained release of these antimicrobials to the headspace allows for inhibiting microorganisms during the storage and transportation of food (Aziz and Karboune, 2018).

Natural antimicrobials can be incorporated into packaging material in three ways: (1) by addition to packaging material (blending or mixing with polymer resin), (2) by coating antimicrobials onto packaging surfaces, and (3) by Covalent chemical grafting (immobilization). Antimicrobial packaging systems can be categorized into two types based on the nature of the antimicrobial agent (migrating or non-migrating). Non-migrating packaging systems comprise an antimicrobial fixed onto the packaging material, whereas migrating packaging systems contain antimicrobial that migrates to the headspace. The latter case becomes effective against the growth of unwanted microbes when food products are irregular in shape, and package-product direct contact is significantly less (Kardam et al., 2021).

4.2.2.1 ADDITION WITH POLYMERIC FILMS

Researchers have looked into embedding and combining natural antimicrobials with polymers in recent years. Antimicrobial compounds in the packaging films change the film's mechanical, moisture barrier, gas barrier, optical properties, and microstructural changes. Hence, the performance of films must be investigated accordingly. Several researchers used this approach for food preservation like Meira et al. (2017) used nisin with corn starch, Uranga, 2019 incorporated citric acid with chitosan, and Zhang et al. (2017) added natural bacteriocin (plantaricin BM-1) with polyethylene for potential application in food preservation.

4.2.2.2 COATING OVER PACKAGING SURFACES

A coating is a layer of material deposited onto a substrate to modify the surface property. Coating methods have been extensively used to create packaging substrates with new characteristics. Numerous approaches, such as nano/micro-encapsulation, have included antimicrobial compounds into the coating while avoiding further thermal and mechanical stress difficulties. When designing an antimicrobial coating, interactions between the substrate, coated layer, active component, and the food must be considered. These antimicrobial coatings should meet the following criteria:

i. Coating must adhere effectively to the substrate and be non-toxic when directly contacting food
ii. The released agent's concentration should be optimized to ensure effective antimicrobial action
iii. Coated packaging material should be suitable for the specific food. This means properties like barriers should be matched with the specific food requirement (El-Saber Batiha et al., 2021).

4.2.2.3 CHEMICAL IMMOBILIZATION

Antimicrobial compounds can be covalently included in the matrix after chemical or physical changes. This integration can theoretically be accomplished in a variety of ways, for as, forming covalent bonds between the polymer and the active ingredient. The presence of functional groups on both the polymer and the antimicrobial molecule, resulting in covalent bonding, is required for covalent immobilization. Antimicrobial substances with a suitable functional group include polyamines, organic acids, peptides, and enzymes (El-Saber Batiha et al., 2021).

Polylactic acid (PLA) was plasma treated to generate carboxylic groups on the film surface. Then, nisin or ε-polylysine (ε-PL) was covalently bonded to the film surface, resulting in effective inhibition of bacteria

(Huang et al., 2020). In another study, lysozyme was used to immobilize calcium alginate and showed good antimicrobial effect (Wang et al., 2018).

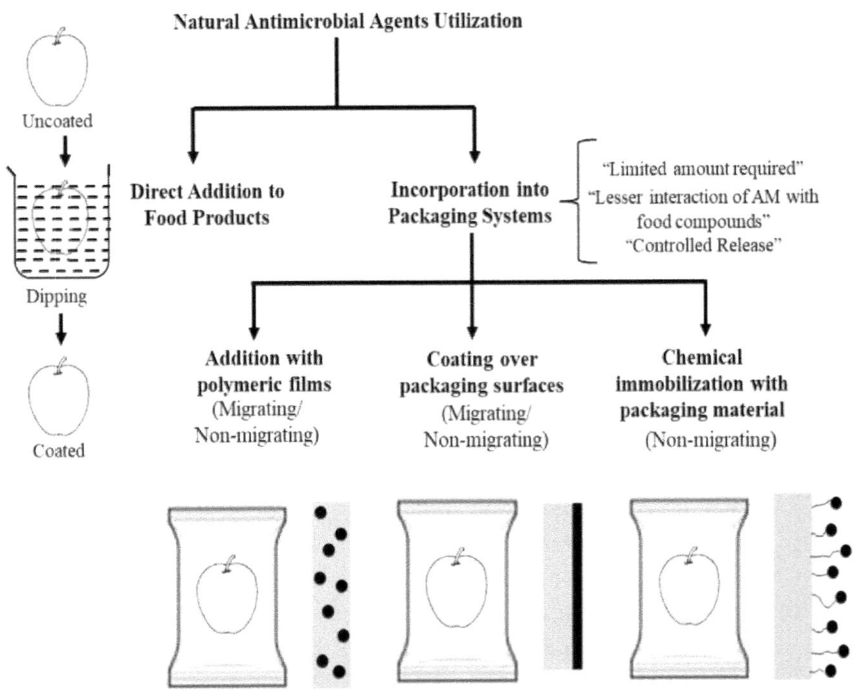

FIGURE 4.2 Natural antimicrobial agents utilization.

4.3 FRUITS AND VEGETABLES: MICROBIOTA AND OUTBREAK

Microbiota of microflora in vegetables and fruits may diverge significantly depending on nutrition, water activity, and pH. Generally, the bacteria found on the surfaces at reduced storage temperatures consist of *Pseudomonadacae*, *Enterobacteriaceae*, and certain *lactic acid bacteria*. After packaging, total mesophile counts on minimally processed vegetables range between 10^3 and 10^6 colonies per gram. However, they vary between 10^3 and 10^9 colonies per gram in various catering and retail outlets. Compared to molds, bacterial growth dominates in fruits and vegetables due to pH lying between acidic and neutral. Most foodborne illnesses are scarce, and most of them go unidentified; they contribute key unseen aspects of morbidity. Between 1996 and 2004, there were 14 outbreaks of foodborne infections, resulting in 859 illness cases reported to the "Food and Drug Administration (FDA)," with lettuce and tomatoes being blamed for the illnesses. *Escherichia coli* O157:H7 and *Salmonella* are particular microbes since they can cause infections at low dosages and develop very quickly (Al-Tayyar et al., 2020). These two human viruses became front-page news in the USA in 2006 and 2007 due to outbreaks related to the intake of various products, such as pre-cooked green vegetables. *Shigella spp, Clostridium botulinum, Yersinia enterocolitica, Staphylococcus spp.,* and *Vibrio cholerae* are other pathogens linked to outbreaks (Prakash et al., 2018). Specific microbes isolated from fresh produce are shown in Figure 4.3 (Balali et al., 2020).

Bacteria	**Parasites**	**Viruses**	**Fungi**
Aeromonas sp.	*Ascaris Spp*	Calicivirus	*Alternaria sp.*
Bacillus cereus	*Cryptosporidium parvum*	Hepatitis A Virus	*Aspergillus niger*
Brucella spp	*Cyclospora*	Norovirus	*Fusarium sp.*
Campylobacter spp	*Giardia sp.*	Norwalk & Norwalk-like	
Enterobacter spp.	*Taxoplasma gondii*	Rotavirus	
Escherichia coli	*Trichinella spp.*	Sapovirus	
Listeria monocytogenes	*Trichuris trichiuria*		
Pseudomonas spp.			
Salmonella spp			
Shigella spp.			
Staphylococcus spp.			
Vibrio spp.			
Yersinia spp.			

FIGURE 4.3 Typical microbes isolated from fresh produce.

Source: Modified from Balali et al. (2020). Copyright © 2020 Gadafi Iddrisu Balali et al. https://creativecommons.org/licenses/by/4.0/

4.4 NATURAL ANTIMICROBIALS IN ACTIVE PACKAGING: RESEARCH AND APPLICATION CHALLENGES

Antimicrobial agents with potential use in food packaging have a long application history in medicine, food, pharmaceuticals, and ethnomedicine. Despite the numerous opportunities for natural antimicrobials in food packaging, various challenges have come into existence during research and development, subsequent in a limited number of applications (Mlalila et al., 2018). An overview of research and application challenges is shown in Figure 4.4.

4.4.1 *IN VIVO STUDIES*

Antimicrobial results obtained using nutritional cultures in the lab are far from when actual food testing (*in vivo*) is used, as only a restricted spectrum of microorganisms may be inactivated. Increased *in vivo* studies under semi-controlled or real-world production circumstances are required. The antimicrobial agent necessary to prevent microbial spoilage in food products is 10 times higher than the quantity needed for *in vitro* (nutrient culture) testing (Mlalila et al., 2018).

For example, Khalid et al. (2018) examined polycaprolactone/starch blend matrix with pomegranate rind for antimicrobial packaging applications; however, *in vivo* experiments were omitted. Zhang et al. (2020) included mangosteen rind powder (MRP) with chitosan film and analyzed *in vitro* antimicrobial analysis but did not include antimicrobial *in vivo* tests (microbial spoilage) for soybean. For genuine efficiency against food microbial spoilage, *in vivo* tests are required (Kardam et al., 2021).

4.4.2 *MECHANISM OF ACTION*

Antimicrobials inhibit microbial growth, and their antimicrobial mechanisms are a complicated concept that has yet to be clarified. In concept, several antimicrobial mechanisms were proposed.

The essential oil can permeate through the bacterial cell due to its hydrophobic nature and disintegrate cell structure. Tannins can block the catalase activity and impose a lethal effect on pathogenic microbes (Aguilar-Veloz et al., 2020).

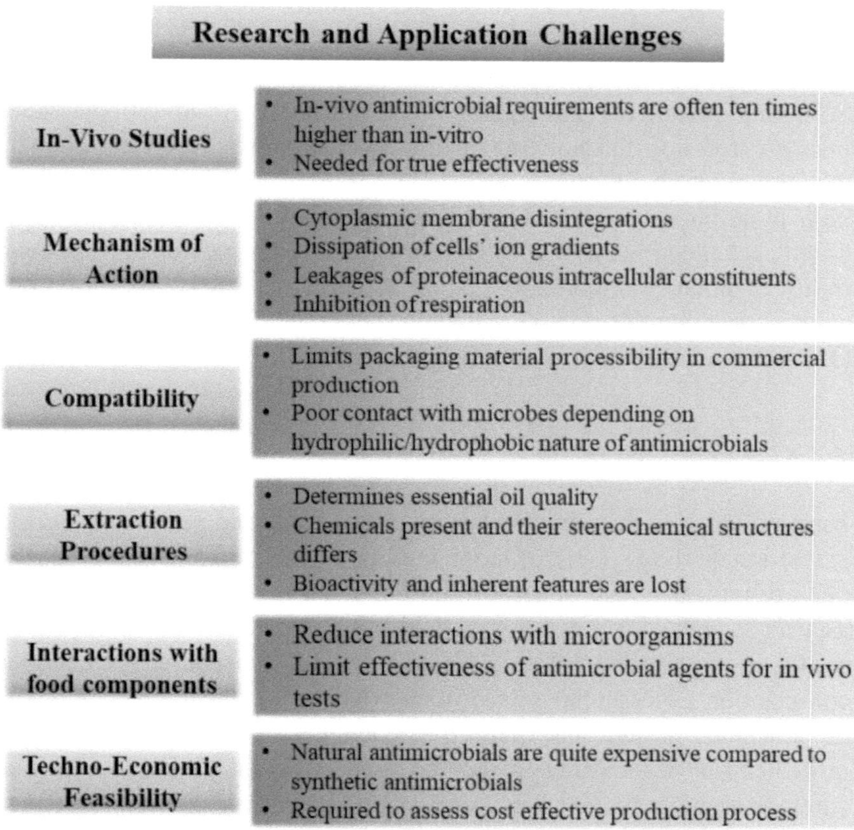

FIGURE 4.4 Overview of research and application challenges.

4.4.3 INTERACTIONS WITH FOOD COMPONENTS

Antimicrobial agent effectiveness in *in vivo* studies and commercial applications is limited due to interactions between the antimicrobial agent and food components. Antimicrobial agents in packaged foods can interact with a food constituent, like carbohydrates, salts, vitamins, and proteins, reducing their antimicrobial action with microbes. As a result, it can be inferred that packaging materials suited for one type of food may not be acceptable for another and that case-by-case research should be used when building a packaging system (Mlalila et al., 2018).

4.4.4 COMPATIBILITY

Engineering packaging materials involves a variety of processing conditions and treatments that aren't straightforward. Antimicrobial agents incorporated into the polymer matrix have compromised the polymers' inherent integrity. Furthermore, some antimicrobial compounds have a high hydrophobicity, which adds to poor microbe interaction in water-based packaged food products. Other considerations include the incompatibility of polar antimicrobial agents like benzoic, sorbic, and propionic acids with nonpolar polymer materials like LDPE due to their polarity (Mlalila et al., 2018).

4.4.5 EXTRACTION PROCEDURES

Essential oils can be extracted from various plants and parts using various extraction methods. The extraction method and the production of essential oils are typically determined by the botanical material employed. Other aspects to consider are the state and species of material. Essential oil quality highly depends on the selected extraction procedure. The chemical composition of essential oils differs in terms of the number and kind of molecules present and their stereochemical structures, which can be influenced by the extraction method used (Aziz et al., 2018). An inaccurate extraction method might damage or change essential oil's chemical signature, bioactivity, and inherent features. Developing non-expensive efficient extraction methods is required, mainly drying procedures used for raw material pre-treatment and extract concentration increment (Aguilar-Veloz et al., 2020).

4.4.6 TECHNO-ECONOMIC FEASIBILITY ANALYSIS

With the comparatively long production cycle, restricted crop availability, and difficulty in purification, natural antimicrobial agents are pretty costly related to artificial antimicrobial agents. Many natural antimicrobials are, in fact, more expensive than equivalent synthetic substances for a variety of reasons. Finally, besides satisfactory laboratory results, a techno-economic viability analysis should evaluate the proposed production process's cost-benefit connection. Infrastructure, local demands, stock availability, and technical facilities play a role in techno-economic viability analysis. This integrated approach should consider for cost-effective control programs for natural antimicrobial-based packaging (Aguilar-Veloz et al., 2020).

4.5 LEGAL REGULATIONS

Antimicrobial packaging can be launched in the market if it correctly and effectively fulfills its application and meets legal criteria. The legal standards for antimicrobial packaging are found in the "European Union's framework regulation (EC) No. 1935/2004." This regulation was enacted to confirm the quality and legal reliability of materials used to actively change food's behavior or improve its condition and materials used to protect it. This regulation is applied to those active materials and objects that contact food so that active agents are gradually released into food and modify the nutritional composition or sensory qualities of food in line with the appropriate standards for food.

"All materials and articles" developed to come into contact with food must be safe, according to article 3 EC No. 1935/2004. Therefore, active packaging must be sufficiently inert or free from those constituents whose migration into food could endanger human well-being, cause unacceptable alterations in food or cause deterioration of the sensory characteristics of the food. While according to article 4, EC No. 1935/2004, active articles and ingredients can alter food composition or sensory properties if these alterations are under the food rules (Wyrwa and Barska, 2017). In compliance with "Regulation (EC) No. 1935/2004," active packaging has been purposefully created to contain active substances that can be progressively migrated into food products or headspace or absorbed from food products.

The "Commission Regulation (EC) No. 450/2009" on active and intelligent packaging and objects designed to contact food covers detailed regulations on the usage of active packaging. This act governs concerns connected to the secure application of active packaging materials and establishes marketing rules for these materials. It also establishes the European Commission's list of approved compounds for use in active articles, the criteria for inclusion in the list, the requirements for labeling, and the requirement for a declaration of conformity and supporting papers.

The "European Food Safety Authority (EFSA)" conducts a safety assessment on the chemicals in the active component utilized in the active material. Since 2002, the EFSA has served as a consultation body to the European Commission on various topics, including toxicological assessments of novel compounds used to manufacture materials and items proposed to contact food (Wyrwa and Barska, 2017).

4.6 CONCLUSION

Natural antimicrobials are attracting attention among scientists and industry experts due to their potential to support quality and safety measures for food and consumer preferences for natural preservation. Natural antimicrobials embedded into packaging systems have excellent microbial inhibition against various spoilage and pathogenic bacteria. Natural antimicrobials and multiple types of packaging have potential applications in extending the shelf life of fruits and vegetables. Fruits and vegetables are vital human diet, but microbial spoilage deteriorates their quality. Studies conclude that natural antimicrobial agents are good for maintaining nutritional quality and organoleptic properties and increasing fruits and vegetables' shelf life. Incorporating these antimicrobials into the packaging materials can inhibit microbial spoilage. Natural antimicrobials have massive potential for commercial applications, but their commercial utilization is limited. However, despite their potential, natural antimicrobials are rarely used in food packaging systems because of adverse organoleptic alterations in food. In some cases, active natural substances may interact with other food components, such as lipids and proteins. Yet society faces a more significant challenge in consuming fresh fruits and vegetables as many pathogenic microbes are isolated from these. Hence, foodborne diseases have been associated with consuming fresh fruits and vegetables. However, there are still many challenges to commercializing these natural antimicrobial packaging. Few challenges that must be rectified for the industrial implementation are the quality of natural products, extraction conditions, and preliminary *in vivo* studies.

KEYWORDS

- **antimicrobial compounds**
- **active packaging**
- **fresh produce**
- **storage**
- **microorganisms**
- **shelf life**

REFERENCES

Aguilar-Veloz, L. M.; Calderón-Santoyo, M.; Vázquez González, Y.; Ragazzo-Sánchez, J. A. Application of Essential Oils and Polyphenols as Natural Antimicrobial Agents in Postharvest Treatments: Advances and Challenges. *Food Sci. Nutr.* **2020,** *8* (6), 2555–2568. https://doi.org/10.1002/fsn3.1437

Al-Tayyar, N. A.; Youssef, A. M.; Al-Hindi, R. R. Edible Coatings and Antimicrobial Nanoemulsions for Enhancing Shelf Life and Reducing Foodborne Pathogens of Fruits and Vegetables: A Review. *Sustain. Mater. Technol.* **2020,** *26*, e00215. https://doi.org/10.1016/j.susmat.2020.e00215

Aziz, Z. A. A.; Ahmad, A.; Setapar, S. H. M.; Karakucuk, A.; Azim, M. M.; Lokhat, D.; … Ashraf, G. M. Essential Oils: Extraction Techniques, Pharmaceutical And Therapeutic Potential - A Review. *Curr. Drug Metab.* **2018,** *19* (13), 1100–1110. https://doi.org/10.2174/1389200219666180723144850

Aziz, M.; Karboune, S. Natural Antimicrobial/Antioxidant Agents in Meat and Poultry Products as well as Fruits and Vegetables: A Review. *Crit. Rev. Food Sci. Nutr.* **2018,** *58* (3), 486–511. https://doi.org/10.1080/10408398.2016.1194256

Balali, G. I.; Yar, D. D.; Afua Dela, V. G.; Adjei-Kusi, P. Microbial Contamination, an Increasing Threat to the Consumption of Fresh Fruits and Vegetables in Today's World. *Int. J. Microbiol.* **2020,** *2020.*

El-Saber Batiha, G.; Hussein, D. E.; Algammal, A. M.; George, T. T.; Jeandet, P.; Al-Snafi, A. E.; … Cruz-Martins, N. Application of Natural Antimicrobials in Food Preservation: Recent Views. *Food Contr.* **2021,** *126.*. https://doi.org/10.1016/j.foodcont.2021.108066

Guo, Q.; Du, G.; Jia, H.; Fan, Q.; Wang, Z.; Gao, Z.; … Yuan, Y. Essential Oils Encapsulated by Biopolymers as Antimicrobials in Fruits and Vegetables: A Review. *Food Biosci.* **2021,** *44* (PA), 101367. https://doi.org/10.1016/j.fbio.2021.101367

Gutiérrez-del-Río, I.; Fernández, J.; Lombó, F. Plant Nutraceuticals as Antimicrobial Agents in Food Preservation: Terpenoids, Polyphenols and Thiols. *Int. J. Antimicrob. Agents* **2018,** *52* (3), 309–315. https://doi.org/10.1016/j.ijantimicag.2018.04.024

Huang, Y.; Wang, Y.; Li, Y.; Luo, C.; Yang, C.; Shi, W.; Li, L. Covalent immobilization of Polypeptides on Polylactic Acid Films and their Application to Fresh Beef Preservation. *J. Agric. Food Chem.* **2020,** *68* (39), 10532–10541.

Kardam, S. K.; Kadam, A. A.; Dutt, D. Retention of Cinnamaldehyde in Poly(vinyl alcohol) Films Intended for Preservation of Faba Beans Through Vapor-Phase Antimicrobial Effect. *Food Packag. Shelf Life* **2021,** *29*, 100704. https://doi.org/10.1016/j.fpsl.2021.100704

Khalid, S.; Yu, L.; Feng, M.; Meng, L.; Bai, Y.; Ali, A.; ... Chen, L. Development and Characterization of Biodegradable Antimicrobial Packaging Films Based on Polycaprolactone, Starch and Pomegranate Rind Hybrids. *Food Packag. Shelf Life* **2018,** *18*, 71–79. https://doi.org/10.1016/j.fpsl.2018.08.008

Meira, S. M. M.; Zehetmeyer, G.; Werner, J. O.; Brandelli, A. A Novel Active Packaging Material Based on Starch-Halloysite Nanocomposites Incorporating Antimicrobial Peptides. *Food Hydrocoll.* **2017,** *63*, 561–570.

Mlalila, N.; Hilonga, A.; Swai, H.; Devlieghere, F.; Ragaert, P. Antimicrobial Packaging Based on Starch, Poly (3-hydroxybutyrate) and Poly (lactic-co-glycolide) Materials and Application Challenges. *Trends Food Sci. Technol.* **2018,** *74*, 1–11.

Pisoschi, A. M.; Pop, A.; Georgescu, C.; Turcuş, V.; Olah, N. K.; Mathe, E. An Overview of Natural Antimicrobials Role in Food. *Eur. J. Med. Chem.* **2018,** *143*, 922–935. https://doi.org/10.1016/j.ejmech.2017.11.095

Prakash, A.; Baskaran, R.; Paramasivam, N.; Vadivel, V. Essential Oil Based Nanoemulsions to Improve the Microbial Quality of Minimally Processed Fruits and Vegetables: A Review. *Food Res. Int.* **2018,** *111*, 509–523.

Salehi, F. Edible Coating of Fruits and Vegetables Using Natural Gums: A Review. *Int. J. Fruit Sci.* **2020,** *20* (sup2), S570–S589.

Uranga, J.; Puertas, A. I.; Etxabide, A.; Dueñas, M. T.; Guerrero, P.; De La Caba, K. Citric Acid-Incorporated Fish Gelatin/Chitosan Composite Films. *Food Hydrocoll.* **2019,** *86*, 95–103.

Wang, D.; Lv, R.; Ma, X.; Zou, M.; Wang, W.; Yan, L.; ... Liu, D. Lysozyme Immobilization on the Calcium Alginate Film Under Sonication: Development of an Antimicrobial Film. *Food Hydrocoll.* **2018,** *83*, 1–8.

Wyrwa, J.; Barska, A. Innovations in the Food Packaging Market: Active Packaging. *Eur. Food Res. Technol.* **2017,** *243* (10), 1681–1692. https://doi.org/10.1007/s00217-017-2878-2

Zhang, M.; Gao, X.; Zhang, H.; Liu, H.; Jin, J.; Yang, W.; Xie, Y. Development and Antilisterial Activity of PE-Based Biological Preservative Films Incorporating Plantaricin BM-1. *FEMS Microbiol. Lett.* **2017,** *364* (7), fnw283.

Zhang, X.; Liu, J.; Yong, H.; Qin, Y.; Liu, J.; Jin, C. Development of Antioxidant and Antimicrobial Packaging Films Based on Chitosan and Mangosteen (Garcinia mangostana L.) Rind Powder. *Int. J. Biol. Macromol.* **2020,** *145*, 1129–1139.

CHAPTER 5

NATURAL OXYGEN SCAVENGERS FOR MAINTAINING THE QUALITY OF FRUITS AND VEGETABLES

DAKURI RAMAKANTH[1], KIRTIRAJ K. GAIKWAD[2], and PRADIP K. MAJI[1]

[1]*Department of Polymer and Process Engineering, Indian Institute of Technology Roorkee, Roorkee, India*

[2]*Department of Paper Technology, Indian Institute of Technology Roorkee, Roorkee, India*

ABSTRACT

Fruits and vegetables are rich sources of nutrients like vitamins, and minerals most of them are highly susceptible to oxidative degradation thereby altering their taste, flavors, odor, texture, color, and additionally prompt microbial growth. The application of oxygen scavengers is a well-known, promising, and result-oriented technique to remove or reduce the concentration of oxygen in the package headspace thereby improving food quality and resulting in their extended shelf life. The focus on the development of natural or bio-based polymers which acts as active material is of prime concern to defend the environment and cut down the generated carbon footprint. This chapter will focus on the present-day literature and developments in natural oxygen-scavenging materials such as films, coatings, and multi-layered structures with novel natural active agents that are employed to uphold the quality and enhance the shelf-life of fruits and vegetables.

Novel Packaging Systems for Fruits and Vegetables. Kirtiraj K. Gaikwad & Suman Singh (Eds.)

5.1 INTRODUCTION

In the current scenario, there is a massive demand for fresh and quality food around the world. The change in consumer behavioral patterns such as food, lifestyle, and growing health concerns accelerates the need for healthy and quality food, despite increased prices. Additionally, consumer acceptance of the use of state-of-the-art technology in food packaging has grown, thus improving the quality of food such as freshness, texture, odor, color, and flavor (Singh et al., 2021).

There are numerous factors like lipid oxidation, aerobic microorganism growth, interactions between food and the packaging, enzymatic browning, presence of oxygen, ethylene, and moisture, which contribute to the spoilage of food. Notably, the presence of residual oxygen inside the packaging creates rancidity in fatty foods, discoloration, and loss of flavor & micronutrients. The reason to contain residual oxygen (RO) inside the package might be due to an inadequate barrier or inefficient vacuum. For foods with low water activity such as nuts, cured meats, breakfast cereals, and smoked fish, vacuum packaging has been widely used to get rid of atmospheric oxygen inside the container. On the other hand, RO can't be removed if the oxygen is permeated through packaging material from the outside.

The oxidation of lipids causes deterioration reactions, which affect the quality, and reduces the shelf-life of the food, which in turn affects human health. Oxygen can react with unsaturated fatty acids rapidly and can deteriorate its quality. Oxygen scavengers are the game changers in the field of food packaging, as their introduction made food packaging more accessible and extended the shelf life up to a certain extent (Ramakanth et al., 2022). Commercially, these come in different forms, such as sachets, labels, caps, and films as a multilayer package. The integration of oxygen-scavenging materials into packages has a significant impact on consumers. Sachets create ambivalence among consumers; thus, incorporating oxygen scavenging or absorbing material directly into the packaging material is gaining importance.

Fruits and vegetables are very essential and extremely perishable foods and comparatively have a shorter shelf life. Fruits and vegetables are living tissues, which undergo continuous changes post-harvest as well. It shows that there will be metabolic reactions that are active in cells. However, fruits and vegetables are living and respiring products even after harvesting. The degradation rate of these foods depends on the rate of respiration. Like humans, fruits and vegetables require oxygen (Makino, 2013). The complete anaerobic condition leads to fermentation and decay of FV which gives off-flavors and odors. The presence of air inside the package contains a high amount of oxygen (21%), which is very huge for storing FV. A very low concentration of oxygen, that is, less than 2% is enough to store the FV for a longer time. Most of the freshly harvested fruits and vegetables comprise water. Due to their extremely perishable property and microbiological instability and the increase in demand, the companies are focusing on processed fruit and vegetable products, and the market for cut fruits and vegetables has been rising. Hence, the usage of oxygen-absorbing materials in the FV packaging has become an essential part in order to protect it from browning and arresting the microbial growth, therefore, maintaining its quality till consumption (Scienchef, 2017).

The global statistics on the processed fruits and vegetable market are estimated at 337 billion USD in 2022 and projected to reach 481.2 billion USD by 2027 at a compounded annual growth rate of 7.4%, which is huge in terms of market size. So, the requirement for keeping it safe and fresh using active packaging technologies such as oxygen scavengers also has huge demand. By 2025, the market demand for oxygen scavengers is projected at USD 3 billion, with a 4.8–5% of CAGR from 2019 to 2025. The demand for oxygen scavengers increases day by day due to the requirement for quality packaged food and rising consumer concern about reducing food wastage. The change in consumer lifestyle, urbanization, and adaptation to ready-to-eat food (RTE), on-the-go (OTG), and the increased demand for advanced packaging is fuelling the growth of the oxygen scavengers' market (Dynamics, 2020; Markets and Markets, 2022).

Most of the oxygen-scavenging packaging materials developed to date are based on conventional polymer materials. The triggering mechanism for initiating oxygen scavenging depends on the type of packaging and scavenging materials used. Some of them are light activated, moisture activated, a few require photosensitizing dye to excite oxygen to its singlet state, and a few are based on transition metal-catalyzed oxidation technique. On the other hand, most commercial oxygen scavengers are based on iron as scavenging material, which has the drawback of being detected in metal detectors and creating ambiguity among consumers as it comes as additional material in the packaging (Sultana et al., 2022).

This has made researchers worldwide think and explore bio and organic materials as oxygen scavengers for food packaging applications to create a sustainable packaging environment.

This book chapter explains the oxygen scavengers, types of oxygen scavengers, and their role in fruit and vegetable packaging. And it discusses the different incorporation technologies and their effects on food shelf life, challenges, recent developments of natural oxygen scavengers along with some regulatory issues, and future scope.

5.2 OXYGEN-SCAVENGING SYSTEMS

Food packaging is referred to as the technique of providing protection against external influencing factors like light, dust, dirt, greases & oils, and providing a barrier against moisture, oxygen, and other gases, chemicals, and microbiological contamination which is also termed passive packaging. On the other hand, smart packaging technology is the fastest-growing packaging technology to safeguard foods but is not only limited to the food packaging sector. It has a wide product portfolio including foods, beverages, pharmaceuticals, and other types of products. The term active packaging refers to "the packaging to which subsidiary constituents or additives have been incorporated purposely in the material of the package or the headspace to improve shelf life and quality of foods packed inside." Even though active packaging is dated back to the twentieth-century beginning, nowadays, active packaging has become more innovative for both researchers and companies. Oxygen scavengers are by far the most successfully commercialized active packaging technology.

5.2.1 OXYGEN SCAVENGERS

Oxygen scavengers are also termed as oxygen absorbers. These are the active materials that can be incorporated or added into the package or its materials in order to remove or eliminate oxygen from the package headspace to preserve the quality and enhance the shelf-life of foods packed inside. The oxygen scavengers can react with oxygen chemically or enzymatically to protect the foods against their oxidative degradation.

The oxygen-scavenging technology depends on the oxidation of active material that is incorporated in or on packaging material/headspace or the combination of active material, photosensitizer, a catalyst, and enzyme, etc. The incorporated active compounds could reduce oxygen levels below 0.01% which is a desirable limit to maintain the quality of food and improve its shelf life. Oxygen scavengers can be selected based on the specific requirement, the oxygen level in the headspace, and the quantity of oxygen that can be transferred from the outer environment to the package along with the nature of food and desired shelf life (Roberta, 2020).

Oxygen scavengers must fulfill certain necessities such as not producing/releasing undesirable gases or products, by-products, or any other toxic substances or foul odor and they should be lightweight, compact, and easy to process. There shouldn't be any leaching/transfer of active material to food, must possess an appropriate oxygen-scavenging rate (OSR) and should be available at a reasonable price, shouldn't be harmful to the human health and the environment as well (Apicella and Incarnato, 2019).

In general, fruits and vegetables are packed in conventional polymeric films such as high-density, low-density polyethylene (PE), linear-low density polyethylene (LLDPE), polypropylene (PP), biaxially oriented polypropylene (BOPP), and polyethylene terephthalate (PET). However, addressing the permeability of gases from the outer environment to the inner side of the package through the packaging materials is a big issue, as these conventional polymeric materials possess low or negligible barriers against gases such as oxygen. The recommended oxygen permeability values for films containing oxygen scavengers should not exceed 20 mL/m^2.d.atm. Without a proper oxygen barrier, the oxygen scavengers can't be used; otherwise, the developed oxygen scavengers can become saturated rapidly and lose their ability to scavenge oxygen inside the package, thereby leading to food deterioration (Busolo and Lagaron, 2012).

The oxygen-scavenging ability can be expressed in the form of oxygen-scavenging capacity (OSC) and rate.

5.2.1.1 OXYGEN-SCAVENGING CAPACITY (OSC)

The OSC can be expressed as the amount of oxygen absorbed per gram of the material used. The capacity of an oxygen-scavenging material can be calculated by using eq 5.1

Oxygen-scavenging capacity (OSC)

$$= \frac{\text{amount of oxygen absorbed}}{\text{oxygen} - \text{scavenging material used}} \sim \frac{[ml]}{[g]} \tag{5.1}$$

5.2.1.2 OXYGEN-SCAVENGING RATE (OSR)

The rate of any oxygen-scavenging material can be expressed as the ratio of the amount of oxygen scavenged per gram of substance used per day. The rate can be calculated by using eq 5.2

Oxygen-scavenging rate (OSR)

$$= \text{Vp} \left[(\%O_2)_i - (\%O_2)_f \right] \frac{1}{m.t} \sim \frac{[ml\ O_2]}{[g]\cdot[day]} \tag{5.2}$$

here, OSR is oxygen-scavenging rate of the material used and is expressed as $\frac{[ml\, O_2]}{[g]\cdot[day]}$, m ~ mass of the oxygen-scavenging material used in (g), Vp ~ volume of the package in (mL), t~ time period expressed in (days) or (hours), $(\%O_2)_i$ ~ initial oxygen content inside the package (%), that is, 20.9% in case of air, $(\%O_2)_f$ ~ final oxygen content inside the package measure at time t (%) (Gaikwad and Lee, 2016).

5.3 ROLE OF OXYGEN SCAVENGERS IN FRUITS AND VEGETABLES

Oxygen scavengers deliberately protect and preserve food products packed inside it thus increasing shelf life. During the process of protecting foods, oxygen scavengers play numerous roles such as simply protecting the foods from oxidation to preventing discoloration due to enzymatic browning. Figure 5.1 illustrates the different roles of oxygen scavengers can play in protecting fruits and vegetables.

5.3.1 PREVENTING OXIDATION

Oxygen scavengers efficiently prevent oxidative deterioration in various kinds of food products such as pigments and flavors in plant and muscle products to arrest discoloration and taste, vitamins to prevent nutritional

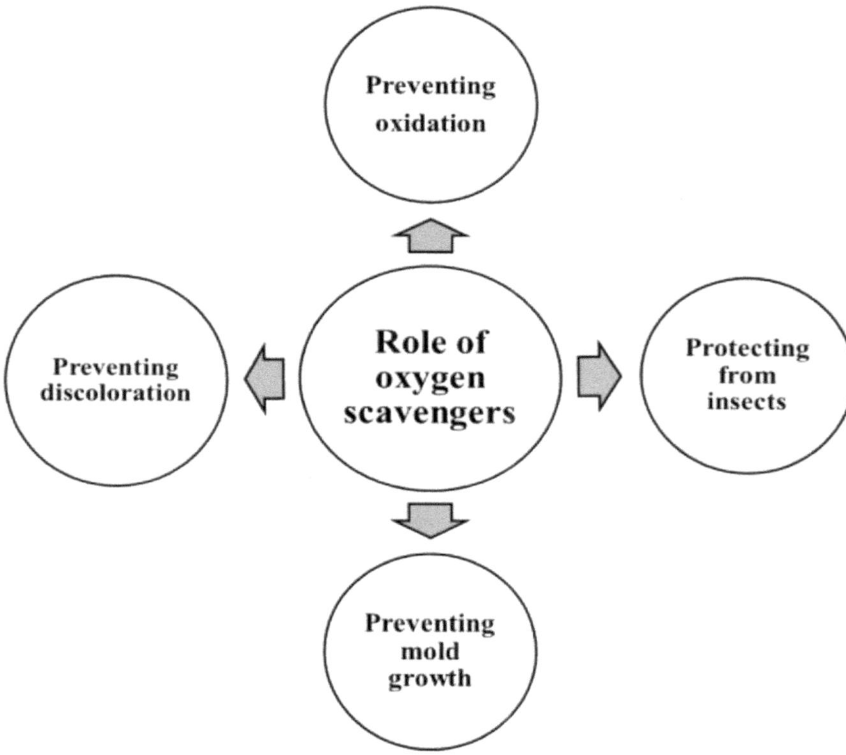

FIGURE 5.1 Representation of various kinds of roles that oxygen scavenger can play in protecting fruits and vegetables.

loss, and fats and oils to prevent rancidity. These are irreversible chemical absorbers that absorb molecular oxygen. These materials are used in controlling or preventing oxygen-induced degradation or damage in packed foods. Oxygen scavengers can mitigate these impacts by eliminating oxygen from a given space, preventing oxygen from entering, or both, resulting in improved product quality and shelf-life. The scavenger function is based on chemical interactions between oxygen and substrates that have been carefully chosen and prepared. A reducing agent (typically the scavenging substrate) chemically binds gaseous or oxygen before reacting with the product. Oxygen scavengers are generally characterized based on their capacity to absorb oxygen, absorption rate, and their mode of activation. Lee et al., 2018 investigated the effect of sulfite-based oxygen scavengers on kimchi quality and shelf-life extension. Kimchi is a mixture of radish, Chinese cabbage, and cucumber along with other spices, as a result, when kimchi is exposed to O_2 on a regular basis, aerobic microorganisms proliferate. Film-forming yeasts might appear as a white spot or thin film on the surface of kimchi. The beauty and value of kimchi are severely harmed as a result of this. Furthermore, the presence of O_2 causes rapid oxidative degradation, which includes softening, taste, discoloration, and nutritional loss. To avoid O_2-mediated kimchi quality degradation, O_2 must be eliminated from kimchi packets. For this, kimchi was packed in sulfite-based oxygen scavengers and stored at 0 and 10°C for a storage period of 12 weeks. The SOS-treated kimchi had considerably lower headspace oxygen inside the packages than the control.

5.3.2 PROTECTING FROM INSECTS

Fruits and vegetables are rich in water content and contain many vitamins, nutrients, carbohydrates, and minerals which are highly susceptible to oxidative degradation. The presence of oxygen by any means can accelerate the growth of microorganisms thereby developing insects inside the package. Earlier, fumigation techniques such as gaseous pesticides (bromides, methyl disulfides) were used to kill or remove insects, and harmful microorganisms. However, it has several adverse effects on food products as their residues can remain on them and they are harmful to human health as well. In order to overcome such issues, researchers have developed oxygen scavengers. Oxygen scavengers can deliberately kill microorganisms, insects, and worms formed/entered inside a package. The use of OSs can effectively kill the insects/aerobic microorganisms by removing/reducing oxygen levels inside the package below 0.01%, which is not sufficient to survive (Cichello, 2015).

5.3.3 PREVENTING MOLD GROWTH

Due to their high-water activity (a_w), fruits and vegetables are highly vulnerable to mold growth and aerobic bacteria in the presence of oxygen. The spoilage and losses due to the mold growth are serious economic concerns in fruits and vegetables packaging industry. It requires at most care to safeguard highly perishable foods. With the help of active oxygen barriers, keeping the oxygen levels inside the package to 0.1% or lower is achievable. However, mycotoxins are very harmful chemicals produced by fungi such as *Aspergillus flavus and Aspergillus parasiticus*. Singh et al., 2019 developed oxygen scavenger based on pyrogallic acid for extending the shelf-life of peeled garlic. Garlic is rich in phenolic compounds and flavonoids; in the presence of oxygen, it undergoes enzymatic/non-enzymatic browning, microbiological deterioration, and surface discoloration which ultimately lead to lower shelf life. The pyrogallic acid-coated sodium carbonate and LDPE films were used to pack peeled garlic cloves and stored at 25 and 5°C for a storage period of 30 days. When compared, oxygen concentration stabilized faster in the OS package at low temperature to the OS package held at 25°C. At both temperatures, it was observed that the package containing OS, weight loss, aerobic microorganisms count, and the fungal decay relative to the control package. The negligible color change values at 5°C show that pyrogallic acid acted as an efficient oxygen-scavenging agent.

5.3.4 PREVENTING DISCOLORATION

The discoloration of food plays a dual role in affecting consumers' decision-making (either positive or negative). For instance, in case of fruits and vegetables, it restricts the consumers from purchasing, thus generating a lot of food wastage; on the other hand, it is essential to impart color and flavor in other foods such as tea, coffee, and chocolate. Fruits and vegetables are extremely vulnerable to discoloration with the presence of oxygen, as they possess enzymes in their tissues. The enzymes come into contact with oxygen in the air when the fruit is sliced or squished, or when the fruit or vegetable begins to break down with age. The fruit will turn brown as a result of this phenomenon. Depending on the mechanism involved during the browning process, it can either be enzymatic browning or non-enzymatic browning. In the latter process, the food doesn't require any enzyme to initiate the change; rather it involves a chemical reaction of individual molecules or elements present in the food itself. Some examples of such kinds of browning include the Maillard process, caramelization, and ascorbic acid oxidation reactions that occur in food as a combined one rather than as independent ones owing to the complexity in food ingredients. The enzymatic browning reaction, unlike non-enzymatic browning, uses a most commonly found enzyme named polyphenol oxidase to carry out the color change. Enzymatic browning is abundantly observed in fruits and vegetables while harvesting, storage & transportation, and during processing. This adversely affects the organoleptic properties and nutritional value of the food. Physical tissue damage in FVs can be caused by mechanical and physical stresses during processing. These stresses include cutting, dicing, slicing, peeling, shredding, and extreme temperature variations during storage. An external damage to the tissue exposes the PPOs and other phenolic compounds to oxygen which in turn leads to their oxidation into quinones. Following that, these quinones and their derivatives are polymerized by a series of processes, resulting in melanin, a rather insoluble brown pigment. The enzymatic activity of PPOs determines the rate of enzymatic browning (Moon et al., 2020).

The increase in demand for ready-to-eat (RTE) and ready-to-cook (RTC) foods is due to the change in consumer attitudes and lifestyles. Providing fresh and quality RTE or RTC fruits and vegetables is the responsibility of the food processing industry. Researchers are working on cut fruits and vegetables, how they deteriorate faster and what is the reason behind their deterioration. Cut fruits and vegetables are more prone to oxidative degradation via browning, faster mold growth, and flavor loss due to their high respiration rate. What happens when a fruit or a vegetable is cut into pieces? As we know, breaking or size reduction creates bigger surface areas which mean increased surface area. The amount of oxygen in headspace has a lot more surface area to expose and reacts with it to impart browning and starts decaying the quality of fruits and vegetables (Scienchef, 2017).

5.4 TYPES OF OXYGEN-SCAVENGING SYSTEMS

There are numerous oxygen-scavenging systems available in the present market. They are divided into metallic and non-metallic oxygen-scavenging categories. Additionally, there are new types of oxygen scavengers developed in the past decade, emphasizing natural materials as oxygen scavengers. Different types of oxygen scavengers and their compositions, activation mechanisms, etc., are described below.

5.4.1 METALLIC OXYGEN SCAVENGERS

These are widely used and the largest market-occupied type of oxygen scavengers. And the most popularly commercialized oxygen scavengers are iron-based. Inspired by nature, that is, rusting of iron in the presence of moisture motivated the researchers to develop iron-based oxygen scavengers. Being highly efficient, low price, availability, and high scavenging rates favored their commercial success. Iron-based oxygen scavengers work on the basic principle, that is, oxidation of iron in the presence of external influencing/accelerating agents like moisture or Lewis's acids ($FeCl_3$ and $AlCl_3$). The basic reactions involved in it are

$$Fe \rightarrow Fe^{2+} + 2e^-$$

$$\tfrac{1}{2} O_2 + H_2O + 2e^- \rightarrow 2OH^-$$

$$Fe^{2+} + 2OH^- \rightarrow Fe(OH)_2$$

$$Fe(OH)_2 + \tfrac{1}{4} O_2 + \tfrac{1}{2} H_2O \rightarrow Fe(OH)_3$$

Iron-based oxygen scavengers were used to come in sachet form. Wołosiak-Hnat et al., 2019 investigated the iron compound with an ascorbic acid-based oxygen-scavenging system coated onto an LDPE/PET laminate packaging system. It was observed that the oxygen content was decreased from 20 to 1% within a storage period of 15 days and suggested these types of oxygen scavengers can be best applicable for high water activity food products such as fruits and vegetables. The most important property of any oxygen scavenger is reaction rate. One of the most influencing factors to improve the rate of reaction is the reduced size of the particle. A 10-times reduction in particle size increases the area of the same particles by 10 folds. Based on this concept, Foltynowicz et al., 2017 developed nanoscale oxygen scavengers consisting of iron particles blended in a silicone base matrix had exhibited excellent oxygen-scavenging properties even in both dry and wet environments. Excellent dispersion and high availability of active surface area, both improved the scavenging capacity.

Metallic oxygen scavengers are not only limited to iron-based, other metals such as platinum, palladium, and zinc have gained importance in the active packaging sector due to their limited toxicity and high catalyzing efficiency. These materials are either impregnated or blended or anchored to polymeric films in order to safeguard the foods from oxidative degradation. Different types of oxygen scavengers along with some examples are shown in Figure 5.2. Even though metallic oxygen scavengers have high efficiency and OSR, and are the most commercialized products, still they have a few drawbacks such as metal contamination with foods, and line disturbances due to metal detector issues thereby creating wastage, premature activation due to temperature reliance.

5.4.2 ORGANIC OXYGEN SCAVENGERS

Researchers and industries have shifted their focus onto organic scavengers due to the limitations in using metallic scavengers. Oxidizable organic substrates or incorporation of oxidizable functional groups in the polymer or oxygen-absorbing polymers in which the backbone or side chain reacts with oxygen. Organic scavengers are versatile in nature due to their high dispersion in polymer matrices; they produce excellent transparency compared to inorganic dispersions. Ascorbic acid (AA)-based scavengers work on ascorbate oxidation and generate dehydroascorbic acid (DHAA). The scavenging reaction alone is slow, to accelerate it light or a metal catalyst is needed. The absorption capacity here depends on the amount of ascorbic acid used, for example, to reduce 1 mole of O_2, it was reported that 2 moles of ascorbic acid are required. Generally, ascorbic acid-based scavengers are available in the form of films. AA-based scavengers with transition metal catalysts require moisture to activate. These kinds of scavenging materials are applicable to high moisture-containing foods such as cucumber, melons, strawberries, and pineapple fruits (Dey and Neogi, 2019). Singh et al., 2020 developed gallic acid (GA)-based oxygen-scavenging films containing chitosan (CH) as base matrix and sodium carbonate (SC) as an activator. It was reported that the developed CH/GA/SC films not only exhibited oxygen-scavenging properties but also provided an excellent barrier to oxygen and water vapor. The oxygen transmission rate (OTR) was observed to be 4.10 ± 1.07 $cm^3/\mu m/m^2$. day. kpa along with an OS capacity and rate of 19.55 mL O_2/g and 2.66 mL O_2/g. day, respectively, at 23 ± 2°C. The presence of moisture catalyzed the reaction, making it suitable for fruits and vegetables packaging. Other examples for organic oxygen scavengers are ascorbic acid/whey protein, isolate films, pyrogallic acid/sodium carbonate/LDPE films, α-tocopherol/ PLA microparticles, and other oxidizable polyphenolic compounds such as catechol and resorcinol. Edible oxygen-scavenging films are the other emerging technologies under the organic scavengers' category which addresses issues like accidental ingestion, leaching/migration of active materials, and recyclability. However, organic scavengers have certain drawbacks like limited OSC and highly expensive compared to metallic oxygen scavengers. Figure 5.2 explains the different types of oxygen scavengers available with few examples under each category.

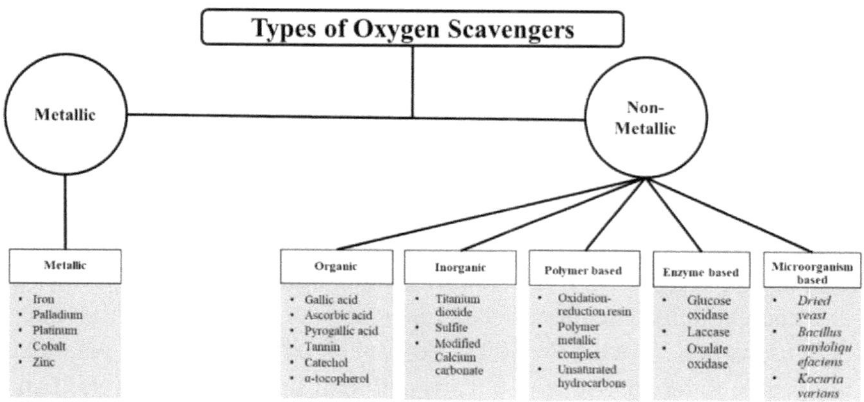

FIGURE 5.2 Illustration of types of oxygen scavengers broadly classified into metallic and non-metallic oxygen scavengers.

5.4.3 INORGANIC OXYGEN SCAVENGERS

Inorganic oxygen scavengers are the materials that scavenge oxygen and have also been included in the bulk of the packaging material. This technology is popularly known as the "active oxygen barrier" and it provides an even scavenging effect on the entire package, as well as intercepts and scavenge oxygen that permeates through the package's walls, resulting in the package's lowest oxygen level. Inorganic powders and/or salts have been used in the package to provide active oxygen barriers. Incorporating such powders and/or salts, on the other hand, can reduce the transparency and mechanical qualities of the packing material, as well as complicate manufacturing, particularly when thin films are required. Furthermore, foods packed inside may absorb these chemicals and by-products, causing the food product to fail to fulfill legislative criteria for human consumption.

5.4.4 POLYMER-BASED OXYGEN SCAVENGERS

The majority of polymeric films manufactured are used in the packaging industry. Applications needing a low quantity of oxygen in a package are the major challenges with those films. This sort of packaging is most commonly used for foods like fruits and vegetables, which are frequently wrapped in a low-oxygen environment. For example, by minimizing the exposure of oxygen to oxygen-sensitive foods in a packaging system, their quality can be preserved and spoilage can be delayed. Furthermore, such packaging keeps the product in stock for a longer period of time, decreasing waste expenses. Dealing with barrier packing and low oxygen levels is an art. This involves the use of oxygen-scavenging films along with inserts to retrieve the majority of any remnant oxygen post-packaging. Compounds that scavenge oxygen for usage in plastic films are fairly well known. Recently, an oxygen-scavenging film containing tea polyphenols-β-cyclodextrin embedded complex was developed by coextrusion with LDPE. The active film was activated by moisture through β- cyclodextrin and had exhibited excellent scavenging capacity at extremely high humid conditions like above 84% RH. The developed films can be potentially applicable to oxygen-sensitive fruits and their processed products (Pan et al., 2020).

5.4.5 ENZYME-BASED OXYGEN SCAVENGERS

Enzyme-based oxygen scavengers are another type of natural scavengers developed to safeguard moist foods including fruits and vegetables. The grouping of glucose oxidase (GOx) and catalase (Cat) is a typical example of enzyme-based scavengers. In this, either glucose should be abundant in the food or can be added to the product. The glucose upon reaction with oxygen in the presence of moisture forms gluconic acid and hydrogen peroxides. To remove or convert objectionable end products, that is, H_2O_2–H_2O and O_2, catalase is required. It breaks down the hydrogen peroxides which are an undesirable by-product of the oxygen-scavenging reaction. The basic reaction mechanism can be as follows

$$2\text{glucose} + 2O_2 + H_2O \rightarrow 2\text{gluconic acid} + 2\,H_2O_2$$

$$2H_2O_2 \rightarrow 2H_2O + O_2 \text{ (in the presence of catalase)}$$

Changes in pH, a_w, temperature, and a variety of other variables can affect coupled enzyme systems. Furthermore, they necessitate the addition of water and hence can be used efficiently for fruit and vegetable packaging. The removal of O_2 from processed fruits and vegetables products such as purees, and ketchups is one application for glucose oxidase.

Different processes, such as adsorption and encapsulation, are used to immobilize the substance. Other enzymes, in addition to glucose oxidase, that have the ability to scavenge O_2 are oxalate oxidase (OOx) and catalase (Cat) as efficient scavengers owing to their scavenging property as well as the ability to produce carbon dioxide. Another example of enzymatic scavengers is laccase. Laccases are copper-containing oxy-reductases found in plants and fungi such as *Trametes Versicolor and Myceliophthora thermophila*. Since the enzymatic activity is dependent on relative humidity, such kind of oxygen scavengers can essentially be applied to fruits and vegetables with high moisture content such as peaches, oranges, and celery.

5.4.6 MICRO-ORGANISMS-BASED OXYGEN SCAVENGERS

Due to its environmentally beneficial and low-cost nature, oxygen-absorbing systems based on entrapped microorganisms in a solid matrix are an intriguing technique. Dried yeast immobilized on solid wax was used to scavenge oxygen in a hermetically sealed package. The inclusion of immobilized yeast onto the bottle closure liner or coating application or contained in a pouch within the container. It is reported that when yeast is moistened, it activates and respires, using oxygen while producing carbon dioxide and alcohol. Any carbon dioxide and alcohol produced in the oxidation of yeast can control the product respiration due to the alterations in O_2 and CO_2 concentrations inside the package without generating noticeable alterations in the product. Other examples of aerobic microorganisms utilized for oxygen scavenging include *Bacillus amyloliquefaciens, Kocuria varians*, and *Pichia subpelliculosa*. It was reported that *Kocuria varians* and *Pichia subpelliculosa* microorganisms were entrapped onto the hydroxyethyl cellulose (HEC) and polyvinyl alcohol (PVA) films and found that they could be stored for 20 days without affecting the viability of the microorganisms. This kind of scavenging mechanism would be highly beneficial for foods with high moisture content like fruits and vegetables.

For modern food tech systems, using natural or bio-oxygen scavengers that work on microbial activity can be an intriguing concept, especially encapsulating them in a polymer matrix to avoid direct contact with food. They are successful in preserving foods, safe to use, acceptable to consumers, economical, and environmentally benign. In reality, the prospect of developing a new packaging with a variety of desirable properties, without neglecting emerging customer demands for moderately processed convenience foods with preserved freshness is quite attractive. For fruits and vegetables, freshness and quality are the key parameters; oxygen-scavenging packaging in combination with other active packaging techniques might resolve the issue.

5.5 INCORPORATION TECHNOLOGIES OF OXYGEN-SCAVENGING MATERIALS

The success of any oxygen-scavenging material depends on its scavenging rate, availability, and processability. The incorporation of active materials into/onto the packaging system plays a crucial role in exhibiting its property. The choice of food application depends on the type of incorporation technique used. Earlier, oxygen scavenger sachets were the most utilized and commercialized products, over time and due to their limitations, new inventions were made to incorporate them into the packaging system. Different types of incorporation techniques are described as follows.

5.5.1 SACHETS

Sachets are the small packets or porous bags with a material inside them and are intended to deliberately interact with the surrounding atmosphere. Sachets are the most common type of oxygen scavengers used for the protection of solid food products. These come in various sizes according to package size. Nowadays tailor-made sachets are gaining importance to cater to the needs of the food packaging industry due to the change in demand for different package sizes and varieties. In general, sachets can be made up of different materials starting from aluminum, paper, polymers, and cloth, etc. When it comes to oxygen scavenger sachets, mostly polymer or paper-based are popular in use. Paper-based sachets can't be used. Sachets contain iron-based oxygen-scavenging system inside it and are placed inside the package. The porous outer layer allows oxygen present in the headspace to pass through so that it comes in contact with iron-based scavenging system and is trapped inside it. A typical form of oxygen scavenger sachet can be seen in Figure 5.3a. Usually, iron-based scavengers are activated through moisture and the minimum requirement for the scavenging process to begin is 65% RH. For moisture levels below 65% RH, there must be an additional compound such as sodium chloride or other salts should be added which acts as a catalyst or activator, and accelerates scavenging reaction at low humidity levels. The requirement of moisture activation in the case of an iron-based oxygen-scavenging system makes it suitable for fruits and vegetable packaging applications. One such example is shown in Figure 5.3, where an iron-based oxygen-scavenging sachet containing metal salt as a catalyst is placed inside date fruit packaging. Even though sachets are widely used and possess higher OSCs, there are certain limitations such as accidental ingestion and migration of the materials to food as they are in direct contact, and these types of sachets can't be used in liquid products.

5.5.2 FILMS

Oxygen-scavenging films have gained popularity over the past few decades due to their ease of processability, light weight, high barrier, and wide range of product applications. When developing an oxygen-scavenging film, the mode of activation for oxygen-scavenging materials must be taken into account based on the desired product application. There are various techniques available to incorporate oxygen-scavenging materials onto the films by coating, multilayer structure by co-extrusion or co-extrusion lamination, or absorption onto base material.

5.5.2.1 COATINGS

Coatings are thin layers of substances applied or deposited on the surface of a substrate. Coatings may be functional or decorative purpose and are done to improve or alter the surface properties and to provide a barrier against the exposing environment. Oxygen-scavenging materials are actively coated on to the polymeric or paper substrates as inner layer so that it can be in direct contact with head space gases to readily oxidize them (Figure 5.3b). However, issues arise when oxygen-scavenging films come in contact with oxygen present in the atmosphere prior to use. In order to overcome this, an activation mechanism such as through moisture, light illumination, catalysts, or reagents on filling lines can be added. Recently, Cherpinski et al., 2018 and co-workers developed palladium nano particles (PdNP)-coated poly (3-hydroxybutyrate) (PHB)-based oxygen scavengers. PdNPs were coated on PHB matrix by using electrospinning technique. The developed OS film showed notable OSC at 100% RH. These can be best suitable for high moisture-containing fruits like water melon and honeydew melon.

5.5.2.2 BLEND FILMS

Blend films are the kind of oxygen-scavenging films containing active compounds and base matrix together. The oxygen-scavenging materials are dispersed into the polymer matrix and developed films by casting, extrusion, etc. The active materials are dispersed completely into the bulk of the polymer matrix so that they can react with oxygen. Mahieu et al., 2015 investigated ascorbic acid (AA) and iron-based thermoplastic starch blend extruded films for oxygen-scavenging activity. The resultant films exhibited excellent oxygen-scavenging activity around 80% RH upon moisture activation. The prepared films can be applied to high-moisture foods such as fresh fruits and vegetables. Although effective, these coatings can lend an unpleasant flavor to foods, especially when iron-based oxygen scavengers are added. One of the major drawbacks of this type of packaging is that heat-sensitive compounds can't be used since they are rendered inactive during the packaging process.

5.5.2.3 MULTILAYER SCAVENGING FILMS

Multilayer scavenging films are commonly used oxygen-scavenging packaging materials due to their lite weight, a high barrier against WVTR and OTR, and wide range of applicability. The oxygen-scavenging active layer is sandwiched in between a barrier layer and an inert porous layer. Even though multilayer films are effective, the main drawback is the high cost. Recently, Pant et al., 2017 and co-workers have developed bio-based multi-layered oxygen-scavenging packaging based on gallic acid. Gallic acid along with sodium carbonate was mixed together and cast extruded with other layers together consisting of BioPE as the inner layer followed by bioPE + oxygen-scavenging layer + adhesive layer + biodegradable PLA. The maximum capacity of oxygen absorbed was 447 mg O_2/g of GA at 21°C and 100% RH. And it was reported that, the developed gallic acid-based multi-layered oxygen-scavenging films can be applicable to high-water activity foods including fruits and vegetables. A high barrier multi-layered oxygen-scavenging film is shown in Figure 5.3d.

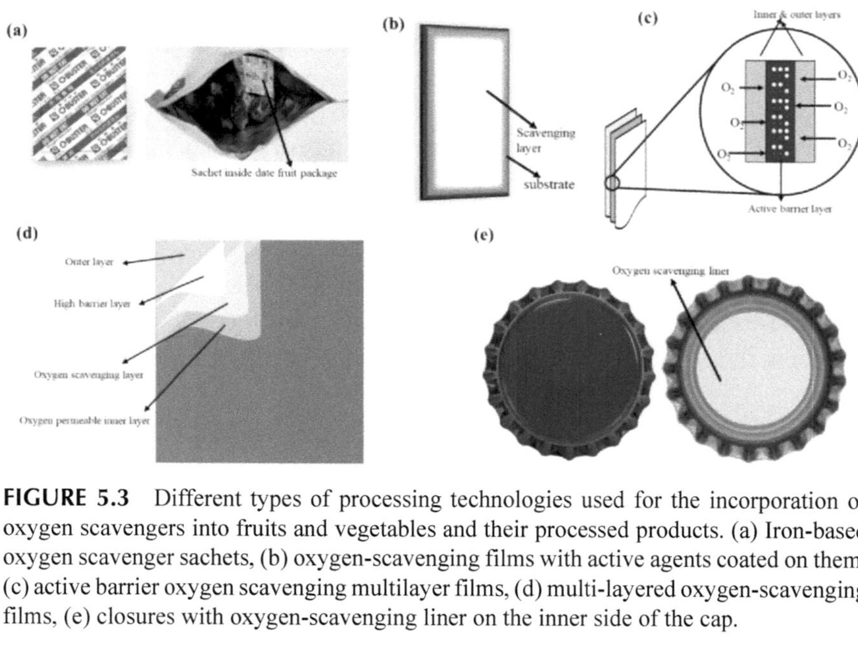

FIGURE 5.3 Different types of processing technologies used for the incorporation of oxygen scavengers into fruits and vegetables and their processed products. (a) Iron-based oxygen scavenger sachets, (b) oxygen-scavenging films with active agents coated on them, (c) active barrier oxygen scavenging multilayer films, (d) multi-layered oxygen-scavenging films, (e) closures with oxygen-scavenging liner on the inner side of the cap.

5.5.2.4 IMMOBILIZATION

Immobilization is another type of incorporation technique for oxygen scavenger films. As the name indicates, it is a process of making active substances immobile on an inert or insoluble substrate. The most commonly used biodegradable and non-biodegradable polymers are PE, PP, PET, PVA, PLA, and PCL, and are all having commercial importance. The precursor to immobilizing active substances on substrates is, that they should be inert and pre-treated to develop reactive sites to the substrate surface to attach active agents. The common surface activation techniques include UV irradiation. Immobilization is the most commonly used technique for incorporating oxygen-scavenging enzymes onto the surface of the polymer used as packaging material. The commercialization of this kind of technique depends on the processability of active enzymes and their retention on the substrate surface, and the exposure to the conditions like temperature, pH, and relative humidity inside the package. Second, the process of immobilizing enzymes onto polymer substrate must be scaled up to get large outputs, via roll-to-roll pressing. Other techniques such as making enzymes soluble in different organic solvents are applied via coating, spray coating, and printing techniques. These are economical and give high throughput as well. Johansson et al., 2012 investigated the oxygen-scavenging activity of enzyme laccase. Laccases are copper-based oxidoreductases obtained from different plants and fungi. The films were prepared by including laccase and lignosulfonates in coating colors and embedded in latex or starch-based films. The developed films started scavenging activity after 3 days due to the insufficient humidity required for its activation. However, there was no scavenging activity reported below 84% RH. So, the developed laccase-based active films can be best suitable for foods that possess high water activity, that is, fruits such as apples and oranges.

5.5.3 CLOSURES

Closures are the critical parts of the packaging system without which the concept of packaging is unimaginable. They simply contain the products inside the package and provide a barrier against moisture, oxygen, dirt, and other gases and chemicals, etc. Nowadays, a closure not only contains the product, but can act as a tamper-proof agent, can avoid pilfering, and also interacts with packages to that of the consumer. The inclusion/incorporation of active agents has opened tremendous opportunities for closures in a variety of food products. When it comes to oxygen-scavenging packaging, the utilization of closures as active agents was dated back to the late 90s. Incorporation of enzymes onto the liners of bottle closures and activating them with moisture has gained popularity (Roberta, 2020). Caps and closures are mostly applicable to liquid-based packaging materials and are very limited to fruits and vegetables. Instead, active closures are popular in processed fruits and vegetable products such as ketchups, purees, and juices. A typical structure of closures with oxygen-scavenging liners is shown in Figure 5.3e.

5.6 EFFECT OF OXYGEN SCAVENGERS ON SHELF LIFE AND QUALITY OF FRUITS AND VEGETABLES

Owing to the high perishability of fruits and vegetables, it is essential to use active packaging techniques such as oxygen scavengers, ethylene absorbers, moisture absorbers, CO_2 emitters, and absorbers. Several parameters such as oxygen, ethylene, moisture, other gases, and light affect the quality of fruits and vegetables. Among all, the presence of oxygen along with moisture critically affects the quality and shelf-life of fresh produce via vitamin C loss, accelerated respiration of fresh FV, growth of aerobic microorganisms, and enzymatic browning and discoloration of fresh and processed FVs. To maintain the quality and to improve the shelf life, several types of oxygen-scavenging materials were developed and applied in the past and still, researchers are focusing on the development of natural and biodegradable materials. Some of the developed oxygen-scavenging materials are discussed here.

Niazmand and Yeganehzad, 2020 evaluated the shelf-life extension and quality of fresh barberry. A commercial oxygen scavenger with a capacity up to 3000 mL has been placed inside the package made up of LDPE/PET films. The barberries packed with OS in LDPE/PET stored at 4°C had no effect on sensory properties and minimal effects on chemical and texture after 4 weeks. Aday and Caner, 2013 demonstrated the usage of commercial oxygen scavengers (ATCO® -100) which are able to scavenge up to 100 mL oxygen inside the package on fresh strawberry fruits and was reported that, the shelf life has been improved to 4 weeks. The quality of strawberries was analyzed and shown as low mold growth, minimal loss of total soluble solids, and sensory properties were retained and color was observed to be similar. A summary of oxygen scavengers developed by several researchers and their effects on different fruits and vegetables are presented in Table 5.1.

5.7 CHALLENGES ASSOCIATED WITH THE OXYGEN SCAVENGERS FOR FRUITS AND VEGETABLES PACKAGING

Even though the use of oxygen scavengers in fruits and vegetables packaging system has numerous advantages, there are few challenges associated with them via material or the processing technology used.

TABLE 5.1 Different Types of Oxygen-Scavenging Materials Incorporated into Different Packaging Systems and their Applications on Various Kinds of Fruits and Vegetables.

Active material	Packaging system	Oxygen-scavenging capacity/content/rate	Food application	References
Copolyester-based polymeric oxygen scavenger	Polyethylene Terephthalate/ Aluminum/ Polyethylene Terephthalate	0.0086 mL O_2/μm, 23 days	Mango fruit	(Leneveu-Jenvrin et al., 2021)
Ascorbic acid-based reactive oxygen scavenger	Dip coating	--	Longan Fruit	(Liu et al., 2021)
Iron nanoparticles	Montmorillonite	0.20 mL O_2/g	High moisture containing foods	(Kombaya-Touckia-Linin et al., 2019)
Gallic acid and sodium carbonate	Chitosan film	19.55 mL O_2/g	High moisture foods	(Singh et al., 2020)
Ascorbic acid and iron compounds	Low density polyethylene	18.3–1% in 15 days	Foods with high water activity	(Wołosiak-Hnat et al., 2019)
Activated carbon and sodium L-ascorbate	Polypropylene	39.1 mL O_2/g, 4 days at 25°C	Foods with high water activity	(Lee et al., 2018)
Gallic acid with Na_2CO_3	Bio polyethylene/polylactic acid	208.6 mL O_2/g, 25 days at 21°C	Food products with water activity >0.86	(Pant et al., 2017)
Nano-TiO_2	Low-density polyethylene	20.9–5%, 8 days	Strawberries	(Dongdong et al., 2016)
Nano scale iron powder	Silicon matrix	22.4 mL O_2/ g	High humidity food	(Foltynowicz, 2017)
Ascorbic acid, Iron powder/ Copper chloride	Thermoplastic starch	20.9–1%, 15 days	High humidity food	(Mahieu et al., 2015)

The application of developed or existing oxygen-scavenging materials on fruits and vegetables packaging has so many hurdles. Either the developed materials can't be incorporated into the FV packaging system such as closures, sachets, or the active materials are not compatible with FV packed inside. The majority of the challenges were posed during processing stages, few of them are discussed below.

First and most important issue is the **retention of oxygen-scavenging activity** and stability. Oxygen scavengers have received huge demand in the past decade and it is continuing due to the versatility and functionality that the OS provides to safeguard foods packed in it. However, the success of OS technology depends on its OSC retention. As most of the OS lose their capacity and efficiency during the processing due to exposure to abnormal temperatures, pressure, and other mechanical forces while making films, and the interactions between adhesive layers, barrier layers when using in multi-layered films. So, the stability evaluation in various storage conditions prior to commercialization is the best way to understand newly developed OS performance (Gaikwad et al., 2018).

Iron-based OS sachets are the most successfully commercialized techniques and are applied to different product portfolios. However, the limitations aroused due to their form of incorporation. As it comes in different packaging form and must be placed inside a package headspace/on the food products, the outer material selection is an important point to be considered while developing it. It creates confusion to the consumers, thereby chances of accidental ingestion. The physical conditions such as mechanical hazards during transportation and storage can rupture or burst the OS sachets placed inside the package. To avoid this kind of contamination, the packaging material of the sachets must be tested and evaluated to sustain physical hazards. In order to aware/warn consumers, the sachets were printed with a warning sign and the manufacturing company and product details. The migration of printing inks to the foods must be evaluated prior to utilization. However, due to their adverse effects and the limited product portfolio of sachets, alternative OS technologies are emerging with much focus on natural and biodegradable materials (Kamaruddin et al., 2021).

Other considerable factors that affect the quality and efficiency of oxygen scavengers in food packaging applications include film properties, different food processing methods, migration of the active compounds into the foods packed, and focus on how to control the by-products generated during scavenging reactions.

Another important factor that shouldn't be ignored while developing any OS is its **saturation level.** Most of the oxygen scavengers reach their saturation level at one point, the majority of the time due to either its limited capacity or the permeation of oxygen from the outer environment to the inner side of the package. In such conditions, the oxygen saturates the headspace thereby oxygen scavenger as well. To overcome these issues, the focus should be shifted to developing slow scavenging films so that shelf life can be extended, use of active barrier scavenging agents so that OS layers can't allow oxygen inside, and at the same time, it scavenges oxygen present in the headspace, and the other factors like hermetic sealing and section of appropriate OS size according to the package volume (Gaikwad et al., 2018).

5.8 REGULATORY CONCERNS AND FUTURE SCOPE

The emergence of active packaging technologies and their rapid commercialization challenges the existing regulatory framework. Improvements or modifications have to be done with respect to the usage of new-age active packaging technologies, products to make sure about the stability, safety, and quality of foods in which the concerned active packaging is used. Continuation of regulations developed for traditional packaging would not suffice for the modern cutting-edge technologies. The technical amendments, environmental impacts and safety of consumers, migration limits, etc., are the factors to be considered for regulatory upgradation. In the case of oxygen scavengers, numerous commercial vendors make oxygen absorbers, and investigations have shown that varied degrees of water activity or activation mechanism type improve the shelf life of packed foods significantly. Apart from the USA, Europe, and Japan, the countries with the highest manufacturing standards for the production of OS, OxySorbTM from Australia falls into this category. The introduction of different types of new programs such as generally recognized as safe (GRAS), food additive petition (FAP), and food contact substances (FCS), provides authorization and can monitor the developments. Oxygen scavengers must comply with food safety regulations such as the US FDA document MAPP 5015.5, "plastic materials for food contact use" by 2016 guidance in continuation to the amendment established by FSANZ in 1999, that is, Australian Standard 2070–1999 (Nielsen and Skaggs, 2017), European Commission Directive 2002/72/EC, and HACCP standards for food contact objects. Furthermore, third-party laboratories accredited by the International Laboratory Accreditation Co-operative (ILAC) are used to conduct the shelf-life trials to find the stability of the oxygen absorbers. The advantages of oxygen absorbers are attributed to their proper application (Dainelli, 2015; Poças and Franz, 2018).

The following are some of the advantages of oxygen absorbers:

- Browning and darkening effect on FV are postponed
- Inhibiting the oxidation of fat-soluble and water-soluble vitamins, such as those found in tomatoes
- Preventing berry oxidation and condensation
- Reducing the requirement of a few additives/preservatives (such as calcium propionate "E282"), which have been related to behavioral changes in children

5.8.1 FUTURE SCOPE

The future scope of the natural oxygen scavengers depends on the acceptance of cutting-edge technologies and innovative packaging concepts by consumers. Active packaging technologies were well established and commercialized since the late '90s, still, researchers are tirelessly working on new developments for catering to the needs of rapidly changing consumer lifestyle, urbanization, and globalization. The usage of traditional oxygen-scavenging materials in different forms such as sachets, films, and closures has become limited, in fact, the modern consumer market demands attractive, affordable, innovative, and multifunctional such as a combination of active and smart packaging systems. Nevertheless, food packaging scientists and young researchers are focusing on the development of natural oxygen scavengers, to make the planet more sustainable and eco-friendlier. The recent publication trends in natural oxygen-scavenging packaging materials shown in Figure 5.4, anticipates that there is a huge demand for such products in the near future.

5.9 CONCLUSION

Oxygen scavengers/absorbers are the most commercially successful active packaging technologies to date in the protection and preservation of foods. The use of the most promising active packaging technology in increasing the quality and shelf-life of fruits and vegetables has been systematically discussed in this chapter. The advantages of new-age oxygen scavengers over conventional scavengers, their drawbacks, and challenges were highlighted. Research related to natural oxygen scavengers has increased in the past decade and is yet to be commercialized and proven. However, the focus has been shifting to the development of natural/organic-based oxygen scavengers for preserving highly perishable fruits and vegetables.

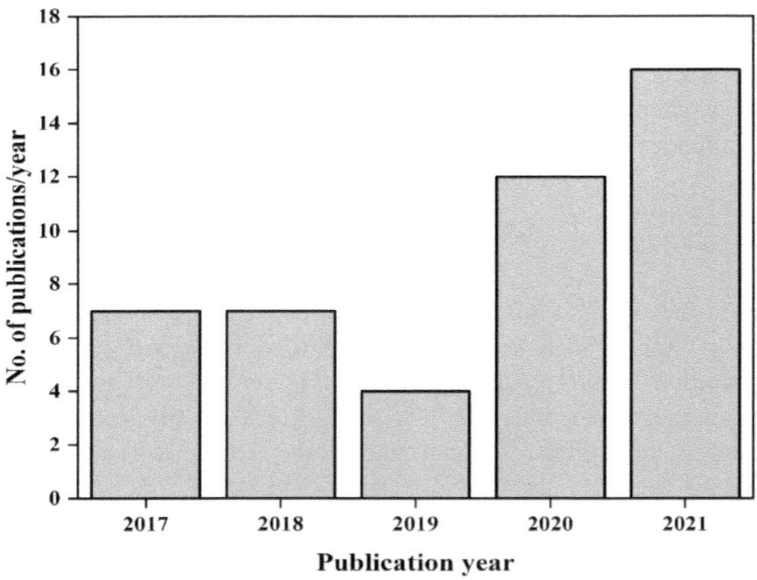

FIGURE 5.4 Recent publication trend in natural oxygen-scavenging packaging materials for food packaging applications from 2017 to 2021. Data were collected from the Web of Science by entering the keyword "natural oxygen scavenging food packaging."

KEYWORDS

- **natural oxygen scavengers**
- **bio-based oxygen scavengers**
- **biopolymers**
- **active barrier**
- **oxygen-scavenging films**

REFERENCES

Aday, M. S.; Caner, C. The Shelf Life Extension of Fresh Strawberries Using an Oxygen Absorber in the Biobased Package. *LWT Food Sci. Technol.* **2013,** *52* (2), 102–109. https://doi.org/10.1016/J.LWT.2012.06.006

Apicella, A.; Incarnato, L. Oxygen Scavengers in Food Packaging. In *Innovative Food Processing Technologies: A Comprehensive Review*, 2019; pp 487–506. https://doi.org/10.1016/B978-0-12-815781-7.22942-9

Busolo, M. A.; Lagaron, J. M. Oxygen Scavenging Polyolefin Nanocomposite Films Containing an Iron Modified Kaolinite of Interest in Active Food Packaging Applications. In *Innovative Food Science and Emerging Technologies*, 2012; vol 16, pp 211–217). https://doi.org/10.1016/j.ifset.2012.06.008

Cichello, S. A. Oxygen Absorbers in Food Preservation: A Review. *J. Food Sci. Technol.* **2015,** *52* (4), 1889. https://doi.org/10.1007/S13197-014-1265-2

Cherpinski, A.; Gozutok, M.; Sasmazel, H. T.; Torres-Giner, S.; Lagaron, J. M. Electrospun Oxygen Scavenging Films of Poly(3-hydroxybutyrate) Containing Palladium Nanoparticles for Active Packaging Applications. *Nanomaterials* **2018,** *8* (7), 469. https://doi.org/10.3390/NANO8070469

Dainelli, D. Global Legislation for Active and Intelligent Packaging Materials. In *Global Legislation for Food Contact Materials*, 2015; pp 184–199. https://doi.org/10.1016/B978-1-78242-014-9.00008-5

Dey, A.; Neogi, S. Oxygen Scavengers for Food Packaging Applications: A Review. *Trends Food Sci. Technol.* **2019,** *90*, 26–34. https://doi.org/10.1016/J.TIFS.2019.05.013

Dongdong, L.; Qingyang, Y.; Jiang, L.; Luo, Z. Effects of Nano-TiO 2 -LDPE Packaging on Postharvest Quality and Antioxidant Capacity of Strawberry. *J. Sci. Food Agric.* **2016,** *97*, 1116–1123.

Dynamics, M. *Oxygen Scavengers Market,* (2020; pp 1–10.

Foltynowicz, Z. Nanoiron-Based Composite Oxygen Scavengers for Food Packaging. In *Composites Materials for Food Packaging*, 2017; pp 209–234. https://doi.org/10.1002/9781119160243.ch6

Foltynowicz, Z.; Bardenshtein, A.; Sängerlaub, S.; Antvorskov, H.; Kozak, W. Nanoscale, Zero Valent Iron Particles for Application as Oxygen Scavenger in food packaging. *Food Packaging and Shelf Life*, **2017,** *11*, 74–83. https://doi.org/10.1016/J.FPSL.2017.01.003

Gaikwad, K. K.; Lee, Y. S. Novel Natural Phenolic Compound-Based Oxygen Scavenging System for Active Packaging Applications. *J. Food Meas. Charact.* **2016,** *10* (3), 533–538. https://doi.org/10.1007/S11694-016-9332-1

Gaikwad, K. K.; Singh, S.; Lee, Y. S. Oxygen Scavenging Films in Food Packaging. *Environ. Chem. Lett.* 2018, *16*(2), 523–538. https://doi.org/10.1007/S10311-018-0705-Z/FIGURES/5

Johansson, K.; Winestrand, S.; Johansson, C.; Järnström, L.; Jönsson, L. J. Oxygen-Scavenging Coatings and Films Based on Lignosulfonates and Laccase. *J. Biotechnol.* **2012,** *161* (1), 14–18. https://doi.org/10.1016/J.JBIOTEC.2012.06.004

Kamaruddin, I.; Dirpan, A.; Bastian, F. In *The Novel Trend of Bacterial Cellulose as Biodegradable and Oxygen Scavenging Films for Food Packaging Application : An Integrative Review*, IOP Conference Series: Earth and Environmental Science, **2021;** vol *807* (2), 022066. https://doi.org/10.1088/1755-1315/807/2/022066

Kombaya-Touckia-Linin, E. M.; Gaucel, S.; Sougrati, M. T.; Stievano, L.; Gontard, N.; Guillard, V. Elaboration and Characterization of Active Films Containing Iron–Montmorillonite Nanocomposites for O2 Scavenging. *Nanomaterials* **2019,** *9* (9), 1193. https://doi.org/10.3390/NANO9091193

Lee, J. S.; Chang, Y.; Lee, E. S.; Song, H. G.; Chang, P. S.; Han, J. Ascorbic Acid-Based Oxygen Scavenger in Active Food Packaging System for Raw Meatloaf. *J. Food Sci.* **2018,** *83* (3), 682–688. https://doi.org/10.1111/1750-3841.14061

Lee, J. S.; Jeong, S.; Lee, H. G.; Cho, C. H.; Yoo, S. R. Development of a Sulfite-Based Oxygen Scavenger and its Application in Kimchi Packaging to Prevent Oxygen-mediated Deterioration of Kimchi Quality. *J. Food Sci.* **2018,** *83* (12), 3009–3018. https://doi.org/10.1111/1750-3841.14374

Leneveu-Jenvrin, C.; Apicella, A.; Bradley, K.; Meile, J. C.; Chillet, M.; Scarfato, P.; Incarnato, L.; Remize, F. Effects of Maturity Level, Steam Treatment, or Active Packaging to Maintain the Quality of Minimally Processed Mango (Mangifera indica cv. José). *J. Food Process. Preserv.* **2021,** *45* (7). https://doi.org/10.1111/JFPP.15600

Liu, J.; Lin, Y.; Lin, H.; Lin, M.; Fan, Z. Impacts of Exogenous ROS Scavenger Ascorbic Acid on the Storability and Quality Attributes of Fresh Longan Fruit. *Food Chem. X* **2021,** *12*, 100167. https://doi.org/10.1016/J.FOCHX.2021.100167

Mahieu, A.; Terrié, C.; Youssef, B. Thermoplastic Starch Films and Thermoplastic Starch/Polycaprolactone Blends with Oxygen-Scavenging Properties: Influence of Water Content. *Ind. Crops Prod.* **2015,** *72*, 192–199. https://doi.org/10.1016/J.INDCROP.2014.11.037

Makino, Y. Oxygen Consumption by Fruits and Vegetables. *Food Sci. Technol. Res.* **2013,** *19* (4), 523–529. https://doi.org/10.3136/FSTR.19.523

Markets and Markets. *Fruit and Vegetable Processing Market Analysis, Growth Trends, and Forecasts to 2027*; *MarketsandMarkets*, 2022. https://www.marketsandmarkets.com/Market-Reports/fruit-vegetable-processing-market-140232885.html

Niazmand, R.; Yeganehzad, S. Capability of Oxygen-Scavenger Sachets and Modified Atmosphere Packaging to Extend Fresh Barberry Shelf Life. *Chem. Biol. Technol. Agric.* **2020,** *7* (1), 1–14. https://doi.org/10.1186/S40538-020-00195-3/TABLES/7

Nielsen, C. R.; Skaggs, K. C. *Food Packaging Regulations in Australia and New Zealand; PackagingLaw.com.*, 2017. https://www.packaginglaw.com/special-focus/food-packaging-regulations-australia-and-new-zealand

Pan, L.; Zhang, M.; Lu, L.; Ou, B.; Chen, X. A Polyethylene Base Moisture Activating Oxygen Scavenging Film Co-Extruded with Tea Polyphenols-β-Cyclodextrin Inclusion Complex. *Materials* **2020,** *13,* 3857. https://doi.org/10.3390/MA13173857

Pant, A. F.; Sangerlaub, S.; Muller, K. Gallic Acid as an Oxygen Scavenger in Bio-Based Multilayer Packaging Films. *Materials* **2017,** *10*, 489. https://doi.org/10.3390/MA10050489

Poças, F.; Franz, R. Overview on European Regulatory Issues, Legislation, and EFSA Evaluations of Nanomaterials. In *Nanomaterials for Food Packaging: Materials, Processing Technologies, and Safety Issues*, 2018; pp 277–300. https://doi.org/10.1016/B978-0-323-51271-8.00010-3

Ramakanth, D.; Akhila, K.; Gaikwad, K. K.; Maji, P. K. UV-Activated Oxygen Scavenging System Based on Natural Rubber Latex from Hevea brasiliensis for Active Packaging Applications. *Ind. Crops Prod.* **2022,** *178*, 114658. https://doi.org/10.1016/J.INDCROP.2022.114658

Roberta, M. A. Oxygen Scavenging Films and Coating of Biopolymers for Food Application. In *Biopolymer Membranes and Films*, 2020; pp 535–551. https://doi.org/10.1016/B978-0-12-818134-8.00022-5
Scienchef. Factors That Influence the Respiration of Fruits and Vegetables, *FoodCrumbles*, 2017. https://foodcrumbles.com/respiration-fruits-vegetables/
Singh, S.; Gaikwad, K. K.; Lee, Y. S. Development and application of a pyrogallic acid-based oxygen scavenging packaging system for shelf life extension of peeled garlic. *Scientia Horticulturae*, **2019,** *256*, 108548. https://doi.org/10.1016/J.SCIENTA.2019.108548
Singh, A. K.; Ramakanth, D.; Kumar, A.; Lee, Y. S.; Gaikwad, K. K. Active Packaging Technologies for Clean Label Food Products: A Review. *J. Food Meas. Charact.* **2021,** *15* (5), 4314–4324. https://doi.org/10.1007/S11694-021-01024-3
Singh, G.; Singh, S.; Kumar, B.; Gaikwad, K. K. Active Barrier Chitosan Films Containing Gallic Acid Based Oxygen Scavenger. *J. Food Meas. Charact.* **2020,** *15* (1), 585–593. https://doi.org/10.1007/S11694-020-00669-W
Sultana, A.; Kathuria, A.; Gaikwad, K. K. Metal–Organic Frameworks for Active Food Packaging. A Review. *Environ. Chem. Lett.* **2022,** *20* (2), 1479–1495. https://doi.org/10.1007/S10311-022-01387-Z
Wołosiak-Hnat, A.; Zych, K.; Mężyńska, M.; Kifonidis, A.; Dajworski, M.; Lisiecki, S.; Bartkowiak, A. LDPE/PET Laminated Films Modified with FeO(OH) × H2O, Fe2O3, and Ascorbic Acid to Develop Oxygen Scavenging System for Food Packaging. *Packag. Technol. Sci.* **2019,** *32* (9), 457–469. https://doi.org/10.1002/PTS.2449

CHAPTER 6

NATURAL CLAY-BASED ETHYLENE SCAVENGERS FOR ACTIVE PACKAGING OF FRUITS AND VEGETABLES

AFREEN SULTANA, ANUSHIKHA, and SAMPAT SINGH BHATI

Department of Paper Technology, Indian Institute of Technology Roorkee, Roorkee, Uttarakhand, India

ABSTRACT

Fruits and vegetables are always in demand due to their nutritional worth and health-promoting properties. The demand for fresh produce that has been preserved without any treatment or addition of additives has increased as consumer awareness has grown. Ethylene is a phytohormone that causes fruits and vegetables to mature faster. Fresh food ripening can be delayed and shelf life extended by adopting an ethylene scavenging packaging technique. One of the strategies for preserving the freshness of fresh fruit and minimizing undesired changes is active packaging. The numerous natural clay-based ethylene scavengers and new technologies recently developed for the effective removal of ethylene are summarized in this chapter. The clays zeolite and halloysite are often employed for ethylene adsorption. Zeolite, halloysite nanotube, sepiolite, smectite, and Japanese oya stone are some of the most often utilized clays for ethylene adsorption. The basic mechanism of ethylene removal by these scavengers is reviewed, along with many applications in the food packaging industry. Several studies have suggested that acid treatment, cation exchange, or

Novel Packaging Systems for Fruits and Vegetables. Kirtiraj K. Gaikwad & Suman Singh (Eds.)

some other approach can be used to improve the ethylene adsorption effectiveness of clays. These scavengers can be included into the packaging system by impregnating the film or in the form of sachets, and spraying is also being examined in certain research. However, using a natural clay-based ethylene scavenger for active packaging has the potential to prevent fresh fruits and vegetables from ripening, hence, preserving quality and prolonging shelf life.

6.1 INTRODUCTION

Consumers demand that the fresh produce they buy stays fresh for long and maintain its quality throughout its shelf life. But, in case of fresh fruits and vegetables, uncontrolled ripening plays the key role in reducing their shelf life. Ripening is a natural ageing phenomenon which causes degradation of plant tissues and is triggered by ethylene. Ethylene is a colourless gas that acts as a plant hormone which is biosynthesized by plants (Shenoy et al., 2022). It is responsible for development, ripening, stress tolerance, and germination of fresh fruits and vegetables. Ethylene is not only associated with ripening, it also leads to over-ripening, which further causes decay and spoilage. Ethylene concentration plays an important role in determining the maturity index of fruits and vegetables, which helps in predicting the time of harvest. It is important to control ethylene concentration in the storage room for avoiding undesirable physiological change in fruits and vegetables. The presence of ethylene concentration in the ripening room depends on the type of fresh commodity and is usually maintained between 10 and 200 ppm (Dan et al., 2018; Sun et al., 2019). Hence, it can be stated that ethylene concentration and ripening are directly proportional, which needs to be controlled during storage and transportation. To minimize fruits and vegetables loss due to ethylene, various strategies are implemented such as controlled atmosphere, pickering, controlled temperature, and usage of LED lights (Wei et al., 2021).

Ethylene present around the fresh produce can be regulated via three pathways: first is by implementing modified atmosphere packaging; second is by using perforated packaging material to allow the passage of gas through the packaging material; and third is by using ethylene scavengers. Ethylene scavengers can be defined as the material capable of removing the ethylene from the surrounding by chemical reactions or physical adsorption. Chemical reaction takes place between the ethylene scavenger and the ethylene, which leads to breakdown of ethylene into carbon dioxide and water; this mechanism is also referred to as catalytic oxidation. Examples of such scavengers include potassium permanganate, titanium dioxide, and sodium permanganate. Scavenging of ethylene by physical adsorption includes van der Waals force interaction between adsorbate and adsorbent. Examples of such scavengers include nano-clays (zeolite, halloysite nanotube), activated carbon, and many others (Gaikwad et al., 2019).

Clay can be defined as the hydrous silicate or alumino-silicate, which dominates in colloidal fraction of rocks, soils, sediments, and water. Since clay is obtained from nature, it is in the heterogenous form consisting of minerals such as quartz, calcite, feldspars, or any other. The structure of clay can be explained as layers composed of silica and alumina sheets. Structural units, which are commonly found in most of the atomic lattice of clay, are octahedral and tetrahedral sheet. Tetrahedral sheet is formed using silicon atom, which is equidistance from the four oxygen or hydroxy groups (in case it is needed to balance the structure). Octahedral sheet is made up of two sheets of hydroxyls or oxygens embedded with iron, aluminium, or magnesium atoms in a way that they are equidistance from oxygens or hydroxyl atoms forming an octahedral coordination (Luckham and Rossi, 1999). Ethylene scavengers can be incorporated into the active packaging system either in the form of sachets or by directly incorporating into the packaging material.

6.2 RIPENING OF FRUITS AND VEGETABLES

Fruit ripening can be defined as an irreversible mechanism which involves alteration to biochemical, organoleptic, and physiological attributes of the fruit leading to softening of edible portion with enhanced quality (Prasanna et al., 2010). Based on regulation of ripening process, fruits are broadly classified into climacteric (such as tomato, apple, and melon) and non-climacteric fruits (such as orange, pineapple, and grape). In climacteric fruits, ripening is dependent on increase in respiration rate and ethylene production. In non-climacteric fruits, ethylene-associated respiratory peak is absent; thus, respiration and production of ethylene is lesser. Non-climacteric fruits do not produce carbon dioxide burst as observed during ripening of climacteric fruits (Bouzayen et al., 2010). However, both the fruits require hormones and gene expression regulated by transcription factors in the ripening process. Hence, the two primary factors involved in the ripening are respiration rate and ethylene. Respiration can be defined as the natural phenomenon of intake of oxygen and producing carbon dioxide. Plants undergo aerobic respiration, which involves conversion of starch, sugars, and organic acids into carbon dioxide and water (Jalali et al., 2020). And, ethylene plays a key role in regulating respiration rate, which is further explained in Section 6.3.

6.3 ETHYLENE BIOSYNTHESIS AND ITS ROLE IN RIPENING

Ethylene plays a crucial role in ripening of fruits and vegetables. Ethylene can be categorized into two groups: endogenous ethylene, which is produced by plant itself and exogenous ethylene, which is present in atmosphere. Climacteric fruits produce higher level of ethylene compared to non-climacteric fruits. Exogeneous ethylene triggers the respiration rate and initiates autocatalysis of endogenous ethylene in climacteric fruits; whereas, for non-climacteric fruits, removal of exogenous ethylene causes respiration to come back to level of processing. Uncontrolled ethylene regulation leads to overripening and then ultimately causes decay (Wei et al., 2021).

In endogenous ethylene biosynthesis, three key catalysts are involved, which are S-adenosyl-l-methionine synthetase (catalyst 1), 1-aminocyclopropane-1-carboxylic acid synthetase (catalyst 2), and 1-aminocyclopropane-1-carboxylic acid oxidase (catalyst 3). This process starts with conversion of l-methionine to S-adenosyl-l-methionine using adenosine triphosphate in the presence of catalyst 1. Then, Schiff base is produced due to reaction between S-adenosyl-l-methionine pyridoxal phosphate. The Schiff base results in formation of cyclopropane ring and 5′-methylthioadenosine, where cyclopropane ring is converted to 1-aminocyclopropane-1-carboxylic acid via catalyst 2. Then, the by-product 5′-methylthioadenosine is converted to 5-methylthioribose, which is recycled back to methionine, and aminocyclopropane-1-carboxylic acid is further converted to ethylene by aminocyclopropane-1-carboxylic acid oxidase (Zhang et al., 2016; Zhang et al., 2020) (Figure 6.1).

6.4 ETHYLENE SCAVENGERS

Ethylene scavengers are the material which are capable of reducing the ethylene concentration from the headspace with the purpose of reducing the ripening process. These materials can be of two types—scavengers which adsorb the ethylene by physical adsorption and scavengers which cause catalytic oxidation of the ethylene. Ethylene can also be controlled by using inhibitors such as 1-methylcyclopropane which binds to the receptor and blocks the binding of ethylene's binding site (Mariah et al., 2022). Natural clays usually adsorb the ethylene due to their property to hold the ethylene molecule in their porous structure by interacting with the active sites. The mechanism of ethylene adsorption of natural clays is further discussed in this section.

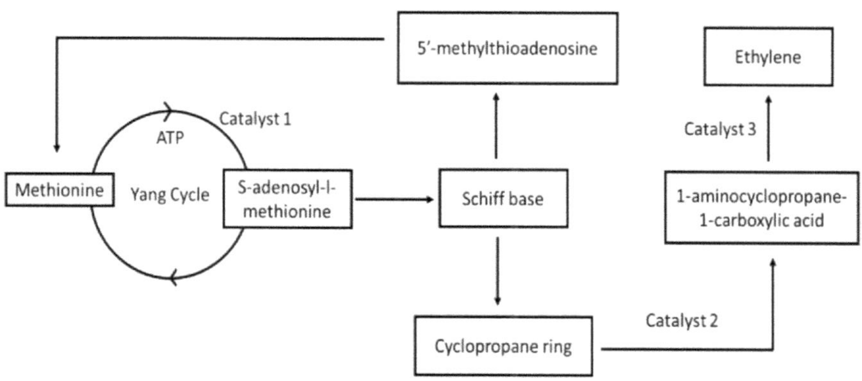

FIGURE 6.1 Schematic representation of ethylene biosynthesis, which involves conversion of methionine to ethylene, using three catalysts, which are S-adenosyl-l-methionine synthetase (catalyst 1), 1-aminocyclopropane-1-carboxylic acid synthetase (catalyst 2), and 1-aminocyclopropane-1-carboxylic acid oxidase (catalyst 3).

6.4.1 *ZEOLITE*

Zeolite is a class of crystalline microporous solids with the structure of an aluminosilicate framework (Mitchell et al., 2015). The word "zeolite" is derived from the Greek words "Zeo," which means boil and "lithos," which means stone (Derbe et al., 2021). The chemical composition of zeolite is represented by a formula: $Mx/n[(AlO_2)x\ (SiO_2)y\ ] \cdot zH_2O$, where n is the valence of cation M; z is the number of water molecules in each unit cell; and x and y are integers (Yang, 2003). Zeolite forms a tetrahedral arrangement by connecting silicon ion and aluminium ion consisting of bridging four oxygen anions (Figure 6.2).

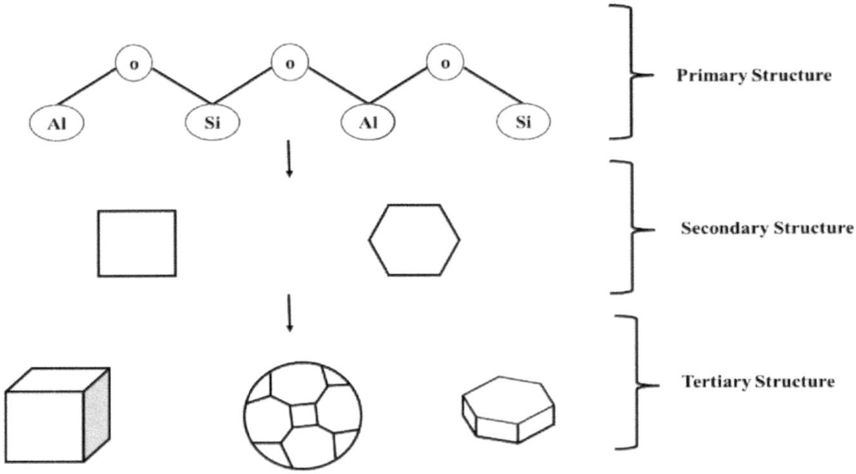

FIGURE 6.2 Schematic representation of the primary, secondary, and tertiary structure of the zeolite framework.

6.4.1.1 PROPERTIES AND CLASSIFICATION OF ZEOLITE

Silicon content plays an important role in zeolite framework; when silicon atom is added to the framework as a substitute for aluminium, it converts the charge zero to negative, which is further balanced by hydroxyl group or metal cations (Abreu et al., 2019). Properties of zeolite contributing to ethylene adsorption are:

1) Ion exchange: Ion exchange is the important attribute of the zeolite framework; it has the aluminium ionic charge which has to be balance by the cation (Moshoeshoe and Nadiye-Tabbiruka, 2017).
2) Cation exchange: According to this property, zeolites have the feature of exchanging the cations present on the zeolite on the active site from the cations present in the surrounding.

Properties of zeolite depends on ion exchange, dehydration, rehydration, and adsorption characteristics. Based on dehydration and rehydration property, zeolites can be broadly classified into natural and synthetic. Exposure of high temperature on natural zeolites causes decrease in weight loss without affecting the structural framework. The other group undergoes structural changes without any continuous weight loss. Natural zeolites have lesser purity compared to synthetic polymer due to presence of components other than Si, Al, and quartz (Tran et al., 2019). Zeolites can also be grouped based on their silicon amount and pore diameter. Based on silicon concentration, zeolites are categorized as low silica zeolite (ratio of silica to aluminium range between 1 and 1.5), intermediate silica zeolite (ratio of silica to aluminium range between 2 and 5), and high silica zeolite (ratio of silica to aluminium is above 10) (Cruciani, 2006). Based on pore diameter, zeolite having pore size between 3 and 4.5 Å are known as small pore zeolite, zeolites with pore size ranging between 4.5 and 6 Å are intermediate pore zeolite, and large pore zeolite have pore diameter above 6 Å (Zarrintaj et al., 2020). These classifications of zeolite based on different factors are represented in Figure 6.3.

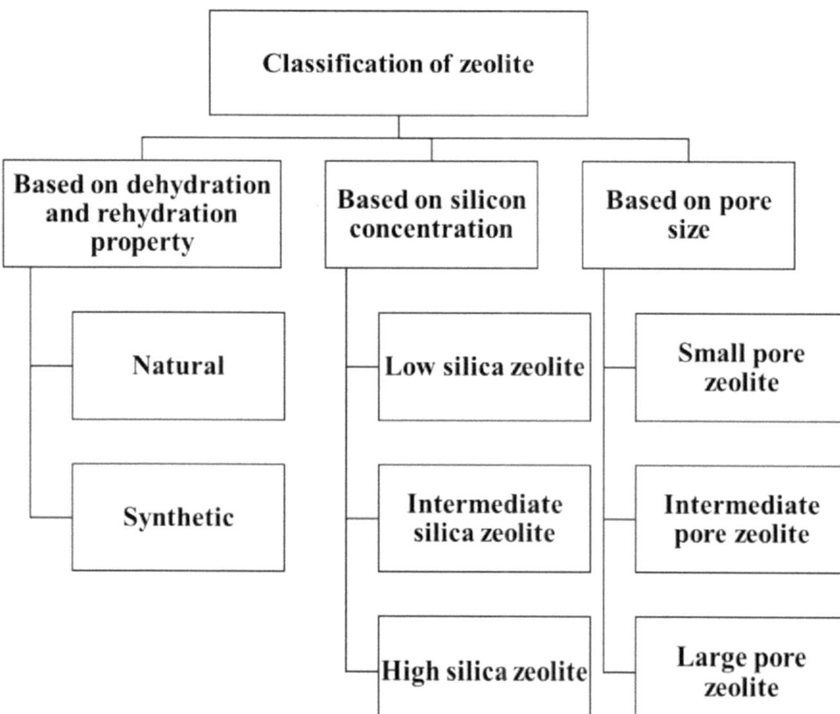

FIGURE 6.3 Graphical representation of the categories of zeolite based on dehydration and rehydration, silicon concentration, and pore size.

6.4.1.2 MECHANISM OF ETHYLENE ADSORPTION BY ZEOLITE AND ITS APPLICATIONS

The catalytic activity of the molecular sieve is influenced by parameters such as acidic property and pore structure. In case of zeolite, acidity and BrØnsted acid site plays a key role for catalytic activity in adsorption process. Depending on the acidity of the zeolite framework, two mechanisms are reported: (1) stepwise mechanism—in this, initially alkoxide is formed, which further reacts with alkene and results in the formation of C–C bond, and (2) concreted mechanism—this involves simultaneous occurrence of two steps, addition of proton and C–C bond formation. In addition, pore diameter also contributes in adsorption of ethylene (Shen, 2017). Ethylene is also adsorbed by interacting with the hydroxyl group of zeolite framework forming a π-complex. In this mechanism, change in chemical bond strength of aluminum–oxygen zeolite framework takes place due to the transformation of donor–acceptor bond to chemical and visa-versa (Senchenya and Kazansky, 1991). Still there is a lack of proper explanation on active sites responsible for ethylene adsorption. Factors affecting ethylene adsorption into zeolite are:

i. Presence of water molecule: It has been reported that water molecules compete with the ethylene molecule by rapidly occupying Si-OH-Al and Si-OH sites and reduces its adsorption into zeolite (Abreu et al., 2019).
ii. Cation exchange: Modification of zeolite framework by exchanging the cations already present in the framework with other cations such as potassium, rubidium, and cesium ions can help in improving the ethylene adsorption efficiency (Sue-aok et al., 2010).

Zeolite has various food applications such as ethylene scavenger, filler, immobilizing stabilizing agent, antimicrobial agent, sieving at molecular level, and also nanoreactor. Additionally, zeolite has ion exchange capability, biocompatible behavior, and adjustable pore framework (Villa et al., 2022). According to USFD, zeolite is considered as generally recognized as safe (GRAS) material for food purpose (USFDA, 2015). Zeolites in food packaging applications can be incorporated into packaging material in three ways, namely, directly into film matrix such as paper or polymer; in the form sachets; or embedded into coating solutions. Some of the studies involving zeolite as an ethylene scavenging material are mentioned in Table 6.1.

TABLE 6.1 Some of the Applications of Zeolite Incorporated into Packaging System for Ethylene Adsorption.

Type of zeolite	Form of incorporation	Results	References
Zeolite 13X	Incorporated into lignocellulosic active paper	About 62% of ethylene was scavenged by paper with 30% of zeolite 13X.	Kumar et al. (2021)
Watercel ZF zeolite	Poly (butylene-co-terephthalate) and cassava starch-based film	Broccoli stored in the fabricated packaging stayed fresh for days at 12°C.	Marzano-Barreda et al. (2021)
Clinoptilolite	Chitosan film	At 1 bar pressure, 25% of the ethylene was adsorbed.	do Nascimento Sousa et al. (2020)
Zeolite	Direct spraying	Results suggests 6% of zeolite preserved the seedless table grapes for 60 days.	Huwei et al. (2021)
Mordenite and mordenite-clinoptilolite	Sachet	Shelf life of Salak Pondoh fruit was extended to 25 days at 10 ± 2°C.	Widayanti et al. (2021)

Studies are also performed on improving the ethylene scavenging activity of zeolite by treating it with other promoters such as potassium permanganate and palladium (Mansourbahmani et al., 2018; Tzeng et al., 2019). Hence, it can be stated that zeolite has the ability to adsorb ethylene, which makes it a potential candidate for developing ethylene scavenging food packaging system.

6.4.2 HALLOYSITE NANOTUBE

Halloysite nanotube are one-dimensional natural nanomaterial with hollow tubular type of aluminosilicates nanostructure which are considered as one of the kaolin family (Duan et al., 2020). It is obtained naturally due to pedagenesis, weathering, and hydrothermal changes of rocks and soils (Yurdacan and Murat Sari, 2021).

6.4.2.1 STRUCTURE AND CLASSIFICATION

It has the chemical structure $Al_2Si_2O_5(OH)_4 \cdot nH_2O$. The atoms were arranged in dioctahedral 1:1 clay structure, the wall of nanotube comprises of 10–15 bilayers with the 0.72 nm of the space between them. The length of halloysite nanotube ranges between 0.2 and 1.5 μm with the inner diameter of 10–30 nm and outer diameter of 40–70 nm (Goda et al., 2019). Figure 6.4 displays the structure of halloysite nanotube showing the two surfaces and atomic arrangement.

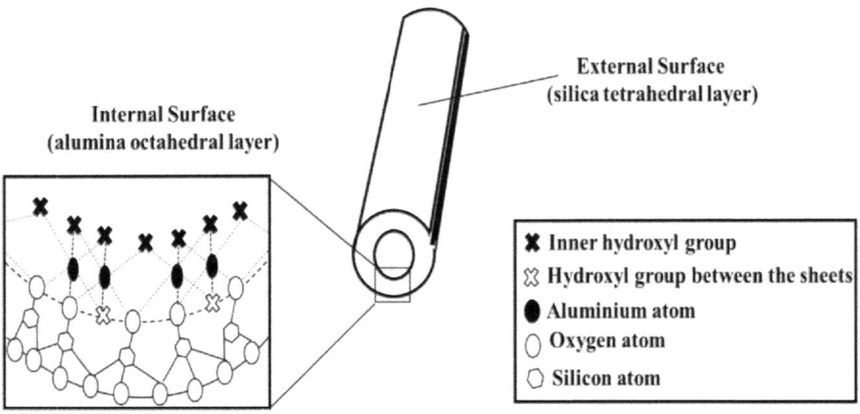

FIGURE 6.4 Diagrammatic representation of structure of halloysite nanotube highlighting the internal surface (alumina octahedral layer) and external surface (silica tetrahedral layer).

The difference between the internal and external surface of halloysite makes it an interesting structure as the inner one has Al–OH groups and the external surface comprises of Si–O–Si. Negative charge is present on the outer layer (pH is less than or equal to 8.5) and positive charge on the inner surface (pH is more than or equal to 1.5) (Fakhruddin et al., 2021). The halloysite nanotube has three types of pores—pores due to dehydration and voids formed due to agglomeration and inherent lumen (Yuan et al., 2015). The factors such as concentration, the pH, and the presence of electrolytes have effect on the crystalline phase behavior of the halloysite nanotube. Halloysite nanotubes have some attractive properties such as surface reactivity, a hollow cavity, easy dispersibility, stability in solvent media, and long aspect ratios (Cavallaro et al., 2020). The low-surface charge property of the halloysite nanotubes make it more convenient to be used with polymers for active packaging (Tas et al., 2017).

Halloysite nanotubes are classified into the following two categories depending on the hydration property (Hanif et al., 2016; Yuan et al., 2015):

i) Hydrated halloysite nanotube: Hydrated halloysite nanotube has the crystalline structure with the spacing of 10 Å d_{001} and consists of one water molecule layer.
ii) Dehydrated halloysite nanotube: Dehydrated nanotube has the crystalline structure with the spacing of 7 Å d_{001} and the layer of water molecule is absent, which might have been lost due to heating.

6.4.2.2 MECHANISM OF ETHYLENE ADSORPTION

Halloysite nanotubes have a unique framework with the capability of adsorb anionic molecule on the outer surface and cation molecule on the inner surface. A study reported that one gram of halloysite nanotube can absorb around 6.36 mL of ethylene gas (Tas et al., 2017). Aspect ratio of hollow structure is an important factor which affects ethylene adsorption by halloysite nanotube. Larger the aspect ratio, higher is the ethylene adsorption capacity of the halloysite nanotube. In various studies, it is shown that halloysite nanotube can be treated with alkali to achieve increase in aspect ratio (Gaikwad et al., 2018). Several studies have reported the application of halloysite nanotube in food packaging matrix for ethylene adsorption (Table 6.2).

6.4.3 SEPIOLITE

Sepiolite is a trioctahedral magnesium-rich silicate clay with the chemical formula $Mg_8[Si_{12}O_{30}](OH)_4(H_2O)_4.nH_2O$. Its structure consists of tetrahedral sheet and octahedral sheet. Both these sheets combine to form a ribbon with a 2:1 phyllosilicate structure, where tetrahedral sheets are continuous and octahedral sheets are discontinuous. The strong ethylene adsorption capacity of sepiolite can be associated to the substantial polarity of the sepiolite surface and the quadrupole moment of ethylene. Ethylene adsorption on sepiolite clay depends on particle size distribution, capillary structure, microporosity, surface area, active sorption sites (especially silanol groups), and negative charge (Sakizci, 2013). Adsorption activity of the sepiolite can be improved by mild acid treatment, which increases the surface area. In this treatment, metal exchange cation and magnesium ions are eliminated from the structure, which leads to increase in porosity. Compared to cation exchange modification of sepiolite, acid-activated samples show better ethylene adsorption (Erdoğan and Esenli, 2021). Sepiolite can act

TABLE 6.2 Some of the Applications of Halloysite Nanotube Incorporated into Packaging System for Ethylene Adsorption.

Treated/untreated halloysite nanotube	Packaging matrix	Incorporation method	Results	References
Halloysite nanotube	Low density polyethylene	Twin screw extruder	Nanocomposite polyethylene with 5% HNT improved the shelf life of strawberry and banana.	Tas et al. (2017)
Alkaline-treated Halloysite nanotube	Low density polyethylene	Twin screw extruder	5.27 ± 0.65 μL ethylene/100 cm^2 was absorbed by the film consisting of 5% of alkaline-treated HNT.	Boonsiriwit et al. (2020)
Potassium permanganate-loaded halloysite nanotube	Low density polyethylene	Twin screw extruder	Tomato wrapped in film with 1% of potassium permanganate-loaded halloysite nanotube showed lesser ethylene production.	Joung et al. (2021)
Halloysite nanotubes were incorporated with essential oil	Cardboard box	Spraying	Decay was less than 5% and firmness remained unchanged for 6 days at 8°C for the samples with halloysite nanotube.	Buendía–Moreno et al. (2020)

as supporting framework for other ethylene scavenging material, such as potassium permanganate, which improves the effectiveness and can also reduce the quantity of material required for adsorption. For example, in a study, it was reported that incorporation of potassium permanganate into the sepiolite reduced the quantity usage by 50% (Álvarez-Hernández et al., 2020). Sepiolite can be added into packaging matrix to develop an active film, such as, corn starch-gum Arabic incorporated with 20% of sepiolite showed ethylene scavenging activity of around 69%. This investigation also revealed that when broccoli was packed in the prepared active film, the ethylene present in the headspace was recorded as approximately 190 ppm which was lesser than control samples (149 ppm approximately) (Upadhyay et al., 2022).

6.4.4 SMECTITE CLAYS

Smectite clay are a group of phyllosilicate mineral possessing variable net negative charge (Odom, 1984). Bentonite comprises of smectite mineral which has two silica tetrahedral sheets with a central octahedral sheet. Intercalation of organometallic, organic or inorganic, molecules in the interlamellar area, as well as heat or acid treatments, can typically increase the adsorption capabilities of smectite clays (Marouf et al., 2021). On the active sites of the smectite clay, there is a surface interaction for ethylene adsorption. But the intraparticle diffusion mechanism is well established to have a major role in the adsorption process. Acid activation is a method for improving adsorptive characteristics, surface crystallinity, functionality, specific surface area, and selectivity for various adsorbates by partially dissolving the adsorbent (Shattar et al., 2020). Modification of bentonite-using metals such as silver, copper, and iron can be a method to increase the specific surface area and it enhances the retention of ethylene molecule (Erdoğan Alver and Günal, 2016). In an investigation, porous clay heterostructures were fabricated using silica and sodium–bentonite and modified with addition of thiol group. Ethylene scavenging activity of the composite improved with addition of thiol group on the surface of clay (Srithammaraj et al., 2018).

Montmorillonite are clay that are derived from the bentonite consisting of sodium or calcium (Uddin, 2018). Montmorillonite embedded with potassium permanganate has been reported to extend the shelf life of banana from 4 days (control) to 20 days at 25°C (Álvarez-Hernández et al., 2018).

In a study, protonated montmorillonite was incorporated with potassium permanganate, which, when combined with modified atmosphere packaging, resulted in increase in shelf life of blueberries from 7 to 21 days at 10°C. Hence, these investigations reveal the applications of halloysite nanotube in food packaging system to preserve the quality by adsorbing ethylene.

6.4.5 *JAPANESE OYA STONE*

The Japanese oya stone is a tuff stone discovered in the Japanese city of Utsunomiya. By finely grinding a little amount of metal oxide, kneading, and heating to 200–900°C, Japanese oya stone may be employed as an ethylene adsorber. The resulting mixture is oxidized using ozone or electromagnetic radiation (Zagory, 1995). Not much literature is available on Japanese oya stone as an ethylene scavenger for food packaging, but it is considered as one of the scavengers for efficient ethylene adsorption.

6.5 CHALLENGES FACED IN COMMERCIALIZATION OF THE NATURAL CLAY-BASED ETHYLENE SCAVENGERS

Although natural clay-based ethylene scavengers have potential scope in the field of preserving the quality of fresh fruits and vegetables, it also undergoes various challenges such as consumer perspective; packaging cost; limited capacity and reusability issue; competitors of ethylene molecule; regulatory compliances; and examination for usability, stability, and conformity.

6.5.1 *CONSUMER PERSPECTIVE*

Manufacturing companies need to make packaging decisions based on the consumer behavior and market trend analyses (Schmid, 2008). Consumer perspective plays an important role due to their sophisticated needs based on convenience, quality, ethics, health, and cost. Mistrust caused between the packaging and consumer has negative impact on the product, which is usually due to lack of transparency. In order to spread awareness among consumers about the innovative technology and packaging systems, non-traditional method should be used for sharing consumer-friendly information, which will build trust.

6.5.2 PACKAGING COST

Addition of ethylene scavenger to the packaging system, increases the additional cost of the product. Efforts need to be made for developing a packaging system with minimal cost of packaging because consumers expect to pay for the product and not the packaging material (Mariah et al., 2022).

6.5.3 LIMITED CAPACITY AND REUSABILITY

Limited adsorption capacity of the ethylene scavenger is yet another significant challenge as it reaches saturation with no further adsorption activity (Sadeghi et al., 2021). Natural clays-based ethylene scavenger loses their adsorption activity after single use, which might be due to the presence of moisture; thus, reusability feature of the scavenger needs further investigation (Chopra et al., 2017).

6.5.4 COMPETITORS OF ETHYLENE MOLECULES

Some studies report that adsorption of ethylene molecule is obstructed in the presence of other molecules with structure similar to ethylene, such as ethylene adsorption is reduced in the presence of oxygen and carbon dioxide molecules (Bailen et al., 2007; Martínez-Romero et al., 2007). These competitors reduce the adsorption efficiency of the ethylene, which can be improved by further modifications in the pore size or cation exchange of the framework.

6.5.5 REGULATORY COMPLIANCES

In the United States, Food and Drug Administration plays the role of regulatory authority for food packaging. It includes an important term, food contact substance (FCS), which means any substance in the contact of food used for manufacturing, holding, packaging, packing, and transporting of food and is not intended to have any technical effect on the food material. The FDA performs the safety assessment by considering resultant dietary exposure, which is determined by the toxicological report of the food contact substances. A proper knowledge of the US regulatory processes for food contact substances helps the industry to decide the adequate regulatory options needed for their product. Food packaging industry can be categorized into releasing packaging systems and immobilized systems. Releasing packaging systems need food additive petition premarket approval; whereas, immobilized packaging system needs food contact substance notification (Werner et al., 2017). Lack of awareness and global gap between the regulations make the commercialization process of active packaging a challenging task.

6.5.6 EXAMINATION FOR USABILITY, STABILITY, AND CONFORMITY

Before commercialization of any innovative idea, it is important to study the compatibility of the production of the product at large scale with traditional equipment and forming procedures. It gives a better understanding about the usability, stability, and conformity with regards to the manufacturing process (Mariah et al., 2022).

6.6 ETHICS AND SAFETY ISSUES

Insufficient investigation and lack of awareness regarding the compatibility of the natural clay-based ethylene scavengers with humans and environment lead to inadequate regulation of the packaging system. It is important to establish a regulatory body to monitor and regulate such packaging system to ensure the safety of consumers and environment. Some of the important factors which need to be understood, such as usage of natural clay-based ethylene scavengers, informative labelling, migration study, and environmental impact, are discussed below.

6.6.1 HOW TO USE MANUAL

Consumers need to be made familiar with the usage of ethylene scavengers by providing them a proper usage manual for better understanding about how to use and dispose the scavenging material (Gaikwad et al., 2019).

6.6.2 LABELING

Food packaging needs to labeled with the description about the ingredients present in the ethylene scavenger to create consumer transparency and build confidence for purchase of the product. Proper labeling provides the consumer with the information regarding the product and avoids misinterpretations (Chopra et al., 2017).

6.6.3 MIGRATION COMPONENT FROM THE PACKAGING MATERIAL

One of the risks that must be avoided is the migration of undesired components from the packaging material into the food, necessitating a thorough examination of the packaging material. The rules for analyzing migration of components in contact with food are established by Directives (82/711/EEC or 85/572/EEC) (Dainelli et al., 2008). Migration of components from the ethylene scavenging packaging system is a critical issue; it needs to investigated well before making it commercially available. Efforts need to be made to avoid the migration of any component into the food item and, if migration is reported, it should not exceed the acceptable level.

6.6.4 ENVIRONMENTAL IMPACT

The packaging product incorporated with ethylene scavenging system should not have adverse effect on the ecosystem. Biodegradability and eco-toxicity of the packaging system needs to be studied well to prevent any undesirable impact on nature. Asia, United States, and Europe are working on designing and establishing new policies to achieve sustainable and eco-friendly approach toward packaging system. One of the issue in recycling arises from the multilayering of the packaging and, in case of incorporated materials, separation is another challenge (Werner et al., 2017).

6.7 CONCLUSIONS

Maintaining the freshness quality of fruits and vegetable is one of the highly focused areas of research and development. Various technology and processing methods, such as modified atmosphere packaging, controlled atmosphere storage, low temperature, edible coatings, and scavengers, are used to preserve quality. Since, the main reason for deterioration of fruits and vegetables is uncontrolled ripening, reducing the respiration rate contributes to increase in the shelf life. Fresh produce continues to respire after harvesting and the presence of ethylene accelerates this process leading to faster deterioration. Incorporation of ethylene scavengers into the packaging system helps to impede the ripening process by either of the three mechanisms: adsorbing the ethylene, catalytic oxidation of ethylene, and inhibiting the ethylene production. Among these mechanisms, adsorptions of ethylene is commonly studied, which can be attained using natural clays such as zeolite, halloysite nanotube, smectite clay, and Japanese oya stone. Apart from serving as a potential candidate for ethylene adsorption, these natural clays also possess challenges such as sophisticated consumer perspective, increase in packaging cost, limited adsorption capacity and reusability issue, and competitors molecules, which hinder the ethylene adsorption, regulatory compliances and examination for usability, stability, and conformity, which in turn limits their usage in food packaging system. Combating with these challenges can lead to the development of efficient natural clay-based ethylene scavenger for preserving the freshness and quality of the fresh produce with satisfactory consumer attention and safety and as well as zero or minimal environmental impact.

KEYWORDS

- **ethylene adsorber**
- **ethylene scavenger**
- **zeolite**
- **halloysite nanotube**
- **sepiolite**
- **fresh produce**
- **packaging**
- **quality**

REFERENCES

Abreu, N. J.; Valdés, H.; Zaror, C. A.; Azzolina-Jury, F.; Meléndrez, M. F. Ethylene Adsorption onto Natural and Transition Metal Modified Chilean Zeolite: An Operando DRIFTS Approach. *Microporous Mesoporous Mater.* **2019,** *274*, 138–148. https://doi.org/10.1016/J.MICROMESO.2018.07.043

Álvarez-Hernández, M. H.; Artés-Hernández, F.; Ávalos-Belmontes, F.; Castillo-Campohermoso, M. A.; Contreras-Esquivel, J. C.; Ventura-Sobrevilla, J. M.; Martínez-Hernández, G. B. Current Scenario of Adsorbent Materials Used in Ethylene Scavenging Systems to Extend Fruit and Vegetable Postharvest Life. *Food Bioproc. Technol.* **2018,** *11* (3), 511–525. https://doi.org/10.1007/S11947-018-2076-7/TABLES/2

Álvarez-Hernández, M. H.; Martínez-Hernández, G. B.; Avalos-Belmontes, F.; Miranda-Molina, F. D.; Artes-Hernandez, F. Postharvest Quality Retention of Apricots by Using a Novel Sepiolite–Loaded Potassium Permanganate Ethylene Scavenger. *Postharvest Biol. Technol.* **2020,** *160*, 111061.

Bailen, G.; Guillen, F.; Castillo, S.; Zapata, P. J.; Serrano, M.; Valero, D.; Martínez-Romero, D. Use of a Palladium Catalyst to Improve the Capacity of Activated Carbon to Absorb Ethylene, and its Effect on Tomato Ripening. *Spanish J. Agric. Res.* **2007,** *5* (4), 579–586. https://doi.org/10.5424/SJAR/2007054-5359

Boonsiriwit, A.; Xiao, Y.; Joung, J.; Kim, M.; Singh, S.; Lee, Y. S. Alkaline Halloysite Nanotubes/Low Density Polyethylene Nanocomposite Films with Increased Ethylene Absorption Capacity: Applications in Cherry Tomato Packaging. *Food Packag. Shelf Life* **2020,** *25*. https://doi.org/10.1016/j.fpsl.2020.100533

Bouzayen, M.; Latché, A.; Nath, P.; Pech, J. C. Mechanism of Fruit Ripening. *Plant Dev. Biol.* **2010,** *1*, 319–339. https://doi.org/10.1007/978-3-642-02301-9_16

Buendía−Moreno, L.; Sánchez−Martínez, M. J.; Antolinos, V.; Ros−Chumillas, M.; Navarro−Segura, L.; Soto−Jover, S.; ... López−Gómez, A. Active Cardboard Box with a Coating Including Essential Oils Entrapped within Cyclodextrins and/or Halloysite Nanotubes. A Case Study for Fresh Tomato Storage. *Food Control* **2020,** *107*, 106763. https://doi.org/10.1016/J.FOODCONT.2019.106763

Cavallaro, G.; Milioto, S.; Lazzara, G. Halloysite Nanotubes: Interfacial Properties and Applications in Cultural Heritage. *Langmuir* **2020,** *36* (14), 3677–3689. https://doi.org/10.1021/ACS.LANGMUIR.0C00573/ASSET/IMAGES/ACS.LANGMUIR.0C00573.SOCIAL.JPEG_V03

Chopra, S.; Dhumal, S.; Abeli, P.; Beaudry, R.; Almenar, E. Metal-organic Frameworks have Utility in Adsorption and Release of Ethylene and 1-methylcyclopropene in Fresh Produce Packaging. *Postharvest Biol. Technol.* **2017,** *130*, 48–55. https://doi.org/10.1016/J.POSTHARVBIO.2017.04.001

Cruciani, G. Zeolites Upon Heating: Factors Governing their Thermal Stability and Structural Changes. *J. Phys. Chem. Solids* **2006,** *67* (9–10), 1973–1994. https://doi.org/10.1016/J.JPCS.2006.05.057

Dainelli, D.; Gontard, N.; Spyropoulos, D.; Zondervan-van den Beuken, E.; Tobback, P. Active and Intelligent Food Packaging: Legal Aspects and Safety Concerns. *Trends Food Sci. Technol.* **2008,** *19* (SUPPL. 1), S103–S112. https://doi.org/10.1016/J.TIFS.2008.09.011

Dan, M.; Huang, M.; Liao, F.; Qin, R.; Liang, X.; Zhang, E.; ... He, Q. Identification of Ethylene Responsive miRNAs and Their Targets from Newly Harvested Banana Fruits Using High-Throughput Sequencing. *J. Agric. Food Chem.* **2018,** *66* (40), 10628–10639. https://doi.org/10.1021/ACS.JAFC.8B01844/SUPPL_FILE/JF8B01844_SI_001.ZIP

Derbe, T.; Temesgen, S.; Bitew, M. A Short Review on Synthesis, Characterization, and Applications of Zeolites. *Adv. Mater. Sci. Eng.* **2021,** *2021.* https://doi.org/10.1155/2021/6637898

do Nascimento Sousa, S. D.; Santiago, R. G.; Soares Maia, D. A.; de Oliveira Silva, E.; Vieira, R. S.; Bastos-Neto, M. Ethylene Adsorption on Chitosan/Zeolite Composite Films for Packaging Applications. *Food Packag. Shelf Life* **2020,** *26,* 100584. https://doi.org/10.1016/J.FPSL.2020.100584

Erdoğan, B.; Esenli, F. Sepiolite as an Efficient Adsorbent for Ethylene Gas. *Clay Miner.* **2021,** *56* (3), 222–228. https://doi.org/10.1180/CLM.2021.36

Erdoğan Alver, B.; Günal, A. Thermal, Structural and Ethylene Adsorption Properties of Ag-, Cu- and Fe-modified Bentonite from Turkey. *J. Therm. Anal. Calorim.* **2016,** *126* (3), 1533–1540. https://doi.org/10.1007/S10973-016-5640-2/FIGURES/7

Fakhruddin, K.; Hassan, R.; Khan, M. U. A.; Allisha, S. N.; Razak, S. I. A.; Zreaqat, M. H.; ... Hassan, A. Halloysite Nanotubes and Halloysite-Based Composites for Biomedical Applications. *Arabian J. Chem.* **2021,** *14* (9), 103294. https://doi.org/10.1016/J.ARABJC.2021.103294

Gaikwad, K. K.; Singh, S.; Lee, Y. S. High Adsorption of Ethylene by Alkali-Treated Halloysite Nanotubes for Food-Packaging Applications. *Environ. Chem. Lett.* **2018,** *16* (3), 1055–1062. https://doi.org/10.1007/s10311-018-0718-7

Gaikwad, K. K.; Singh, S.; Negi, Y. S. Ethylene Scavengers for Active Packaging of Fresh Food Produce. *Environ. Chem. Lett.* **2019,** *18* (2), 269–284. https://doi.org/10.1007/S10311-019-00938-1

Hanif, M.; Jabbar, F.; Sharif, S.; Abbas, G.; Farooq, A.; Aziz, M. Halloysite Nanotubes as a New Drug-Delivery System: A Review. *Clay Miner.* **2016,** *51* (3), 469–477. https://doi.org/10.1180/CLAYMIN.2016.051.3.03

Huwei, S.; Asghari, M.; Zahedipour-Sheshglani, P.; Alizadeh, M. Modeling and Optimizing the Changes in Physical and Biochemical Properties of Table Grapes in Response to Natural Zeolite Treatment. *LWT* **2021,** *141,* 110854. https://doi.org/10.1016/J.LWT.2021.110854

Jalali, A.; Linke, M.; Geyer, M.; Mahajan, P. V. Shelf Life Prediction Model for Strawberry Based on Respiration and Transpiration Processes. *Food Packag. Shelf Life* **2020,** *25,* 100525. https://doi.org/10.1016/J.FPSL.2020.100525

Joung, J.; Boonsiriwit, A.; Kim, M.; Lee, Y. S. Application of Ethylene Scavenging Nanocomposite Film Prepared by Loading Potassium Permanganate-Impregnated Halloysite Nanotubes into Low-Density Polyethylene as Active Packaging Material for Fresh Produce. *LWT* **2021,** *145,* 111309. https://doi.org/10.1016/J.LWT.2021.111309

Kumar, A.; Gupta, V.; Singh, S.; Saini, S.; Gaikwad, K. K. Pine Needles Lignocellulosic Ethylene Scavenging Paper Impregnated with Nanozeolite for Active Packaging Applications. *Ind. Crops Prod.* **2021,** *170,* 113752.

Luckham, P. F.; Rossi, S. The Colloidal and Rheological Properties of Bentonite Suspensions. *Adv. Coll. Interface Sci.* **1999,** *82* (1–3), 43–92. https://doi.org/10.1016/S0001-8686(99)00005-6

Mansourbahmani, S.; Ghareyazie, B.; Zarinnia, V.; Kalatejari, S.; Mohammadi, R. S. Study on the Efficiency of Ethylene Scavengers on the Maintenance of Postharvest Quality of Tomato Fruit. *J. Food Meas. Charact.* **2018,** *12* (2), 691–701. https://doi.org/10.1007/s11694-017-9682-3

Mariah, M. A. A.; Vonnie, J. M.; Erna, K. H.; Nur'Aqilah, N. M.; Huda, N.; Wahab, R. A.; Rovina, K. The Emergence and Impact of Ethylene Scavengers Techniques in Delaying the Ripening of Fruits and Vegetables. *Membranes* **2022,** *12* (2), 117. https://doi.org/10.3390/membranes12020117

Marouf, R.; Dali, N.; Boudouara, N.; Ouadjenia, F.; Zahaf, F. Study of Adsorption Properties of Bentonite Clay. In *Montmorillonite Clay*, 2021. https://doi.org/10.5772/INTECHOPEN.96524

Martínez-Romero, D.; Bailén, G.; Serrano, M.; Guillén, F.; Valverde, J. M.; Zapata, P.; Castillo, S.; Valero, D. Tools to Maintain Postharvest Fruit and Vegetable Quality Through the Inhibition of Ethylene Action: A Review. *Crit. Rev. Food Sci. Nutr.* **2007,** *47* (6), 543–560. https://doi.org/10.1080/10408390600846390

Marzano-Barreda, L. A.; Yamashita, F.; Bilck, A. P. Effect of Biodegradable Active Packaging with Zeolites on Fresh Broccoli Florets. *J. Food Sci. Technol.* **2021,** *58* (1), 197–204. https://doi.org/10.1007/S13197-020-04529-9/FIGURES/5

Mitchell, S.; Pinar, A. B.; Kenvin, J.; Crivelli, P.; Kärger, J.; Pérez-Ramírez, J. Structural Analysis of Hierarchically Organized Zeolites. *Nat. Commun.* **2015,** *6* (1), 1–14. https://doi.org/10.1038/ncomms9633

Moshoeshoe, M.; Nadiye-Tabbiruka, M. S.; Obuseng, V. A Review of the Chemistry, Structure, Properties and Applications of Zeolites. *Am. J. Mater. Sci.* **2017,** *7*, 196–221

Odom, I. E. Smectite Clay Minerals: Properties and Uses. *Philos. Trans. Royal Soc. A Math. Phys. Sci.* **1984,** (A311), 391–409. https://doi.org/10.1098/RSTA.1984.0036

Prasanna, V.; Prabha, T. N.; Tharanathan, R. N. Fruit Ripening Phenomena–An Overview. *Crit. Rev. FoodSci. Nutri.* **2010,** *47* (1), 1–19. https://doi.org/10.1080/10408390600976841

Sadeghi, K.; Lee, Y. S., Seo, J. Ethylene Scavenging Systems in Packaging of Fresh Produce: A Review. *Food Rev. Int.* **2021,** *37* (2), 155–176. https://doi.org/10.1080/87559129.2019.1695836

Sakizci, M. Effect of Salt Modification and Acid Activation on Ethylene Adsorption Properties of Sepiolite. *Adsorption* **2013,** *19* (6), 1083–1091. https://doi.org/10.1007/S10450-013-9531-6/FIGURES/4

Schmid, G. Nanotechnology, 2008.

Senchenya, I. N.; Kazansky, V. B. Quantum Chemical Studies of Ethylene Interaction with Zeolite OH-Groups. *Catal. Lett.* **1991,** *8* (5), 317–325. https://doi.org/10.1007/BF00764193

Shattar, S. F. A.; Zakaria, N. A.; Foo, K. Y. One Step Acid Activation of Bentonite Derived Adsorbent for the Effective Remediation of the New Generation of Industrial Pesticides. *Sci. Rep.* **2020,** *10* (1), 1–13. https://doi.org/10.1038/s41598-020-76723-w

Shen, W. A Theoretical Study of Confinement Effect of Zeolite on the Ethylene Dimerization Reaction. *Microporous Mesoporous Mater.* **2017,** *247*, 136–144. https://doi.org/10.1016/J.MICROMESO.2017.03.062

Shenoy, S.; Pathak, N.; Molins, A.; Toncheva, A.; Schouw, T.; Hemberg, A.; … Mahajan, P. V. Impact of Relative Humidity on Ethylene Removal Kinetics of Different Scavenging Materials for Fresh Produce Industry. *Postharvest Biol. Technol.* **2022,** *188*, 111881. https://doi.org/10.1016/J.POSTHARVBIO.2022.111881

Srithammaraj, K.; Magaraphan, R.; Manuspiya, H. Influence of Thiol Groups on the Ethylene Adsorption and Conductivity Properties of the Modified Porous Clay Heterostructures (PCHS) using as Ethylene Scavenger in Smart Packaging. *Polym. Bull.* **2018,** *75* (9), 3951–3969. https://doi.org/10.1007/S00289-017-2243-4/FIGURES/11

Sue-aok, N.; Srithanratana, T.; Rangsriwatananon, K.; Hengrasmee, S. Study of Ethylene Adsorption on Zeolite NaY Modified with Group I Metal Ions. *Appl. Surf. Sci.* **2010,** *256* (12), 3997–4002. https://doi.org/10.1016/J.APSUSC.2010.01.065

Sun, M.; Yang, X.; Zhang, Y.; Wang, S.; Wong, M. W.; Ni, R.; Huang, D. Rapid and Visual Detection and Quantitation of Ethylene Released from Ripening Fruits: The New Use of Grubbs Catalyst. *J. Agric. Food Chem.* **2019,** *67* (1), 507–513. https://doi.org/10.1021/ACS.JAFC.8B05874/SUPPL_FILE/JF8B05874_SI_001.PDF

Tas, C. E.; Hendessi, S.; Baysal, M.; Unal, S.; Cebeci, F. C.; Menceloglu, Y. Z.; Unal, H. Halloysite Nanotubes/Polyethylene Nanocomposites for Active Food Packaging Materials with Ethylene Scavenging and Gas Barrier Properties. *Food Bioproc. Technol.* **2017,** *10* (4), 789–798. https://doi.org/10.1007/s11947-017-1860-0

Tran, Y. T.; Lee, J.; Kumar, P.; Kim, K. H.; Lee, S. S. Natural Zeolite and its Application in Concrete Composite Production. *Compos. B Eng.* **2019,** *165*, 354–364. https://doi.org/10.1016/J.COMPOSITESB.2018.12.084

Tzeng, J. H.; Weng, C. H.; Huang, J. W.; Shiesh, C. C.; Lin, Y. H.; Lin, Y. T. Application of Palladium-Modified Zeolite for Prolonging Post-Harvest Shelf Life of Banana. *J. Sci. Food Agric.* **2019,** *99* (7), 3467–3474. https://doi.org/10.1002/JSFA.9565

Uddin, F. Montmorillonite: An Introduction to Properties and Utilization, In *Current Topics in the Utilization of Clay in Industrial and Medical Applications,* 2018. https://doi.org/10.5772/INTECHOPEN.77987

Upadhyay, A.; Kumar, P.; Kardam, S. K.; Gaikwad, K. K. Ethylene Scavenging Film Based on Corn Starch-Gum Acacia Impregnated with Sepiolite Clay and its Effect on Quality of Fresh Broccoli Florets. *Food Biosci.* **2022,** 101556.

USFDA. Food: Drug Administration Code of Federal Regulations, Google Scholar, 2015.

Villa, C. C.; Valencia, G. A.; López Córdoba, A.; Ortega-Toro, R.; Ahmed, S.; Gutiérrez, T. J. Zeolites for Food Applications: A Review. *Food Biosci.* **2022,** *46*, 101577. https://doi.org/10.1016/J.FBIO.2022.101577

Wei, H.; Seidi, F.; Zhang, T.; Jin, Y.; Xiao, H. Ethylene Scavengers for the Preservation of Fruits and Vegetables: A Review. *Food Chem.* **2021,** *337*, 127750. https://doi.org/10.1016/j.foodchem.2020.127750

Werner, B. G.; Koontz, J. L.; Goddard, J. M. Hurdles to Commercial Translation of Next Generation Active Food Packaging Technologies. *Curr. Opin. Food Sci.* **2017,** *16*, 40–48. https://doi.org/10.1016/J.COFS.2017.07.007

Widayanti, S. M.; Hoerudin; Andes, I. In *Characteristics and Postharvest Life of Snake Fruit (Salacca edulis Reinw) During Storage as Influenced by Application of Activated Nanostructured Natural Zeolites*, IOP Conference Series: Earth and Environmental Science, 2021; vol *803* (1), 012029. https://doi.org/10.1088/1755-1315/803/1/012029

Yang, R. T. In *Adsorption: Fundamentals and Fundamentals and Applications*, 2003; p 410.

Yuan, P.; Tan, D.; Annabi-Bergaya, F. Properties and Applications of Halloysite Nanotubes: Recent Research Advances and Future Prospects. *Appl. Clay Sci.* **2015,** *112*, 75–93.

Zarrintaj, P.; Mahmodi, G.; Manouchehri, S.; Mashhadzadeh, A. H.; Khodadadi, M.; Servatan, M.; … Mozafari, M. Zeolite in Tissue Engineering: Opportunities and Challenges. *MedComm* **2020,** *1* (1), 5–34. https://doi.org/10.1002/MCO2.5
Zhang, W.; Hu, Y.; Liu, J.; Wang, H.; Wei, J.; Sun, P.; … Zheng, H. Progress of Ethylene Action Mechanism and its Application on Plant Type Formation in Crops. *Saudi J. Biol. Sci.* **2020,** *27* (6), 1667–1673. https://doi.org/10.1016/J.SJBS.2019.12.038
Zhang, M.; Smith, J. A. C., Harberd, N. P.; Jiang, C. The Regulatory Roles of Ethylene and Reactive Oxygen Species (ROS) in Plant Salt Stress Responses. *Plant Mol. Biol.* **2016,** *91* (6), 651–659. https://doi.org/10.1007/S11103-016-0488-1

CHAPTER 7

NATURAL ANTIOXIDANTS FOR ACTIVE PACKAGING OF FRESH-CUT FRUITS

RAM KUMAR DESHMUKH and KIRTIRAJ K. GAIKWAD

Department of Paper Technology, Indian Institute of Technology, Roorkee, Uttarakhand, India

ABSTRACT

The fresh-cut fruits and vegetables are susceptible to different consequences after postharvest operation till consumption. Fruits are a great source of nutritional content, which can be degraded due to different reactions such as over-ripening, decaying, and oxidation reactions. The active antioxidant packaging could bring a sustainable approach to preserving these minimally processed fruits and vegetables. The synthetic antioxidant compounds are harmful to consumers and are expensive investments. The natural active compounds from the plant and herbs have great potential as an alternative antioxidant to synthetic compounds. In this chapter, the authors attempted to explain the physiochemical variation of fresh-cut fruits. The classification of antioxidant compounds and methods of incorporation into the packaging systems has been thoroughly presented. The application of natural antioxidants on different fresh-cut fruits and vegetables is explained. The authors also highlighted the limitation of utilizing natural antioxidant agents for minimally processed produce packaging.

Novel Packaging Systems for Fruits and Vegetables. Kirtiraj K. Gaikwad & Suman Singh (Eds.)

7.1 INTRODUCTION

Consumers have been attracted to the fresh-cut fruits and vegetables (FC&V) of convenience and safety in recent years. FCF&V with high-quality attributes such as flavors, texture, color, and appearance are the preference of these products (Perera, 2007). Any FCF&V or any combined form that has been physically altered by peeling, slicing, or chopping with or without washing from its original form without further processing (blanching or cooking) remains fresh. Fresh-cut produce could be a single commodity or in combination with two or more fruits/vegetables in the same package like salad. Fresh-cut fruits do not include products that have undergone processing like freezing, cooling, canning, or packing as juice, syrup, or dressing (Bintsis, 2018). It is a minimally processed product that provides consumers with ready-to-eat or ready-to-use convenience in the present scenario. FCF&V renders convenient to consumers with health-promoting micronutrients such as vitamins, minerals, fibers, etc. Fresh-cut fruits' processing involves harvesting, collecting, sorting, cleaning, washing, peeling, grading, and cutting (chopping, slicing, shredding, chunking, and dicing). The FCF&V are processed in a manner where they remain active after the postharvest, and it undergoes ripening and finally senescence. FCF&V is meant to be consumed in a short time after harvesting. They are delicate and vulnerable after minimal processing operations like cutting, peeling, and slicing, even at refrigerated storage and controlled humidity (Abadias et al., 2008). Processing operations of FCF&V accelerate the physical-chemical and microbial alteration, limiting the longer shelf life, nutritional loss, and sometimes the formation of harmful compounds. FCF&V are more easily spoiled than whole produce, as they physically change during minimal processing operations. FCF&V reduces meal preparation time and volume of kitchen-generated waste, but it is very delicate due to minimal processing, resulting in biological destabilization. Minimal processing of fresh fruits and vegetables includes breaking down or separating the outer protective layer of the produces and dismantling the natural firm phenomenon of the produces that leads to the release of intracellular products that accelerate bacterial growth (Oliveira et al., 2015). Since the fresh produce is consumed raw or after minimal processing, contamination can cause potential health and safety risks through different means. Several factors can compromise fresh produce's chemical and microbiological integrity, some of which have a potentially fatal outcome.

At the farm level, preharvest hazards have been recognized as necessary because once pathogens contamination is established in the field, it is challenging to disinfect the produce later (Li et al., 2017). Fresh-cut produce is usually associated with the incidents of pathogen growth through various sources, either from processing equipment or from personnel handling. Pathogens such as *Escherichia coli* and *Salmonella spp* can enter the water when washing fresh produce. Cross-contamination of pathogens can occur via manure, irrigation, water, soil, processing equipment, or personnel handling (Alegbeleye et al., 2018). Therefore, to ensure product safety and consumer health, it is vital to correctly understand the nature of fresh produce and the contamination caused by different routes. Although, washing fresh produce with a disinfectant solution enhances microbial inhibition and skips cross-contamination. Sterilizers, ozone treatments, ultraviolet (UV) treatment, and chemical intervention are prevalent methods to extend the shelf life and maintain the wholesomeness of fresh produce in the food industry. But these preservation methods cause a loss of flavor and colors, eliminate the product's natural condition, reduce the sensory and nutritional compositions, and, most importantly, affect the consumers' acceptability (Brody et al., 2011). Therefore, packaging could be an effective method to preserve the quality and freshness of fresh-cut produce for a longer duration. Among all packaging, modified atmosphere packaging (MAP) and controlled atmosphere packaging (CAP) are the most studied packaging method. The basic principle of these methods is to modify fruits' surrounding environment by removing or manipulating the level of gases necessary for the respiration of fruits, such as oxygen, carbon dioxide, and ethylene. This modified atmosphere delays the ripening process, which slows down the metabolic activity of fruits. As a result, it inhibits microbial growth, which slows down the fruit's decaying process (Maringgal et al., 2020). Active packaging with antioxidant and antimicrobial effects could be an effective solution to preserve the quality of fresh-cut produces. Active packaging performs as conventional packaging by preventing outside contamination, moisture, water vapor, and gas barrier. Simultaneously, active packaging has secondary functional protection such as antimicrobial, antioxidant, moisture absorber, oxygen scavenger, carbon dioxide emitter, and ethylene scavenger. Moreover, FCF&V requires further evolution for better preservation, simultaneously assuring quality, safety, and sustainable food processing. Oxidation is one of the most detrimental causes of flavor, color, and nutritional loss due to enzymatic browning (Khan et al., 2021). The food industry is constantly searching for new methods to reduce the economic impact due to the effect of oxidation. However, controlling the level of oxygen availability and MAP has effective control over the oxidative reactions. Therefore, the food industry has long recognized the importance of antioxidants in packaging to retard or delay the oxidation of fruits and vegetables. An antioxidant is a novel active packaging technique that restricts the oxidation reaction through the controlled release of antioxidants inside the package environment. There are numerous artificial antioxidants, such as butylated anisole (BHA), butylated hydroxytoluene (BHT), and propyl gallate (PG), that have been approved by the European Union under directive 2006/52/EC (Union, 2006). The antioxidant packaging can be of different types, such as self-adhesive labels, membranes, sachets, and sheets of paper. These can be incorporated inside the packaging environment. Now researchers have extended to grafting and immobilizing active antioxidant compounds into the packaging matrixes of primary layer packaging. The presence of antioxidant compounds in the primary layer will work on the controlled release mechanism from the packaging matrix to the headspace of packaging and inhibit the oxidation initiation by scavenging free radicals. Several studies suggest that the incorporation of AOs into the packaging matrix provides an extensive improvement in the inhibition of oxidation reactions. Artificial active compounds contain different chemicals that can cause serious health-related issues such as allergy, asthma, hypersensitivity, neurological damage, and even life-threatening diseases like cancer. In recent years, consumers awareness increased for food safety and health, and they now prefer or demand the naturally sourced additives in food and food packaging applications (Tanwar et al., 2021).

Therefore, the search for natural-derived antioxidant compounds is attracting more attention from the academic and industrial community due to their low cost, high availability, and lower safety concern. From this perspective, the researchers from the different fields have put their efforts into extracting, optimizing, and incorporating those natural compounds into the development of active packaging fabrication. The active antioxidant components from plant extract, essential oils of herbs and spices, and the polyphenolic extract from waste bioresources widely exist. The available investigations have demonstrated that the packaging matrix can release active AO at controlled rates, thereby inhibiting lipid oxidation and preserving nutritional values (Rangaraj et al., 2021). In this chapter, the authors will provide insights into the investigation, classification, functions, benefits, limitations, and future perspective on the natural-based antioxidants for packaging of FCF&V.

7.2 QUALITY CHANGES OF FRUITS AND VEGETABLES DURING MINIMAL PROCESSING

Fruits and vegetables that have been minimally processed are an important and rapidly growing food group. These convenient foods are made possible by various novel scientific and technological applications. FCF&V has piqued the interest of diverse food industries, including food manufacturers, retail food markets, restaurants, catering services, and commissary units (Wiley and Yildiz, 2017). Several quality attributes of minimally processed produce are affected by different factors during the processing. When leaves senesce, they undergo a series of cellular breakdowns, macromolecule degradation, and efficient nutrient transport from decaying tissues to developing ones. The control of dell decaying and cell membrane peroxidation is very closely related to the shelf-life extension of FCF&V (Wiley and Yildiz, 2017). The FCF&V tissue integrity changed and susceptible than the raw intact form. The minimally processed fruits can be easily contaminated or deteriorated due to physiological aging, chemical intervention, and microbial invasion. Processing causes injury stress, leading to cellular decompartmentalization or delocalization of enzymes and enzymatic substances, resulting in biochemical deterioration such as browning, off-flavor, and texture deformation. Furthermore, washing, sorting, peeling, cutting, shredding, and slicing allow epiphytic and phytopathogenic microbes to infect fresh fruit and vegetable tissues (Bobo-García et al., 2020). Figure 7.1 illustrates the different physicochemical, sensorial changes over the period of harvesting to consumption of fresh produce.

7.2.1 PHYSIOLOGICAL AND BIOCHEMICAL CHANGES

The main physiological phenomenon is increased respiration and, in some cases, ethylene production; however, the response is dependent on the severity of the stress. The fruits and vegetables are prone to desiccation, chilling injury, CO_2 injury, and wound stress that causes metabolic loss.

The respiration rate of FCF&V is higher than intact; the consumption of O_2 by shredded fruits and vegetables increased more than that of the whole one. Shredded endive has 1.2 times the O_2 consumption of broccoli, 1.4 times the O_2 consumption of shredded lettuce. Plant tissues that have been damaged breathe three to seven times faster than intact tissue, such as grated carrots, which breath seven times faster (Sharma et al., 2019). The damaged plant's metabolism accelerated, resulting in rapid oxygen consumption in the packaging atmosphere. Compared to an intact head of lettuce, MAP reduces the rate of respiration in shredded lettuce by up to four times after 16 days of storage at 2°C (Tudela et al., 2016). Even in trace amounts, ethylene as a ripening hormone can harm vegetable growth, development, and shelf-life. In fruits, mainly climacteric fruits, ethylene accelerates ripening, affecting the storage life of fresh produces. Ethylene is involved in the neo-synthesis of enzymes, which assists in maturation and is also responsible for the physiological disorder in sliced fruits and vegetables. Ethylene biosynthesis can be inhibited by maintaining the CO_2 and O_2 concentration, low temperature, vacuum packaging, blocking ethylene generating sites and using active packaging such as ethylene absorber and active antioxidant packaging.

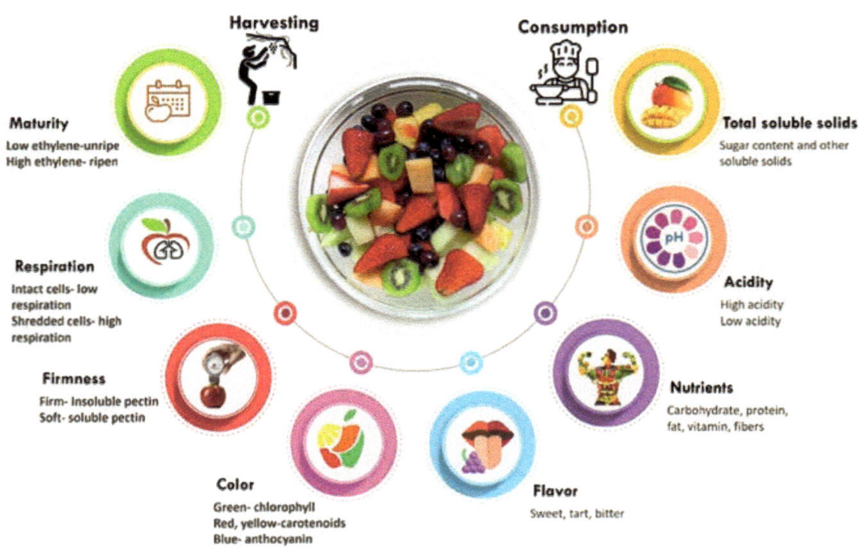

FIGURE 7.1 Physical, chemical, sensorial, nutritional variation in fresh produce during the storage period.

7.2.2 MICROBIOLOGICAL CHANGES

Various bacterial species and fungi are responsible for the spoilage of minimally processed fresh produce. Gram-negative bacteria such as Pseudomonas, Erwinia, and Enterobacteriaceae are common in fresh fruits and vegetables, while Gram-positive bacteria such as lactic acid bacteria (LAB) and yeast are common. Fresh lightly processed produce's surface is exposed to the air, and microbial contamination occurs through different means during the processing. The large surface area is exposed to an open environment where the acidity is low or mild provides a suitable condition for microbial growth (Kaczmarek et al., 2019). The initial high population of microorganisms makes it challenging to set a threshold cell number where the product can be considered fresh or spoiled. All spoilage mechanisms of fresh produce are interdependent and contribute to the quality compromise of the minimally processed horticulture produce.

7.2.3 NUTRITIONAL CHANGES

Fruits and vegetables, either fresh or processed or minimally processed, are the primary dietary source for human health. The FCF&V industry claims that the minimally processed products are as nutritious as fresh, intact ones to fulfill consumers' satiety and diet requirements. However, many nutritional variations occur in fresh produce during minimal processing. Little information about the nutritional changes in fresh-cut produces during minimal processing is available. Klein (1987) thoroughly reviewed the nutritional aspects of minimally processed produces and suggested that the conditions can maintain desirable sensory characteristics that prevent nutrients. Minimal processing like washing, grated, peeling, and shredded does not affect the macronutrients such as protein, carbohydrates, and lipid content of fruits and vegetables significantly but vitamin content (vitamin C). Ascorbic acid is a water-soluble organic acid similar to vitamin C and is commonly found in fruits and vegetables. The loss of ascorbic acid in minimally processed produces is an indicator of oxidative degradation. Ascorbic acid can be oxidized due to mechanical damage caused by slicing, cutting, and shredding. Fresh raw produce's cell is bursted due to washing, grading, peelin, and cutting, and thus chemical and biological reactions rate elevated, which lead to reducing the storage life of the cut product.

Ascorbic acid oxidizes ascorbic acid to dehydroascorbic acid (DHAA) after releasing it from the cell walls, rendering the product vitamin C deficient (Klein, 1987).

7.2.4 OFF FLAVOR AND DISCOLORATION

The plant tissue's compositions are constantly altering during growing and ripening, resulting in either positive or negative quality attributes. Along with these compositional variations, the total soluble solids (TSS) and acidity are known as quality indexes. Climacteric fruits and vegetables are more susceptible to these changes since they continue to ripen after harvesting. Flavors of fresh produce are greatly influenced by the sugar content, organic acid (acidity), phenolic compounds (astringency), and specific volatile compounds. Flavor loss or off-flavor can be correlated to the improper postharvesting process, ripening, unsuitable storage, or enzymatic action such as peroxidase and lipoxygenase. These reactions form free reactive radicals and hydroperoxides that may catalyze the oxidation of lipid compounds, which leads to the development of unpleasant flavors as rancid and oxidized. For example, n-hexanal, a byproduct of hydroperoxide degradation, has been linked to off-flavor in peas and typical flavor in tomatoes (Barrett et al., 2010).

Discoloration (pink, reddish-brown, blackening), particularly for FCF&V, is an essential in quality parameters. Polyphenol oxidase (PPO) caused enzymatic browning in the presence of O_2 converts phenolic components in fresh-cut produces into dark-colored pigments. Usually, the discoloration encountered in vegetable products such as onion, garlic, and potatoes, in which pink, red, green, and blackish colors can develop due to cell disruption due to minimal processing. Discoloration occurs more often in cut produce than in fresh commodities, but it can develop during the minimal processing if the operations lead to significant tissue damage (Salehi, 2020). For example, pink coloration is a significant deterioration index of fresh lettuce and carrots or tomatoes; a whitish color is attributed to an enzymatic reaction to dehydration of the surface.

7.3 ACTIVE PACKAGING OF FC&V

The main deterioration factor of each food product must be understood before using active or intelligent packaging technologies. Active packaging should be designed for each product to overcome the main deterioration factor. FCF&V are sensitive products with limited shelf life due to their susceptibility to physical, chemical, and microbiological deterioration. Therefore, fresh-cut produce poses unique challenges to packaging. Packaging systems have evolved considerably in the last decades. Extensive research has been put into novel techniques such as CAP, MAP, oxygen scavenging, ethylene scavenging, CO_2 emitter, and antimicrobial and antioxidant packaging (Mehyar and Han, 2011). The standard function of fresh produce is its postharvest metabolism, which results in respiration, ripening, and senescence. Fruits and vegetables have their optimum maturity stage with all the desired quality characteristics like firmness, color, flavor, and nutritional value. Minimally processed fresh produce exhibits a considerably higher respiration rate due to mechanical processing stress.

All these points have been predetermined with the potential of packaging design to create a gas composition that fits the individual product requirement. The gas composition effect on the quality of FCF&V has been thoroughly described in many previous investigations, which suggested the optimum level of O_2 and CO_2 concentration inside the package environment (Oliveira et al., 2015). O concentrations of 0.5–5% are required to maintain aerobic respiration when storing commodities. Low O_2 levels cause anaerobic metabolism, which causes off-flavors and rapid product deterioration. Compared to air (20.95% v/v O_2), a lower O_2 concentration can extend the shelf life of fresh produce by retarding ripening. A lower respiration rate is generally seen as a positive effect of increased CO_2 levels, although fruit and vegetables have varying CO_2 tolerance levels. For instance, grapes and tomatoes should have no more than a 2% of CO_2 level in the packaging atmosphere, whereas melons and berries can tolerate CO_2 up to 15%. The storage of FCF&V in altered gas composition finds broad application in commercial packaging.

MAP establishes a headspace gas atmosphere that differs from the air. The non-climacteric fresh produce package headspace is flushed with desired gas composition to create the required storage condition for product-specific. In comparison, climacteric-type products are continuously modifying the package headspace atmosphere with a natural respiration process, leading to lower O_2 and higher CO_2 content compared to air (Mehyar and Han, 2011). MAP with the gas composition of 2–5% of CO_2 and 2–5% of O_2 was traditionally applied to keep fresh fruits and vegetables, but higher gas composition could be essential in the case of FCF&V. Fresh-cut pineapple found increased shelf life for seven days at 10°C with the MAP having 6% O_2 and 14% CO_2 concentration, and minimally processed carrot packed in high O_2 with 80% low CO_2 with 10% was found better keep quality (Oliveira et al., 2015).

Fruits and vegetables are a rich source of many bioactive components with antioxidant properties apart from a good reservoir of vitamins, minerals, and fibers. These compounds are carotenoids, anthocyanins, malic acid, citric acid, polyphenolic compounds, indole, isothiocyanates, and vitamins. Additionally, the external quality attributes such as color, shape, size, and surface become important quality parameters. Antioxidants in fruits and vegetables can be degraded by various factors, including improper handling, autoxidation, fungal decay, chilling injury, irradiation, improper temperature and relative humidity, and other stresses. The antioxidant activity of FCF&V was significantly lower than that of whole fresh produce, which could be attributed to the larger surface area of FCF&V and contact with oxygen, sunlight, and other factors (Ioannou, 2013). With various postharvest treatments, the quality of fruits can be preserved for an extended period. These include ultraviolet radiation, controlled atmospheric packaging (CAP), heating treatment, edible coatings, and active packaging. The incorporation of different natural antioxidant compounds into the packaging matrix can be accomplished by different cutting-edge techniques such as extrusion, casting, coating, and encapsulation. An antioxidant packaging that slowly releases antioxidants or incorporates oxygen or radical scavengers into the packaging is a novel alternative to these methods. The European Union has approved several antioxidant compounds under Directive 2006/52/EC. Antioxidant embedded packaging is a novel alternative to MAP and CAP. It restricts or stops oxidation by releasing antioxidant compounds from the packaging material to the food material inside the package over time. Previously, synthetic antioxidants were widely used in active packaging applications; however, due to health concerns and resource constraints, natural antioxidants are increasingly replacing synthetic antioxidants (P. Sharma et al., 2019).

7.4 ANTIOXIDANT PACKAGING

Active antioxidant packaging is an emerging area of active packaging. Incorporating bioactive antioxidant potential compounds into the packaging matrix restricts the undesirable consequences of oxidative reactions in food products. There is literature on incorporating synthetic or artificial antioxidants, such as butylated hydroxytoluene (BHT), propyl gallate (PG), and butylated hydroxyanisole. Their implementation primarily focuses on processed food such as cereal-based bakery, meat, and dairy products. In recent years, the fresh fruits and vegetable market attracted researchers to find an alternative to these synthetic antioxidant compounds. The natural antioxidant compounds from the plant, herbs, and spices such as caffeic acid, tocopherol, curcumin, menthol, quercetin, and catechins have been explored in recent years for antioxidant packaging applications for different food products. A controlled diffusion mechanism releases these antioxidant compounds in the packaging matrix (Khan et al., 2021). It is widely accepted that essential oils (EO) derived from various plants and animals contain many antioxidants found in synthetic antioxidants. Natural antioxidants decompose in the environment and are biodegradable, biocompatible, antimicrobial, and easily decomposable. While most previous research has concentrated on the antioxidant packaging effect of lipid oxidation, primarily on high lipid foods like fish, meats, and dairy products, there is still a shortage of data on the antioxidant packaging effect on the quality of slightly processed produce during storage. Researchers have recently focused their attention on biodegradable materials such as biopolymers (carbohydrate, protein, polysaccharides, and lipids) for their nutritional preservation and sensory quality of fresh-cut fruits by employing natural antioxidants in the primary layer of edible or biodegradable films or coatings (Kadzińska et al., 2019).

7.5 TYPES OF ANTIOXIDANTS

Various antioxidant-based active packaging developments have been recently investigated to optimize the storing and storage quality of fresh-cut fruits. The active compounds from the packaging matrix are released through the controlled mechanism to neutralize the free radicals from the food; thus, the oxidation reaction is inhibited, and shelf life is improved. Antioxidant compounds are classified based on their source, such as synthetic and natural antioxidant. The use of natural antioxidants as an alternative to synthetic antioxidants is possible because synthetic antioxidants are harmful to human health. The aqueous and alcoholic plant extract, essential oils derived from herbs and spices, and the polyphenolic components from various waste bio-resources are the natural antioxidant agents. Several investigations revealed that extracting natural antioxidant compounds from different natural sources could scavenge the oxidative free radicals by slow and controlled release mechanism of the antioxidant agent from packaging film to package headspace. Figure 7.2 depicts the classification of antioxidants. The antioxidants are classified into two categories based on their reaction mechanism in the active packaging system: release-type antioxidant and scavenging-type antioxidant packaging. Release-type antioxidant packaging is effective and long-lasting, which delivers the antioxidant compounds at a controlled and consistent rate. The uniform concentration of antioxidant compounds in the matrix can be maintained throughout the store to ensure the more prolonged effect of antioxidant packaging. In comparison, the scavenger type of antioxidant packaging works as the preventive mechanism for slowing down or inhibiting the oxidation reaction in the food. Another classification could be based on solubility in the solvent, hydrophilic active antioxidant compounds like uric acid and ascorbic acid, and lipophilic such as carotene and ubiquinol. Lipophilic AOs inhibit lipid peroxidation and protect cell membranes, whereas hydrophilic AOs react with oxidants in blood plasma (Islam et al., 2017).

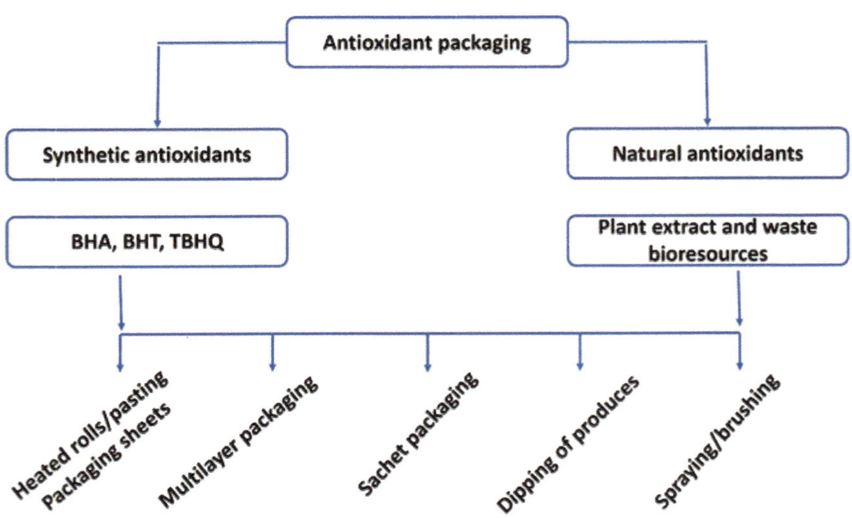

FIGURE 7.2 Classification of antioxidant agent and different incorporation methods for development of different active packaging systems.

7.6 NATURAL ANTIOXIDANTS

Natural antioxidant compounds are complex molecules found primarily in plants, spices, and herb extracts. These can help prevent different oxidation reactions that occur, which degrade the food quality, such as peroxidation, oxidative intermediates, free radicals, and secondary metabolites. These compounds also have an antimicrobial and antifungal activity that protects from the microbial invasion of fresh and cut produce. Therefore, a lot of attention to developing antioxidant-based packaging systems from natural resources for fresh commodities gained popularity. Antioxidant molecules are divided into three categories: vitamins (ascorbic acid, tocopherol), carotenoids (anthocyanins, carotene, and xanthophyll), and polyphenols (flavonoids, phenols, lignin, stilbenes). These compounds' antioxidant efficiency depends on their chemical structure, extraction methods, and implementation methods. Simultaneously, their polyphenolic content determines the antioxidant activity of phenolic concentrates and EOs. Therefore, natural antioxidant agents are adequate for developing active antioxidant packaging films and coatings for fresh-cut fruits and vegetables. These act as potent free radical reducing agents on the exposed area of the fresh-cut fruits to inhibit the oxidation reactions. The most studied natural antioxidants are carnosol, rosmanol, rosemary, quinone, and rosemary diphenol, which have a high capacity for radical oxygen absorption (ORAC). Essential oils with significant free radical scavenging activity include lemongrass, oregano, ginger, cinnamon, clove, mint, and citrus (Deshmukh and Gaikwad, 2022).

7.7 INCORPORATION OF NATURAL ANTIOXIDANT COMPOUNDS INTO PACKAGING MATRIX

Antioxidant packaging for FCF&V is fabricated using mainly three approaches, that is, non-package bound (sachet, pad, labels) and package bound (directly incorporated onto the packaging film layer). The fresh produce has a delicate surface area, which is exposed to the active compounds in the packaging system. The form of active packaging for different products has their pro and cons effect on the quality of fresh-cut produces. Figure 7.2 explains the different types of incorporation method for active packaging solutions.

7.7.1 NON-PACKAGE BOUND SYSTEM

The inclusion sachet, label, or pad, which contains the active antioxidant compounds act as an oxidation radical scavenging system. The non-package or loosen bound system for active packaging has a vast application in the fresh, processed, and semiprocessed food industry. Incorporating natural antioxidant compounds in the loosen bound system does not help protect the produces from oxidation. The oxygen scavenging mechanism will lead to the inhibition of the oxidative reaction in the food products with the limiting oxygen availability in the package atmosphere. Sachets or pads are commonly used for the oxygen scavenging packaging system. Fine ferrous and iron oxides are the most widely used method of oxygen scavenging activity. However, polyphenols, ascorbic acid, and some enzymes have also been used in the sachet form for active packaging. The sachet or pads are made up of highly oxygen permeable material. These oxygen-scavenging systems have the potential to eliminate the oxygen up to the 0.01% level as per the requirement of food products. To optimize the oxygen scavenging action of the active sachet, a specific mechanism can be adapted to initiate the absorbing reaction. For example, the humid condition is vital for the initiation of oxygen removal for metal-based oxygen scavenging systems. However, fresh produce requires oxygen for respiration. Still, many fruits are susceptible to the high level of oxygen, i.e., peeled garlic, cut apple, potato, and lettuces are sensitive to enzymatic browning by polyphenolic oxidation; hence low oxygen (<1%) is essential to inhibit the oxidation incidences. Moreover, it is also a critical point that oxygen content should not be lower than the tolerance limit. The anaerobic condition will lead to anaerobiosis, and toxic compounds will be produced (Gaikwad et al., 2018).

7.7.2 PACKAGE BOUND SYSTEM

The antioxidant compounds can be incorporated inside the container or packaging layer in which the FCF&V are packed or onto the packaging film walls. The active layers on the packaging film employed their activity by absorbing the undesirable compounds such as free radicals and releasing the antioxidant agents into the packaging headspace. Additionally, the polymeric matrix serves as a barrier that prevents oxidation-accelerating factors (mainly oxygen) from penetrating the package, thus increasing the antioxidant activity of the food product and its headspace. Polymer type and the antioxidant's properties (i.e., mechanism of action and heat tolerance) are critical to the development process of antioxidant packaging material. This method may produce active films by casting or extrusion for practically any antioxidant. An antioxidant compound's mode of action is dependent on the principle of migration into food products; therefore, any antioxidant compounds released need not only comply with relevant rules on their maximum permissible level but also be approved as food additives. Antioxidant agents or active compounds are thoroughly mixed with polymer matrix when producing an antioxidant packaging (Deshmukh and Gaikwad, 2022).

7.7.3 COATINGS

Adsorption or coating of antioxidants onto food packaging may also inhibit oxidative responses of foodstuffs due to antioxidant migration from the container to food. While keeping respiration requirements in mind, an active solution with antioxidant compounds can be used as a coating material for direct application on the fruits and vegetable surfaces. Active substances can transmit antioxidant activity in the headspace through direct interaction with the surface of minimally processed fresh produce or by forming a vapor phase. It can act as a gas or vapor barrier and transporter of active chemicals like antioxidants and antimicrobial agents. The application of the edible coating with an active antioxidant agent has been proposed as a successful food packaging solution. Chemical interaction between the coating ingredients and the surrounding atmosphere results in the release of active compounds for antioxidant packaging. Moreover, the active antioxidant component concentration carried by the coating is influenced by the final thickness of the coating on the surface of fresh-cut produces, which depends upon the rheological properties of the coating solution (Ncama et al., 2018).

7.7.4 MULTILAYER ACTIVE FILM

Multilayer packaging is an active antioxidant agent-containing layer sandwiched between inert layers. It can be an alternate solution to overcome concerns relating to active incorporated film but at a high cost. It was found that natural compounds such as tocopherol and quercetin were sandwiched between two layers of EVOH and other common packaging materials to make them more antioxidant-friendly. Thymol, carvacrol, and eugenol have been incorporated into two layers to develop multilayer films. Due to their higher cost and more involved production process, multilayer active films are not as widely used as sachets, coatings, and integrated methods (Basumatary et al., 2022).

7.7.5 COVALENT IMMOBILIZATION

Traditional techniques for active packaging development, including active antioxidant packaging, include coating or blending the active chemical with the packaging material. Such procedures rely on the migration or transfer of active chemicals to food items to be successful. Not only would the packets lose their activity over time, but the antioxidants would undoubtedly infiltrate the food product as additives, causing customer dissatisfaction and posing a regulatory challenge. Therefore, immobilizing the active agent within a polymer matrix would give better protection against the oxidative causing factors in FCF&V. There has been growing interest in the covalent immobilization of active components onto the food contact surface of different packaging films in recent times. Covalent bonding may offer the most stable connection between the surface of the substrate film and the active substances, which may have regulatory implications. Because the active agents are unlikely to transfer from the box to the meal, they may not need to be labeled as food additives (S. M. B. Hashemi and Jafarpour, 2020).

7.8 APPLICATION OF ANTIOXIDANT PACKAGING FOR FRESH-CUT FRUITS AND VEGETABLES

Active antioxidant packaging works differently for different polymer/film/coating. The conjugation, composition and structure, chelating agents, extraction method, and chemical modifications could be the possible reasons for such diversity in the mechanism of antioxidant packaging (Deshmukh and Gaikwad, 2022). The oxidation of fresh produce can be prevented with the polyphenols that undergo single-electron transport and hydrogen atom transfer. It consists of giving away the unpaired electron atom to the oxidant to neutralize the oxidative compounds, and food results in the formulation of safe compounds. The antioxidant compounds greatly influence the physical and mechanical properties of the developed films. Those compounds contain many viable hydroxyl groups, which have intermolecular hydrogen bonding with the base material such as polysaccharide, lipid, and proteins with the addition of plasticizers. This plasticizer increases void space between the matrix structure and results in flexible films. The storage of trimmed cucumber was evaluated by incorporating active saffron petal extract (1–4%) into novel konjac glucomannan (KGM) edible film. The variation in the physical, chemical, and microbiological factors was evaluated during the packaging study. They observed that the addition of natural extract significantly reduces water vapor permeability and transparency. The antimicrobial activity of the film significantly improved against the Escherichia coli, Staphylococcus aureus, and Bacillus cereus till the fifth day of storage (S. M. B. Hashemi and Jafarpour, 2020). Sodium alginate with thyme oil-based edible coatings was applied to the fresh-cut apple and found that the coating could preserve the sensory attributes of the fresh-cut apple for 16 days at 4 C (Hu et al., 2018). Chitosan-based active coating with a different organic acid such as acetic and lactic acid was examined for the preservation of fresh-cut apple, and it demonstrated that the coating significantly reduced the microorganism's proliferation (1< 1log CFU) for 35 days. Fresh-cut mangoes were coated with the denatured protein/guar gum with mango puree significantly reduced the sensory and color degradation of fresh-cut produce (L. Sharma et al., 2019). Pectin-based dip coating was applied for the fresh-cut carrot, found to have a storage life extension of up to 12 days, and alginate with thyme and oregano oil incorporated dip-coating onto the fresh-cut papaya significantly reduced the weight loss and showed inhibition in color degradation at 4 C for 12 days (Yousuf et al., 2020). There are plenty of investigations on the development and application of active antioxidant packaging from natural compounds for fresh-cut products, as given in Table 7.1.

7.8.1 CHALLENGES AND FUTURE PROSPECTUS

Consumers' awareness increased due to consistent food science and processing development. It does not demand the protection of food alone but the expectation of being eco-friendly and sustainable at the same time. Therefore, active packaging with natural bioactive compounds has

TABLE 7.1 Application of Naturally Derived Antioxidant-Based Active Packaging of FCF&V for Extending the Quality and Shelf Life.

S. No.	Natural active functional compound	Form of packaging	Base material	FCF&V application	Output	Reference
1.	Thymol	Composite film	Sodium alginate	Apple	Browning action delayed	(Chen et al., 2021)
2.	Malic acid and cinnamon oil, palmarosa oil, lemongrass oil	Coating	Alginate film	Lemon	Improve shelf-life of a cut lemon for ten days	
3.	α-tocopherol acetate	Coating	Sodium alginate	Potato	The stability in the total phenolic compounds even after 15 days	(Keshari et al., 2022)
4.	Curcuma longa L. (Curcumin)	Membrane	Cellulose acetate	-	Antioxidant activity up to 86 to 93% was observed	(Baldino et al., 2017)
5.	Ascorbic acid	coating	Carboxymethyl cellulose/$CaCl_2$	Fresh-cut apple	The coating formulation suppresses the browning in the fresh-cut apple and stabilizes the firmness during the storage period.	(Saba and Sogvar, 2016)
6.	Tea polyphenol	Film	Gelatin/sodium alginate	-	DPPH and ABTS radical shows that the highest value of 90.62% and 53.36% of Trolox (mg Trolox equivalent/g film)	(Dou, Li, Zhang, Chu, and Hou, 2018)
7.	Aloe vera gel, green and black tea extract	Coating	gelatin	Fresh-cut orange	Gelatin-based active components coating reduces the weight loss, ascorbic acid degradation, and color darkening in the fresh-cut orange for 17 days at 4°C.	(Radi et al., 2017)
8.	Shirazi thyme essential oil	Coating	Alginate	Pistachio	The coating solution reduces the pistachio color browning during storage	(M. Hashemi, et al., 2021)

TABLE 7.1 *(Continued)*

S. No.	Natural active functional compound	Form of packaging	Base material	FCF&V application	Output	Reference
9.	Tannic acid	Film	Chitosan/gelatin	Fresh-cut apple	The developed film decreased the weight loss and suppressed the color browning of fresh-cut apples.	(C. Zhang et al., 2021)
10.	Tween 80, ascorbic acid, and vanillin	Coating	Alginate/carboxymethyl-cellulose	Fresh-cut kiwi fruits	The coating significantly delayed physiological weight loss from 8–14%, and decay was reduced by 3–59% in the control sample.	(Manzoor et al., 2021)
11.	Transglutaminase	Film	Whey protein/pectin	Fresh-cut apple, potato, and carrots	The developed film reduces microbial growth and preserves the phenolic compound, and the carotenoid is carrot.	(Marquez et al., 2017)
12.	Galacturonic acid	Film	Pectin/sodium alginate/xanthan gum	Fresh-cut potato	The developed film showed the anti-browning effect on the fresh-cut potato.	(Fan et al., 2021)
13.	Pterostilbene	Film	Citrus protein/sodium alginate	–	The developed film improved the barrier properties with high antioxidant activity.	(Y. Li et al., 2021)
14.	Ferulic acid	Coating	soy protein	Fresh-cut apple	The coating solution reduced the weight loss and firmness of the apple and inhibited color browning.	(Alves et al., 2017)
15.	Urtica dioica leaf extract	Film	Chitosan	Guava	The developed film with high antioxidant and antimicrobial agent improved the storage life of guava by one week.	(Kalia et al., 2021)
16.	Moringa leaf extract	Film	Papaya	Pears	The developed film with an antioxidation reduced browning in minimally processed pears.	(Rodríguez et al., 2020)

TABLE 7.1 *(Continued)*

S. No.	Natural active functional compound	Form of packaging	Base material	FCF&V application	Output	Reference
17.	Pomegranate peel extract	Film	Chitosan/alginate	Guava	The developed film/coating formation successfully retards the phytochemical loss.	(Nair et al., 2018b)
18.	Pomegranate peel extract	Film	Chitosan/alginate	Capsicum	The application of plant peel extract on the chitosan and alginate coating significantly reduces the physiological losses in weight, firmness, color, and ascorbic acid than the control sample.	(Nair et al., 2018a)
19.	Procyanidin	Coating	Chitosan	Blueberry	A decrease in luminosity and an increase in blue hue color improve shelf life.	(Mannozzi et al., 2018)
20.	Banana peel extract	Film	Chitosan	Apple	Exhibited excellent antioxidant activity in different food simulants	(W. Zhang, Li, and Jiang, 2020)

recently emerged as a high-impact alternative to chemical and petroleum-based packaging solutions. However, food packaging contributes to solid waste due to the high demand for ready-to-eat and served food items. Biopolymers are considered a choice over petrochemical-based plastic food packaging solutions. Unlikely conventional plastic packaging materials degrade gradually within a short period.

Nevertheless, a comprehensive evaluation of the environmental benefits of edible or biodegradable packaging is complex. As a result, there is an urgent need to disseminate and commercialize the active packaging to promote its benefits to society and the environment and gain consumer acceptance. The use of natural antioxidants in developing packaging solutions for FCF&V could be an innovative way to preserve the quality of fresh produce. The perishability of the FCF&V is detrimental to oxidative and enzymatic degradation, which requires inhibiting storage and distribution throughout the supply chain. The antioxidant compound incorporated in packaging films and coating can preserve the quality over a long period. However, the chemical formulation of the extract that is being applied in the development of the packaging of fresh-cut produces needs to be specified because all the plant or animal-derived components could not be beneficial to consumers' health. The published reports in the past five years have suggested the recent advancement in the development of films and coating with essential oils and polyphenolic compounds had groundbreaking revealing to the preservation of fresh-cut produces. The extraction of natural compounds from plant or animal sources has great economic concern because it may be inexpensive at the lab scale. Still, for the commercial application, it would not be that cost-effective (Deshmukh and Gaikwad, 2022; Khan et al., 2021).

Nevertheless, each innovation has their shortcoming and limitations. The main challenge today is the active compound-loaded packaging film or coating that can affect the sensory and nutritional characteristics, which demands attention for future studies. Edible or biodegradable biopolymer development is expected to emerge based on a detailed understanding of biochemistry and interactions with the physical, chemical, and biological aspects of different packaging matrices with fresh produces. Fundamentally, further progress is required in biofilm development, incorporation, migration of components from the packaging films, extraction of active components, application on the fresh produces, and evaluation of physicochemical quality aspects for long-term study.

7.9 CONCLUSION

This chapter explains the active antioxidant-based packaging solution for fresh-cut produce and the incorporation methods for developing functional coating, film, or sachet for the packaging of FCF&V. The active compounds with packaging films significantly contribute to the inhibition of lipid and enzymatic oxidations. The coatings involve the antimicrobial effect and the antioxidant effect, which helps parallelly eliminate the possible microorganism invasion in the fresh produces. Moreover, the emerging demands of research and development of sustainable packaging solutions with low-cost input through bio-resources to incorporate the natural components could bring a great choice to explore this direction. These biopolymers-based edible and biodegradable packaging solutions with natural compounds can fulfill the expectations of global consumers.

KEYWORDS

- **antioxidant packaging**
- **natural antioxidant**
- **fresh-cut produce**
- **oxidation**
- **browning**
- **coating**
- **biopolymer**

REFERENCES

Abadias, M.; Usall, J.; Anguera, M.; Solsona, C.; Viñas, I. Microbiological Quality of Fresh, Minimally-Processed Fruit and Vegetables, and Sprouts From Retail Establishments. *Int. J. Food Microbiol.* **2008,** *123* (1–2), 121–129.

Alegbeleye, O. O.; Singleton, I.; Sant'Ana, A. S. Sources and Contamination Routes of Microbial Pathogens to Fresh Produce During Field Cultivation: A Review. *Food Microbiol.* **2018,** *73,* 177–208.

Alves, M. M.; Gonçalves, M. P.; Rocha, C. M. R. Effect of Ferulic Acid on the Performance of Soy Protein Isolate-Based Edible Coatings Applied to Fresh-Cut Apples. *LWT* **2017,** *80,* 409–415.

Baldino, L.; Cardea, S.; Reverchon, E. Biodegradable Membranes Loaded with Curcumin to be Used as Engineered Independent Devices in Active Packaging. *J. Taiwan Inst. Chem. Eng.* **2017,** *71,* 518–526.

Barrett, D. M.; Beaulieu, J. C.; Shewfelt, R. Color, Flavor, Texture, and Nutritional Quality of Fresh-Cut Fruits and Vegetables: Desirable Levels, Instrumental and Sensory Measurement, and the Effects of Processing. *Crit. Rev. Food Sci. Nutr.* **2010,** *50* (5), 369–389.

Basumatary, I. B.; Mukherjee, A.; Katiyar, V.; Kumar, S. Biopolymer-Based Nanocomposite Films and Coatings: Recent Advances in Shelf-Life Improvement of Fruits and Vegetables. *Crit. Rev. Food Sci. Nutr.* **2022,** *62* (7), 1912–1935.

Bintsis, T. Microbial Pollution and Food Safety. *AIMS Microbiol.* **2018,** *4* (3), 377.

Bobo-García, G.; Arroqui, C.; Merino, G.; Vírseda, P. Antibrowning Compounds for Minimally Processed Potatoes: A Review. *Food Rev. Int.* **2020,** *36* (5), 529–546.

Brody, A. L.; Zhuang, H.; Han, J. H. Modified Atmosphere Packaging for Fresh-Cut Fruits and Vegetables, **2011,** *302.*

Chen, J.; Wu, A.; Yang, M.; Ge, Y.; Pristijono, P.; Li, J.; ... Mi, H. Characterization of Sodium Alginate-Based Films Incorporated with Thymol for Fresh-Cut Apple Packaging. *Food Cont.* **2021,** *126,* 108063.

Deshmukh, R. K.; Akhila, K.; Ramakanth, D.; Gaikwad, K. K. Guar Gum/Carboxymethyl Cellulose Based Antioxidant Film Incorporated with Halloysite Nanotubes and Litchi Shell Waste Extract for Active Packaging. *Int. J. Biol. Macromol.* **2022,** *201,* 1–13.

Deshmukh, R. K.; Gaikwad, K. K. Natural Antimicrobial and Antioxidant Compounds for Active Food Packaging Applications. *Biomass Convers. Biorefin.* **2022,** 1–22.

Dou, L.; Li, B.; Zhang, K.; Chu, X.; Hou, H. Physical Properties and Antioxidant Activity of Gelatin-Sodium Alginate Edible Films with Tea Polyphenols. *Int. J. Biol. Macromol.* **2018,** *118*, 1377–1383.

Fan, Y.; Yang, J.; Duan, A.; Li, X. Pectin/Sodium Alginate/Xanthan Gum Edible Composite Films as the Fresh-Cut Package. *Int. J. Biol. Macromol.* **2021,** *181,* 1003–1009.

Gaikwad, K. K.; Singh, S.; Lee, Y. S. Oxygen Scavenging Films in Food Packaging. *Environ. Chem. Lett.* **2018,** *16* (2), 523–538.

Hashemi, M.; Shakerardekani, A.; Mirzaalian Dastjerdi, A.; Mirdehghan, S. Effect of Sodium Alginate in Combination with Zataria multiflora Boiss. on Phenolic Compounds, Antioxidant Activity, and Browning Enzymes of Fresh In-Hull Pistachio (Pistacia vera L.). *J. Food Qual.* **2021,** *2021.*

Hashemi, S. M. B.; Jafarpour, D. The Efficacy of Edible Film from Konjac Glucomannan and Saffron Petal Extract to Improve Shelf Life of Fresh-Cut Cucumber. *Food Sci. Nutr.* **2020,** *8* (7), 3128–3137.

Hu, W.; Jiang, A.; Xiu, Z.; Feng, K. Effect of Thyme Oil–Alginate-Based Coating on Quality and Microbial Safety of Fresh-Cut Apples. *J. Sci. Food Agric.* **2018,** *98* (6), 2302–2311.

Ioannou, I. Prevention of Enzymatic Browning in Fruit and Vegetables. *Eur. Sci. J.* **2013,** *9* (30).

Islam, R. U.; Khan, M. A.; Islam, S. U. Plant Derivatives as Promising Materials for Processing and Packaging of Meat-Based Products–Focus on Antioxidant and Antimicrobial Effects. *J. Food Process. Preserv.* **2017,** *41* (2), e12862.

Kaczmarek, M.; Avery, S. V.; Singleton, I. Microbes Associated with Fresh Produce: Sources, Types and Methods to Reduce Spoilage and Contamination. In *Advances in Applied Microbiology,* Vol. 107; Elsevier, 2019; pp 29–82.

Kadzińska, J.; Janowicz, M.; Kalisz, S.; Bryś, J.; Lenart, A. An Overview of Fruit and Vegetable Edible Packaging Materials. *Packag. Technol. Sci.* **2019,** *32* (10), 483–495.

Kalia, A.; Kaur, M.; Shami, A.; Jawandha, S. K.; Alghuthaymi, M. A.; Thakur, A.; Abd-Elsalam, K. A. Nettle-Leaf Extract Derived ZnO/CuO Nanoparticle-Biopolymer-Based Antioxidant and Antimicrobial Nanocomposite Packaging Films and their Impact on Extending the Post-Harvest Shelf Life of Guava Fruit. *Biomolecules* **2021,** *11* (2), 224.

Keshari, D.; Tripathi, A. D.; Agarwal, A.; Rai, S.; Srivastava, S. K.; Kumar, P. Effect of α-dl Tocopherol Acetate (Antioxidant) Enriched Edible Coating on the Physicochemical, Functional Properties and Shelf Life of Minimally Processed Carrots (Daucus carota subsp. Sativus). *Future Foods* **2022,** 100116.

Khan, M. R.; Di Giuseppe, F. A.; Torrieri, E.; Sadiq, M. B. Recent Advances in Biopolymeric Antioxidant Films and Coatings for Preservation of Nutritional Quality of Minimally Processed Fruits and Vegetables. *Food Packag. Shelf Life* **2021,** *30*, 100752.

Klein, B. P. Nutritional Consequences of Minimal Processing of Fruits and Vegetables. *J. Food Qual.* **1987,** *10* (3), 179–193.

Li, K. W.; Weidhaas, J.; Lemonakis, L.; Khouryieh, H.; Stone, M.; Jones, L.; Shen, C. Microbiological Quality and Safety of Fresh Produce in West Virginia and Kentucky Farmers' Markets and Validation of a Post-Harvest Washing Practice with Antimicrobials to Inactivate Salmonella and Listeria Monocytogenes. *Food Control* **2017,** *79,* 101–108.

Li, Y.; Lu, J.; Tian, X.; Xu, Z.; Huang, L.; Xiao, H.; … Kong, Q. Alginate with Citrus Pectin and Pterostilbene as Healthy Food Packaging with Antioxidant Property. *Int. J. Biol. Macromol.* **2021,** *193,* 2093–2102.

Mannozzi, C.; Tylewicz, U.; Chinnici, F.; Siroli, L.; Rocculi, P.; Dalla Rosa, M.; Romani, S. Effects of Chitosan Based Coatings Enriched with Procyanidin by-Product on Quality of Fresh Blueberries During Storage. *Food Chem.* **2018,** *251,* 18–24.

Manzoor, S.; Gull, A.; Wani, S. M.; Ganaie, T. A.; Masoodi, F. A.; Bashir, K.; … Dar, B. N. Improving the Shelf Life of Fresh Cut Kiwi Using Nanoemulsion Coatings with Antioxidant and Antimicrobial Agents. *Food Biosci.* **2021,** *41,* 101015.

Maringgal, B.; Hashim, N.; Mohamed Amin Tawakkal, I. S.; Muda Mohamed, M. T. Recent Advance in Edible Coating and Its Effect on Fresh/Fresh-Cut Fruits Quality. *Trends Food Sci. Technol.* **2020,** *96,* 253–267.

Marquez, G. R.; Di Pierro, P.; Mariniello, L.; Esposito, M.; Giosafatto, C. V. L.; Porta, R. Fresh-Cut Fruit and Vegetable Coatings by Transglutaminase-Crosslinked Whey Protein/Pectin Edible Films. *LWT* **2017,** *75,* 124–130.

Mehyar, G. F.; Han, J. H. 14 Active Packaging for Fresh-Cut Fruits and Vegetables. In *Modified Atmosphere Packaging for Fresh-Cut Fruits and Vegetables*, 2011, p 267.

Nair, M. S.; Saxena, A.; Kaur, C. Characterization and Antifungal Activity of Pomegranate Peel Extract and Its Use in Polysaccharide-Based Edible Coatings to Extend the Shelf-Life of Capsicum (Capsicum Annuum L.). *Food Bioproc. Tech.* **2018a,** *11* (7), 1317–1327.

Nair, M. S.; Saxena, A.; Kaur, C. Effect of Chitosan and Alginate Based Coatings Enriched with Pomegranate Peel Extract to Extend the Postharvest Quality of Guava (Psidium Guajava L.). *Food Chem.* **2018b,** *240,* 245–252.

Ncama, K.; Magwaza, L. S.; Mditshwa, A.; Tesfay, S. Z. Plant-Based Edible Coatings for Managing Postharvest Quality of Fresh Horticultural Produce: A Review. *Food Packag. Shelf Life* **2018,** *16,* 157–167.

Oliveira, M.; Abadias, M.; Usall, J.; Torres, R.; Teixidó, N.; Viñas, I. Application of Modified Atmosphere Packaging as a Safety Approach to Fresh-Cut Fruits and Vegetables—A Review. *Trends Food Sci. Technol.* **2015,** *46* (1), 13–26.

Perera, C. O. Minimal Processing of Fruits and Vegetables. In *Handbook of Food Preservation.* CRC Press, 2007; pp 155–168.

Radi, M.; Firouzi, E.; Akhavan, H.; Amiri, S. Effect of Gelatin-Based Edible Coatings Incorporated with Aloe Vera and Black and Green Tea Extracts on the Shelf Life of Fresh-Cut Oranges. *J. Food Qual.* **2017,** *2017.*

Rangaraj, V. M.; Rambabu, K.; Banat, F.; Mittal, V. Natural Antioxidants-Based Edible Active Food Packaging: An Overview of Current Advancements. *Food Biosci.* **2021,** *43,* 101251.

Rodríguez, G. M.; Sibaja, J. C.; Espitia, P. J. P.; Otoni, C. G. Antioxidant Active Packaging Based on Papaya Edible Films Incorporated with Moringa Oleifera and Ascorbic Acid for Food Preservation. *Food Hydrocoll.* **2020,** *103,* 105630.

Saba, M. K.; Sogvar, O. B. Combination of Carboxymethyl Cellulose-Based Coatings with Calcium and Ascorbic Acid Impacts in Browning and Quality of Fresh-Cut Apples. *LWT Food Sci. Technol.* **2016,** *66,* 165–171.

Salehi, F. Edible Coating of Fruits and Vegetables Using Natural Gums: A Review. *Int. J. Fruit Sci.* **2020,** *20* (sup 2), S570–S589.

Sharma, L.; Saini, C. S.; Sharma, H. K.; Sandhu, K. S. Biocomposite Edible Coatings Based on Cross Linked-Sesame Protein and Mango Puree for the Shelf Life Stability of Fresh-Cut Mango Fruit. *J. Food Proc. Eng.* **2019,** *42* (1), e12938.

Sharma, P.; Shehin, V. P.; Kaur, N.; Vyas, P. Application of Edible Coatings on Fresh and Minimally Processed Vegetables: A Review. *Int. J. Veget. Sci.* **2019,** *25* (3), 295–314.

Tanwar, R.; Gupta, V.; Kumar, P.; Kumar, A.; Singh, S.; Gaikwad, K. K. Development and Characterization of PVA-Starch Incorporated with Coconut Shell Extract and Sepiolite Clay as an Antioxidant Film for Active Food Packaging Applications. *Int. J. Biol. Macromol.* **2021,** *185,* 451–461.

Tudela, J. A.; Hernández, N.; Pérez-Vicente, A.; Gil, M. I. Comprehensive Evaluation of Different Storage Conditions for the Varietal Screening of Lettuce for Fresh-Cut Performance. *Postharvest Biol. Technol.* **2016,** *120,* 36–44.

Union, E. Directive 2006/52/EC of the European Parliament and of the Council of 5 July 2006 Amending Directive 95/2/EC on Food Additives Other Than Colours and Sweeteners and Directive 94/35/EC on Sweeteners for Use in Foodstuffs. *OJEU* **2006,** *204,* 10–22.

Wiley, R. C.; Yildiz, F. Introduction to Minimally Processed Refrigerated (MPR) Fruits and Vegetables. In *Minimally Processed Refrigerated Fruits and Vegetables.* Springer, 2017; pp 3–15.

Yousuf, B.; Srivastava, A. K.; Ahmad, S. Application of Natural Fruit Extract and Hydrocolloid-Based Coating to Retain Quality of Fresh-Cut Melon. *J. Food Sci. Technol.* **2020,** *57* (10), 3647–3658.

Zhang, C.; Yang, Z.; Shi, J.; Zou, X.; Zhai, X.; Huang, X.; … Xiao, J. Physical Properties and Bioactivities of Chitosan/Gelatin-Based Films Loaded with Tannic Acid and Its Application on the Preservation of Fresh-Cut Apples. *LWT* **2021,** *144,* 111223.

Zhang, W.; Li, X.; Jiang, W. Development of Antioxidant Chitosan Film with Banana Peels Extract and Its Application as Coating in Maintaining the Storage Quality of Apple. *Int. J. Biol. Macromol.* **2020,** *154,* 1205–1214.

CHAPTER 8

NEW SMART PACKAGING TECHNOLOGIES FOR MONITORING QUALITY OF AGRICULTURE PRODUCE

KONALA AKHILA[1], SUMAN SINGH[2], and KIRTIRAJ K. GAIKWAD[1]

[1]*Department of Paper Technology, Indian Institute of Technology, Roorkee, Uttarakhand*

[2]*Department of Food Engineering, Institute of Food Science and Technology, VCSG Uttarakhand University of Horticulture and Forestry, Majri Grant, Dehradun, Uttarakhand, India*

ABSTRACT

It is well known that fruits and vegetables are highly perishable. They are not only vulnerable to smallest changes in their environment but also susceptible to their own mechanism of respiration and transpiration. When packed in conventional packaging, it is nearly impossible to understand their quality without actually opening the package. Since consumers nowadays started to prefer fresh produce in their everyday meal, there is a high chance that most of the cut fruits and vegetables end in trash because of their stale appearance despite their intact nutritional values. This kind of wastage can be reduced by upgrading the traditional packaging with smart packaging techniques. Intelligent packaging is a class of smart packaging and involves noninvasive quality assessment of packed product without coming in contact with the product. Their mechanism is based on the

Novel Packaging Systems for Fruits and Vegetables. Kirtiraj K. Gaikwad & Suman Singh (Eds.)

changes taking place in the head space that are the result of qualitative changes in fresh produce. There are many commercial examples available in the market. In this chapter, a detailed explanation about different types of intelligent packaging systems was given along with the mechanism involved and their use in fruits and vegetable packaging was focused. Advancements in intelligent packaging techniques were briefed and the challenges it is facing were discussed.

8.1 INTRODUCTION

A packaging is designed to protect a product from external and internal hazards caused during storage, handling, and transportation. External hazards include temperature, light, humidity, oxygen whereas microbial growth, product's water activity, head space gases come under internal hazards. These internal hazards are inextricably dependent on external factors and on packaging material. Especially in the case of agricultural produce, which tends to breath postharvest, internal factors play a very crucial role. Unfortunately, a consumer cannot predict the changes happening inside the packaging, i.e., in the product's atmosphere, with the current knowledge of package labeling. This mere knowledge of state of the product inside the packaging leads to discarding the food after purchase. It may be either at retail level or at household level (Singh et al., 2021).

According to the world statistics, 1.3 billion tons of food produced is being wasted annually. In developed countries likes the USA and Europe, the discarded food costs them up to $161 billion in 2010 and 143 billion euros in 2016. A developing country like India has produced farm waste worth Rs. 92,651 crores in 2016, and it was estimated that almost 16% of produced fruits and vegetables end in dump yard. Of all the food waste generated, 61% is wasted at household level ("Food Loss and Waste | FDA," n.d.; Pnuma, 2021; USDA, 2017). The European Commission has stated that the reasons for food waste are widely varied and can be segment specific. Some of those reasons that contribute prominently are as follows ("Food Waste," n.d.)

- Lack of knowledge about consequences of food waste
- Low perceived value of food
- Inadequate storage and transportation throughout its shelf life
- Misunderstanding of "use by" and "best before" labeling on package

- ❖ Standardized portions in restaurants and canteens
- ❖ Insufficient meal planning and stock management at house hold level
- ❖ High-quality standards for produce sold at retail stores
- ❖ Production errors like over production during low-demand seasons
- ❖ Packaging damage

A severe food shortage will be faced by the humanity unless an appropriate strategy in terms of packaging and storage is designed and followed. Moreover, with busy life style and conflicting priorities, consumers nowadays started to prefer ready-to-eat foods. These foods include fresh agricultural produce in the form of salads, cut fruits and vegetables, natural flavor packed fruit juices, read to cook meals etc. But the postharvest produce is highly susceptible to spoilage and damage with smallest change in its environment. After all, once packed, the conventional packaging makes it impossible to monitor the quality of the food inside ultimately leading to discarding them with some reasons like aesthetics of the food, confusion between use by and best before date despite their nutrition.

Advanced smart packaging techniques need to be introduced to replace conventional packaging in terms of better communication. Intelligent packaging is a type of smart packaging which monitors the quality of the packed food using various mechanisms and visually communicates it to the consumers via different signals and facilitates decision-making. Intelligent packaging (IP) can broadly be categorized into three types: (a) Data Carriers, (b) Indicators, (c) Sensors. IP has gained its popularity in the past decade due to increased use of active components in food packaging which needs monitoring for unprecedented leakage or their ineffective working. The broadly divided IP systems can further be subclassified based on their working principle and mechanism as shown in Figure 8.1.

According to the European Food Safety Authority (EFSA), intelligent packaging is defined as materials and articles that monitor the condition of packaged food or the environment surrounding the food and have the ability to communicate the condition of the product by not interacting with the product directly (Müller and Schmid, 2019).

A recent report on growth of smart packaging market revealed that in 2021 it has a market potential of USD 19.33 billion and is expected to grow at a CAGR of 6.2% to reach USD 29.45 billion by 2028. With Asia Pacific being the fastest growing region, IP is expected to lead the smart packaging market especially in North America and Europe. Although these smart packaging techniques were first commercialized in Japan, in the present-day scenario the US contributes a whopping USD 3.6 billion to the smart packaging market followed by Japan ($ 2.36B), Australia, United Kingdom, and Germany. Some of the leading companies that are huge players in this sector include Sealed Air® corporation, Amcor Plc (Australia), Multisorb Technologies (U.S.), Ball Corporation (U.S.), Sysco Corporation (U.S.), Timestrip PLC (U.K.), Paksense Incorporated (U.S.), and M&G USA Corporation. With its forecasted market growth, it is understood that IP is used in every sector of packaging and has a huge role to play in futuristic packaging of fresh agricultural produce for improved food safety.

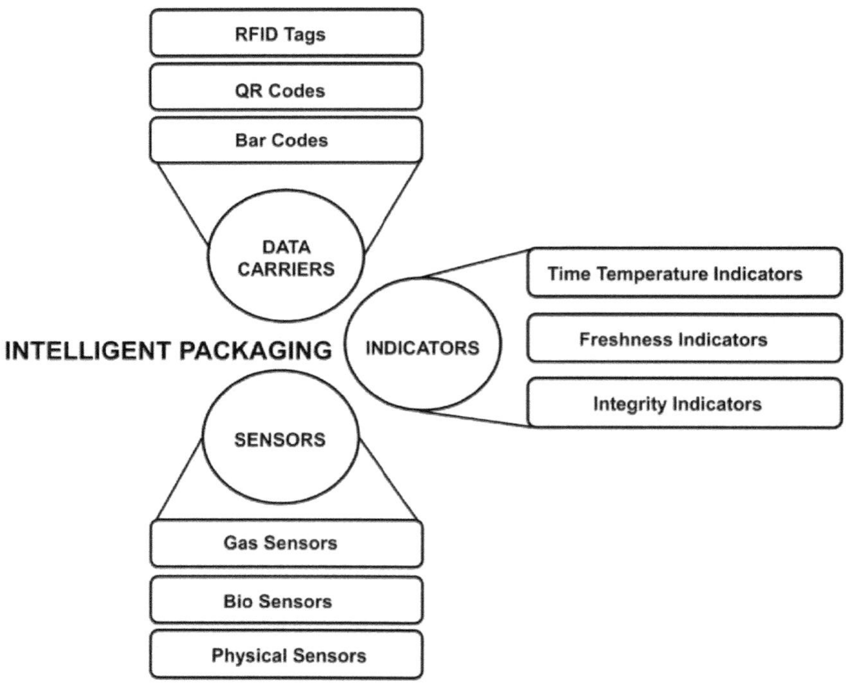

FIGURE 8.1 Classification of intelligent packaging.

8.2 NEED FOR INTELLIGENT PACKAGING IN FRESH AGRICULTURAL PRODUCE

Maintaining sensorial and nutritional quality of the agricultural produce postharvest is a serious challenge. It is realized that managing the quality without compromising on safety and legal aspects cannot be achieved alone by using any single technique. In the process of applying smart packaging techniques to the afore-mentioned problem, one should first know about factors that affect the shelf life of fresh produce postharvesting, after all they are still living tissues after harvesting. They continue to respire and transpire throughout their life. Common deterioration mechanisms that fruit and vegetables suffer are oxidative degradation, microbial spoilage, enzymatic reactions, and moisture losses (Ma et al., 2017; Xanthopoulos et al., 2017). Production and build-up of ethylene gas during ripening of fruit is also a major reason for fruits to deteriorate faster than expected in a closed package. Till date, there are some traditional methods to keep them fresh like cold storage, irradiation preservation technology, chemical-based preservation, and bioactive preservation.

Cold storage is a physical method to constantly expose the fruits and vegetables to significantly lower temperature which lowers their respiration rate, even a micro-organism cannot sustain at those temperatures. But different fruits and vegetables have different filling temperatures, and moreover maintaining these superficial temperatures in a longer run could be expensive, energy demanding (Oyom et al., 2022). So, in recent days, irradiation preservation has been seen using frequently, in which the produce is exposed to ultraviolet-C radiation. Nevertheless, it has its drawbacks like radiation dosage limits, difficulty in controlling the operational conditions and of course high investment. So, comparatively cheaper methods can be chemical and bioactive preservation. The chemical method involves disinfection of surfaces using chlorine, coating with waxes to reduce microbial spoilage and respiration rate. But these chemicals are harmful for human health if consumed leading to food safety regulation issues. Luckily, bioactive preservation techniques do not involve any chemicals (Liu et al., 2020).

Bio preservation technology involves using active agents like carbon dioxide emitters, antimicrobial agents, oxygen scavengers, ethylene absorbers, and modified atmosphere. Involving these kinds of active agents inside the packaging controls the respiration by decreasing O_2, increasing CO_2, and absorbing ethylene produced because of ripening (Qu et al., 2022). Controlling this release and absorption of gases inside the packaging can be more complicated because of their diffusion rate through the film and depends on film's permeability, thickness etc. It is not totally impossible that the active agents added to the packaging might negatively affect the shelf life. But even after extensive efforts made to preserve the postharvest produce, it was reported that 20-30% of losses occurred (Mariah et al., 2022).

These losses can be reduced to some extent by taking simple precautions like storing in right temperatures, responsible purchase, and consumption. But how does a consumer buying any fresh produce know whether or not the product has been stored in desired temperatures so far? Or about the qualitative changes that took place so far because of incorporated active agents? Here comes the need of IP to improve reliable communication function of a package and provide information about the dynamic quality of the product packed. Since all the postharvest produces definitely undergo degradation because of any of the factors mentioned above, there is a need for different systems to monitor different changes. The appropriate intelligent packaging technique to be adopted depends on the analyte that one intends to measure. Factors that can be considered include, whether the produce is climacteric or non-climacteric in nature, water activity, their indigenous shelf life, and storage temperatures. With the increased research interest in smart and intelligent packaging technology, the number of research articles published in the last five years has also shown increasing trend in the data collected from Web of Science till date. It is a clear indication that intelligent packaging techniques have the potential to use in fruits and vegetable packaging.

8.3 INTELLIGENT PACKAGING TECHNIQUES

Intelligent packaging is considered to be a dynamic field which is undergoing continuous advancements. They have the potential to decrease the postharvest losses as discussed above. There are several commercial examples available in the market to act as intelligent system when applied on a variety of products mostly on fruits and vegetables or meat products. Some of them are listed in Table 8.1, and they are briefed in the following discussion.

8.3.1 INDICATORS

8.3.1.1 FRESHNESS INDICATORS

Fruits and vegetables are packed with micro- and macronutrients and are preferred to be consumed fresh and immediately after harvesting. A slightest change in external environment or their respiration rate could alter their appearance which gives an impression that they are not suitable for consumption despite their persistent nutritional quality. Fruits and vegetables do not lose their nutrition unless they undergo internal structural decay due to microbial activity or increased respiration rate. Hence, it is highly recommended to use freshness indicators to detect the quality of produce with the help of the changes in the package head space.

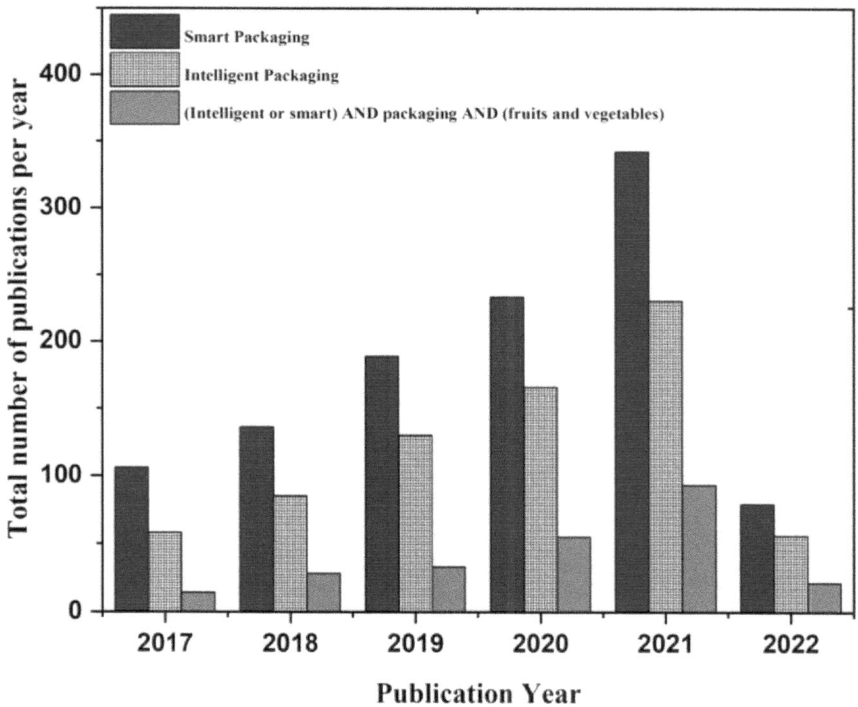

FIGURE 8.2 Recent research publication trend in smart and intelligent packaging technology for fruits and vegetables. Data has been extracted from web of science (till date i.e., 26/04/2022) using keywords such as smart packaging*, Intelligent packaging*, (Intelligent or Smart) AND Packaging AND (Fruits and Vegetables).

Freshness indicators (FI) is a nondestructive method to monitor the quality although it interacts with the gases or metabolites released during the degradation of postharvest produce. The type of freshness indicator that has to be applied is dependent on type of fruit, that is, climacteric or non-climacteric. The former one ripens by releasing ethylene whereas

TABLE 8.1 Examples of Commercial Intelligent Packaging Systems Along with Their Manufacturer.

Intelligent packaging system	Trade name	Manufacturer
Freshness indicators	SensorQ®	DSM NV FQSI Inc.
	RipeSense	Ripesense Ltd.
	Toxin Guard	Toxin Alert Inc.
	Fresh Tag®	COX Technologies
	Raflatac	VTT and UPM Raflatac
	Food Sentinel System	SIRA Technologies
Time-temperature indicator	VITSAB®	VITSAB International AB
	OnVu™/Coolvu™	Freshpoint/BASF
	Best-by®	FreshPoint Lab
	TopCryo®	Cryolog
	Traceo® & eO®	
	Keep -it®	Keep-it Technologies
	Novas®	Insignia Technologies Ltd.
	CheckPoint®	VITSAB
	3M™ Monitor Mark®	3M Company
	FreshCode®	Varcode Ltd.
	Timestrip®	Timestrip Plc
	PLUS	Timestrip Plc
	Colour-Therm	Colour-Therm
	Tempix®	Temptix AB
	TempDot®	Delta Trak
	Cook-Chex	Pymah Corp.
	Thermax®	Thermografic Measurements Ltd.
	WarmMark®	Delta Track
	SmartDot	EVIGENCE SENSORS
	Insignia Deli Intelligent Labels™	Insignia Technologies
	Fresh-Check®	Temptime Corp.
RFID tags	Intelligo	Schiassi Diamo Scaltole Alle Tue Idee
	easy2log®	CAEN RFID
	CS8304	Convergence Systems Ltd.
	Intelligent fish box	Craemer Group GmbH

TABLE 8.1 *(Continued)*

Intelligent packaging system	Trade name	Manufacturer
	Intelligent box	Mondi Plc
	TempTRIP	TempTRIP LLC
Sensors	OxyDot- O2xyDot	OxySense
	Food Sentinel System	SIRA Technologies Inc.
Integrity indicators	Ageless Eye®	Mitsubishi Gas Chemical Co.
	Tufflex GS	Sealed Air Ltd.
	Shelf Life Gaurd	UPM
	Tell-Tab	IMPAK

the latter one spoils mostly due to humidity, temperature, and time. In the course of ripening and aging process, fruits and vegetables release certain analytes that act as base for sensing. These analytes include alkalinity, carbon dioxide, volatile base nitrogen, hydrogen sulfides, ATP, ADP, AMP, lactic acid specifically in the case of fresh produce that indicate their maturity and freshness. For example, the durian fruit releases volatile sulfide, apple releases volatile aldehydes due to maturity. So, the materials that can react with these analytes and show the quality change visually are incorporated in freshness indicators and applied readily on the package. The freshness sensing parameters in FI usually include ripening signs like color, pH, and gases released.

For the ease in communication, visual indicators that change their color by sensing the analytes are used. The main ingredient in FI is a chromogenic agent. They can be synthetic dyes or natural pigments that have the ability to change color while sensing pH, volatile compounds, and gases. The synthetic chromogenic agents like bromocresol green, methyl red, bromophenol red, bromocresol violet, phenol red, chlorophenol red, bromophenol blue are most frequently used as they undergo irreversible reaction with analytes to change their color. A combination of two agents can also be put into use. For example, chlorophenol red based indicator was developed to detect the maturity of grapes. On the other side, combination of methyl red and bromocresol blue was investigated for freshness detection of cut bell peppers (Shao et al., 2021). Such kinds of indicators in the form of labels can be sticked in the package.

Due to the migration of these toxic chemicals into food, their usage has been limited. Natural chromogenic indicators such as anthocyanins, betaines, curcumin, and chlorophyll are the popular ones besides alizarin and shikonin and are approved to be safe to use in freshness indicators. For example, anthocyanins are an electron-deficient compound which readily connects with reactive oxygen or free electron and changes its structure thus changing color. These compounds can be obtained by simple extraction from sources like flowers, fruit peels, roots, and leaves. Till date, these natural indicators are not commercially applied as freshness indicators because of their high instability when exposed to temperature and light.

In the case of climacteric fruits, major parameter that indicates their freshness is ripeness. The amount of ethylene released can be detected and quantified by sticking the label directly to the fruit. Ripesense® is the first freshness indicator that monitors the ripeness of a fruit. This kind of indicator is applied on mangoes, kiwi, pears, avocado, and melons. Synthetic chemical, bromocresol blue, was incorporated in a cellulose membrane and was employed to detect the acidic volatile compounds released from ripening guava. All the above-mentioned examples are direct freshness sensors since they are associated with the internal chemical and physical changes of postharvest produce. But it was discussed earlier that storage temperature, humidity, and time also affect their freshness. FI that work on the changes because of these external factors are called indirect freshness indicators (Alam et al., 2021).

8.3.1.2 TIME-TEMPERATURE INDICATORS

Time-Temperature Indicators (TTI) are the most popular and extensively used indicators and are also indirect freshness indicators because they determine whether the product has gone through adverse storage conditions which in turn affect their quality. They can be classified into three categories: critical temperature indicators, partial history, and full history indicators (Kuswandi and Siddiqui, 2020). On the basis of mechanism involved, TTIs are again enzymatic based, microbial growth based, physical based, and chemical reaction based. They are most successful in commercialization and the examples are given in Table 8.1.

Enzymatic TTIs work on the principle of hydrolysis of chosen enzyme which is dependent on time and temperature that the indicator was exposed to. Enzyme is a biocatalyst that is specific for a biological reaction. This type is uniquely customizable according to the nature of the analyte produced from fruits and vegetables by varying the enzymes or concentration of enzymes or even the polymer substrate. The commercial TTI, Checkpoint® (VITSAB, Sweden) is an example for enzymatic TTI which is made of lipase along with bromothymol blue solution. The color diffuses across the plane and its extent indicates on how long the package has been exposed to temperature. They are very accurate in nature. So far in the literature, this type of TTIs has been used on a variety of foods including fruits and veggies such as lettuce, tomato, fruit juices, salads, mushrooms, peas, and strawberries along with dairy and meat products. Some simple diffusion based commercial TTIs are Monitor Mark by 3M, and Timestrip®, which are simple synthetic dyes attached to a wick. Once they are pressure activated, the ink gradually diffuses through the wick and changes its color. This indicates the quantity of time and temperature that package was exposed to (Siddiqui et al., 2022).

The response of microbial TTIs is based on growth of microbes and their metabolites. This helps in monitoring the food quality depending on the amount of metabolic activity and rate of microbes growing corresponding to the changes in pH. Since in fruits and vegetables, microbes play a very important role in structural damage leading to changes in pH, microbial TTIs will help to effectively communicate both pH and microbes' status within the package by irreversible color change. Since these two parameters rely on temperature and type of microbe used, they are pronounced as microbial TTI's. Although these TTIs are advantageous over other TTIs as they synchronize perfectly with the food packed, it has fewer drawbacks such as they have to be stored in refrigeration temperatures till use to keep active microbe inactive; they take little longer to activate once taken out. Despite the drawbacks, these TTIs are very safe to use. Commercial examples are Traceo® & eO® which are simple films sticked on the barcodes. If the product is exposed to undesired temperatures, then the film turns opaque because of the buildup microbes and makes the barcode unreadable indicating that the food is no longer fit for consumption. Fresh-cut fruits and vegetables can be applied with such TTIs as they ought to be stored in chilling temperatures. But they have found their application more prominently in meat and seafood packaging (Mataragas et al., 2019).

Chemical TTIs usually involve a chemical compound and a reaction to change its color. A photochromic compound is incorporated into a solid substrate and placed/sticked to the package. When light with particular wavelength irradiated this compound, it becomes unstable and changes its structure which results in fading color. This reverse color change indicates the food freshness. OnVu™/Coolvu™ are the commercially available TTIs that work on irradiation of photochromic pigment to indicate the exposure of package to light. Similarly, a typical redox reaction based TTIs also come under this category. A mere oxidation reaction will lead to color change. It is made up of redox dye, reducing agent and alkaline medium. The oxygen diffuses through the barrier film and oxidizes the redox dye to change its color. The rate of this mechanism is linked with temperature exposed.

Physical TTIs involve usage of nanoparticles incorporated into a polymer substrate. Generally, nanoparticles exhibit different shape and color when exposed to light of different wavelengths. The absorption band that comes from nanoparticles depends on the surface plasmon resonance, and it is unique for every metal nanoparticle although its intensity can be changed with shape and size. Using this mechanism, silver and gold nanoparticles exhibit optical properties which can be taken advantage off to employ them as indicators. Sometimes even the polarization effect may contribute to this phenomenon. Unfortunately, using nanoparticles directly in the packaging may face regulatory issues due to their migration over permissible limits. So, this kind of TTI is seldom found in postharvest produce application (Siddiqui et al., 2022).

8.3.1.3 INTEGRITY INDICATORS

Integrity indicators are usually gas and humidity indicators which detect the changes in the composition of head space gases and presence of moisture in vegetable and salads packaging assisted with MAP. Changes in gas composition could mean leaked package or increased microbial activity and increased moisture means an increased respiration rate. The detection of these gases is redox dye based, pH based, luminescence based, and lumophore based (Yousefi et al., 2019). A redox dye like methylene blue is incorporated in a polymer matrix with some strong reducing agent. When this system comes in contact with oxygen the blue color changes to pink rapidly and indicates the presence of oxygen. In the case of carbon dioxide detection, pH-sensitive dyes like cresol red are incorporated in a porous membrane. When it comes in contact with undesired amounts of CO_2, its color changes to yellow. But these indicators are reversible in nature and can be misguiding regarding the quality of product. Hence, in a recent study, a combination of lysine and anthocyanins was applied to indicate the presence of CO_2.

Humidity indicators on the other hand are found rare. A company named IMPAK has developed one such humidity indicator which is made up of cobalt chloride solution. It changes its color from blue to purple with increasing humidity. More recently, copper nanoparticles coated with carbon were made in the form of a strip and used as a humidity indicator. With the increase in humidity, the water condenses over the strip interfering the reflection of light's angle of incidence. This results in visualization of metallic colors on the strip indicating the presence of humidity (Beshai et al., 2020). Such types of indicators can be formed as labels with nanoparticles incorporated in them and employed inside the packaging. Integrity indicators are seldom seen in market because sensors can execute the same job with more precision.

8.3.2 SENSORS

Sensors are an embedded electronic system with the combination of receptor and a transducer. Based on the type of analyte that the receptor is specifically designed for, sensors are classified as chemical, bio, and gas sensors. In a chemical sensor, the receptor is designed for sensing volatile components released by the fruits and vegetables during aging such as terpenes, carboxylic acids, alcohols. In biosensors the analyte will be enzymes, antibodies, aptamers, and cells. Transducer is simply used to convert the signal provided by receptor into electrical signal, so it operates using electrochemical and optical methods to generate qualitative or quantitative output of the sensed analyte.

8.3.2.1 GAS SENSORS

Gas sensors have an important role to play in modified atmospheric packaging for fresh-cut fruits and vegetables. They also act as leakage detectors. Carbon dioxide sensors are famous than oxygen sensors because they serve two purposes of being a leakage detector and microbial growth detector.

The increase in concentration of CO_2 in the fruit and vegetables MAP indicates the growth of microbes, whereas decrease in its concentration detects the leakage since an exact amount of CO_2 is flushed into MAP's. Most recently, metallic organic framework (MOF) based gas sensors are being investigated for their application in postharvest produce as they can efficiently act as ethylene gas detectors. They have elevated sensors for various analytes better than zeolites and other inorganic materials and also possess good regenerating ability. These MOF sensors work on the principle of molecular recognition followed by transduction. Despite the advantages, the usage of MOF-based gas sensors is still in its infancy to apply in fruits and vegetable sector (Yuan et al., 2022).

Electronic nose (e-nose) is the recent, successful, simple, and highly accurate type of gas sensor which can be easily applied for fresh produce. It consists of a gas sensor array which collects the released gases in the package that can sense versatile range of inputs and a pattern reorganization system with algorithms (PRS) to generate output. They can even detect growing molds, volatile metabolites, and flavor components. In the literature, e-nose was used to sense the freshness of apples, bell peppers, strawberries, fresh-cut pineapples, broccoli, peaches, etc. Electronic noses are classified not only on the sensor array but also on the signal processor used before PRS. There are four main algorithms that are used in e-nose for fresh produce. They are (i) super vector machine model for bananas, blueberries, (ii) back propagation neural networks on lemons, cherries, and watermelons, and (iii) principal component analysis for maturity of grapes, fushi apples etc. It has to be noted that e-nose is a complex system to mimic the human olfactory senses involving a sensor array made of different metal oxides and different pattern algorithms. But they are highly sensitive in detecting the smallest changes in postharvest produce (Jia et al., 2019).

8.3.2.2 BIOSENSORS

Biosensors, which are used especially in monitoring the freshness of postharvest produce, should be able to detect pathogens, allergens, pesticides, and mycotoxins even at very slightest concentrations. The detection of pathogens grown on food systems is sensed with the help of their DNA and immune receptors. Since they both need extensive sampling and labors, synthetic antimicrobial peptide receptor has been introduced for easy detection and quantification of four different bacterial strains. The antimicrobial agents used include silver nanoparticles, zinc oxide nanorods, quantum dots, MWCNTs, and graphene. An optical-based transducer is used which relies on luminescence, fluorescence, and light scattering spectroscopy. For example, an enzyme-based biosensor was developed using multiwalled carbon nanotubes that can sense malic acid concentration in tomatoes to detect the ripening of the fruit. Biosensors can also be used to detect food adulteration and authenticity (Mustafa and Andreescu, 2018).

8.3.3 DATA CARRIERS

Data carriers are also called automatic identification devices. Unlike remaining IP systems, they do not necessarily provide information about the quality of the product, rather they provide different features like theft protection, anticounterfeiting, traceability, and automatization to the package and it involves symbology. Hence, they are generally placed on tertiary packaging. Based on the simplicity of their functioning they are classified as: 1) barcodes, 2) radio frequency identification tags.

8.3.3.1 BARCODES

Barcodes are 1-D data carriers that are the first of its kind and were invented dating back to the 1970s. They are the combination of bars and spaces which are read by a laser light. They store very limited information like the details of the manufacturer and item number. The basic principle involved is that the projected laser beam spends relative time on dark bars and spaces which are used to decode individual characters. Due to their small capacity to store data, their successors are designed to store more information in limited space. They are called reduced space symbology (RSS) barcodes or PDF 417 barcodes. These contain an increased number of dots and spaces arranged in a specified area. They should be scanned in omni directions. Hence, they are referred as 2-D barcodes and they can store up to 1.1 KB of information.

The more advanced form of 2-D barcodes is quick response (QR) codes. They store larger amount of data compared with the previous two. In this type, dots and spaces are arranged in an array form and hold information like packaging date, weight, manufacturer, nutritional information, and preparational instructions if necessary. This has provided greater convenience to the customers so far in providing information with a simple scan (Fang et al., 2017). Since these types of data carriers are limited in functioning, they cannot be applied to monitor the freshness of agricultural produce. They can only be printed on the packaging to give very basic details. In order to make data carriers efficient with improved functionalities, RFID Tags were put into use.

8.3.3.2 RADIO FREQUENCY IDENTIFICATION TAGS

RFID tags are similar to electronic sensors and are the most advanced form of data carriers. Unlike barcodes, they are capable of both storing and monitoring the real time data. They can store up to 1 MB of data which is rewritable. They are employed as antitheft, anticounterfeiting, and tracking agents. They also improve the automatic identification process. So, RFIDs have promising qualities to function as an intelligent packaging system. They are essentially sticked to either on primary or tertiary packaging.

Typical components of RFID include microcircuit with tiny transponder and antenna. They are read with the help of high-power radio frequency waves starting from ~125 kHz. A self-powered RFID is called active RFID and the one that depends on the energy supplied by the reader is called passive RFID. The active ones are capable of being read at a maximum distance of 100 m, although operating frequency of the reader often determines the distance. Since, radio waves are interrupted with the presence of metal objects, that can also obstruct the line of reading. Present-day applications of RFIDs are cold chain monitoring, shelf-life prediction, and tracing the product. With forwarding technology, RFIDs are now seen in the form of printed electronics (Sohail et al., 2018). This kind of technology is highly recommended to use in postharvest produce packaging. It prompts the traceability of the product thus decreasing the wastage.

8.4 RECENT ADVANCEMENTS IN INTELLIGENT PACKAGING

The development of IP systems that can be easily incorporated into a polymer matrix by solution casting, extrusion, and attaching in the form of coated labels is very well explored. More recently, production of these systems using sol-gel method, electro spinning, 3D foams and pads, aerogels, electro chemical printing is often focussed since they offer superior color changing properties and strong interactions with concerned polymer matrix for enhanced functionalities of IP. Along with this, using advanced nanomaterials to prepare nanocomposites, nanosensors have also gained its importance.

Sol-gel method is a well-known technique to produce nanoparticles involving simple steps like sol preparation, gelation followed by aging and drying. In the case of indicators, the indicating dye is generally subjected to the mentioned steps and is deposited on the substrate mainly using dip coating. The indicators manufactured in such a method exhibit improved physical and chemical properties like high surface area, porosity, strong adhesion between layers (Dalmoro et al., 2017). The process of electro spinning is efficient and versatile because it can produce ultrafine fibers with different synthetic and biopolymers. This technique is suitable for adding antioxidants, indicator dyes, antimicrobial agents as well. These fibers provide high sensitivity to any changes in food with shorter detection time than regular solvent casted films (Topuz and Uyar, 2020).

In the manufacturing freshness indicators that work on the pH changes of food, common dyes used are bromocresol blue, methyl red, methyl orange, and bromocresol red. These are highly carcinogenic and its migration into the food is dangerous even in smaller quantities. Hence, natural colorants obtained from plants, agricultural waste are more recently explored for packaging applications. These include curcumin, anthocyanins, betalains, shikonin, and alizarin (Yong and Liu, 2020). So far, they are incorporated into a solid support matrix using solvent casting method or any nanocomposites. They have found their application mostly in poultry and meat industry rather than in postharvest agricultural produce.

8.4.1 THREE-DIMENSIONAL PRINTING OF INDICATOR SYSTEMS

Since the last five years, there has been a significant focus on developing 3D materials for freshness indicators. This includes 3D printing on polymer matrix. There are different types of 3D printing which include pressure-assisted micro syringe printing (PAM), electro chemical writing, and inkjet printing. Materials opted for this type printing are called functional inks and possess versatile characteristics like luminescence, electrical conductivity, semiconductivity, electrochromic, pH sensitive, UV light sensitive, and gas sensitive. A praised advantage of printing indicator system is they are flexible along with the substrate. Also, they are lightweight, wearable, foldable, and bendable.

Electro chemical writing involves generation of hydroxyl ions from the cathodic plate which hold a dye-containing film. A platinum pen is connected to anode and moved on the film to produce H^+ where the color of the film changes completely. The movement of Pt pen on the cathode generates hydroxyl ions and pattern is engraved in accordance. An immediate drying of the film has to be done in order to protect the engraved pattern. But the color change during the process itself is undesirable. Hence, instead of using a single layer, another supportive layer was added to reduce the effect of charge (Yang et al., 2021).

Inkjet printing is a regular printing process where ink solution is jetted with a print head onto the substrate. It can either be thermal driven or piezoelectric type to form a droplet. Inkjet printing can also be used to print a QR code for real-time freshness monitoring. Depending on the ink ratio, the printed substrate changes its color. PAM is also a similar deposition system based on nozzle. With the help of air piston, the solution containing indicator dye and polymer matrix are extruded through the nozzle (Mohapatra et al., 2022). Coming to printed electronics, RFIDs are the first systems that are commercialized as printed electronics instead of sticking them separately on the package. A company named ThinFilm has introduced a printed temperature sensor in 2014 to the market and is used to monitor perishable goods like fruits and vegetables.

8.4.2 NANO BIOSENSORS

In the coming years, nanotechnology is predicted to be playing a versatile role in improving the quality of human life. The similar way the same nanotechnology can be used to maintain and monitor the quality of food we consume. Research on implementing this nanoscale technology has been started a decade or couple ago in the form of nanocomposites and nanosensors. Nano biosensors are an integration of nano materials with biosensors. They provide rapid response to the contaminants in food industry. These are noninvasive and very specific to the application.

Some nanomaterials that can be employed as sensors include nanotubes, nanocrystals, nanocomposites, and nanowires to detect allergens, pathogens, heavy metal residues, pesticide residues on postharvest produce (Calicioglu et al., 2019). Nano biosensors are highly sensitive and cannot detect smallest concentration with high energy lasers for example quantum dots. Carbon-based nanomaterials provide better integration of bioactive agents. Agricultural waste can also be used in producing nano biosensors like nano fibers which are actually inexpensive, toxin free when compared to its synthetic counterparts.

Compared to biosensors, nano biosensors require less amounts of analyte to detect the quality. The complexity and cost of the assay can be reduced using nano biosensors (Caon et al., 2017). Since all the agricultural produce degrades because of either respiration or microbial growth, using nano biosensors as an integral part of their packaging can improve the accuracy of measuring their quality. On the basis of transduction mechanism to generate an output nano biosensor used in fruits and vegetables are classified as electro chemical sensors and optical sensors. The latter one uses metallic nanoparticles and quantum dots whereas the former one uses carbon nano tube-based electrodes. Nano barcode technology is also developed for the detection of specific DNA of an organism (Srivastava et al., 2017).

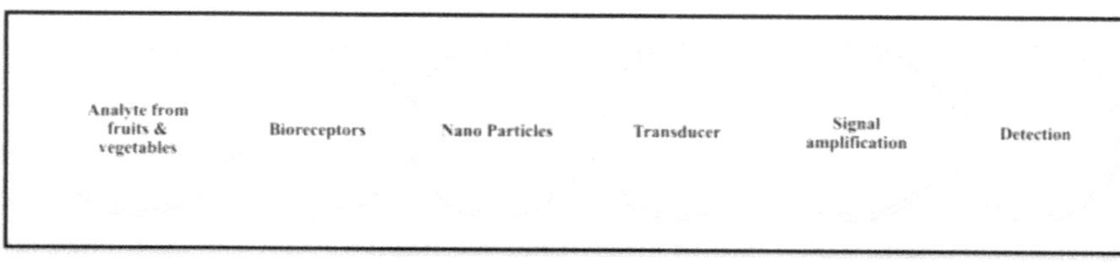

FIGURE 8.3 Schematic representation of working principle of nano biosensors.

Source: Adapted and modified from Srivastava et al. (2017)

8.5 CHALLENGES

It is highly recommended to use intelligent packaging systems for efficient dynamic information of the product. But at what cost? Unlike any other commercialized products, after extensive research on it IPS still suffers to make its place in the market. There are several challenges that IPS face to replace conventional labeling on packaging. First and foremost, the waste generated by intelligent packaging is considered unsustainable for recycling (Schaefer and Cheung, 2018). Hence, improved research on recycling of the materials used needs to be carried or develop any biodegradable solution. Next challenge can be complying with the regulations set by food safety authorities to incorporate IPS into the packaging. But in the case of using nanomaterials to develop advanced IPS, there is a chance that nanomaterials might leach into the product in contact which is undesired from the aspect of food safety. Therefore, full study on migration of nanoparticles into the food has to be established using real food (Bumbudsanpharoke and Ko, 2019).

Considering the safety regulations on migration of indicating system into the package, natural colorants are brought into picture to act as freshness indicators, but after years of research, they are still undermined in terms of performance. They face problems related to stability during processing, sensitivity to bioactivity of pathogen (Poyatos-Racionero et al., 2018). In the case of retailers and manufacturers, the possible drawback could be their apprehension for unsold food products. Even the customers have difficulty in trusting the new IPS because of their accuracy in predicting the quality. The sensors used are fast and reliable but biosensors cannot detect pathogenic disruption at low concentrations, fluorescence dyes cannot be detected with naked eye (Gregor-Svetec, 2018). The printed RFID can be affected by electromagnetic wave. So, it is important to provide awareness among all the consumers, retailers, and manufacturers on how the IPS works and how to read them.

There is no need to mention about the cost limitations of using any kind of IPS in packaging. Even a value is added to the package after the integration of IPS, consumers are unable to accept them since they add extra cost to the product which will be more than the cost of fruits and vegetable itself. Fortunately, if an IPS can be developed in such a way that is both economically cheaper, performs with higher accuracy, complies with regulations and ideally disposable; then there is a higher chance of its acceptance from consumers and retailers (Almasi et al., 2022). And last and most important one is to develop new business models to predict the consumer behavior. Big data and digitalization of innovative models employing newest strategies play an important role in commercializing the IPS. It can also be prompted that combined usage of active and intelligent systems will provide synergistic results in enhancing food safety and security.

8.6 CONCLUSION

Intelligent packaging is a revolution in packaging industry. It provides the dynamic shelf-life monitoring of packed products. Such kinds of systems are highly recommended to be used in perishable food products like fruits and vegetables. Agricultural produce is a living tissue that respires postharvesting also. The climacteric fruits deteriorate due to increased respiration and ripening, whereas the non-climacteric fruits degrade because of external factors like time, storage temperature, light, humidity, and microbial growth. Considering the complexity of mechanisms that involve in the shelf life of fruits and vegetables, an appropriate intelligent packaging system needs to be introduced into the packaging for monitoring their quality. The available types of intelligent packaging systems are data carriers, indicators, and sensors. The latter two are crucial and highly appropriate for use in fruits and vegetable packaging. The freshness indicators technically interact with the product's head space gases, vapors, and microbial metabolites to change their color indicating the freshness of the produce. Time–temperature indicators are dependent on external factors like storage temperature and time. A chemical reaction takes place within the indicator because of exposed temperature and a color change is observed indicating the undesired temperature exposure. Sensor plays a vital role by detecting minute quantities of gases, enzymes, metabolites produced by fruits and vegetables using receptors and transducers. Data carriers like RFIDs are not designed to be an integral part of the product rather they are applied on primary and tertiary packaging to maintain the traceability leading to reduced food waste. Advancements in intelligent packaging system for fresh produce packaging involve printed RFID tags, nano biosensors, metallic organic framework gas sensors, and electronic nose to detect and communicate their quality. Nonetheless, there are existing challenges like high cost, created confusion among consumers due to false readings and operating procedures, economical losses for the industries, and increased plastic wastage. Application of the nanotechnology in freshness indicators needs to be reviewed for the rules and regulations. These factors limit their use giving an enormous scope for the researchers to work on and improve the quality of intelligent packaging systems so that they can be brought into every-day usage.

KEYWORDS

- **intelligent packaging**
- **nano biosensors**
- **freshness indicators**
- **food security**
- **post harvest losses**
- **visual communication**

REFERENCES

Alam, A. U.; Rathi, P.; Beshai, H.; Sarabha, gurusimran k.; Jamal Deen, M. Fruit Quality Monitoring with Smart Packaging. In *Sensors*. Multidisciplinary Digital Publishing Institute, 2021. DOI: 10.3390/s21041509

Almasi, H.; Forghani, S.; Moradi, M. Recent Advances on Intelligent Food Freshness Indicators; an Update on Natural Colorants and Methods Of Preparation. *Food Packag. Shelf Life* **2022,** DOI: 10.1016/j.fpsl.2022.100839

Beshai, H.; Sarabha, G. K.; Rathi, P.; Alam, A. U.; Jamal Deen, M. Freshness Monitoring of Packaged Vegetables. *Appl. Sci.* **2020,** *10* (21), 7937.

Bumbudsanpharoke, N.; Ko, S. Nanomaterial-Based Optical Indicators: Promise, Opportunities, and Challenges in the Development of Colorimetric Systems for Intelligent Packaging. *Nano Res.* **2019,** DOI: 10.1007/s12274-018-2237-z

Calicioglu, O.; Flammini, A.; Bracco, S.; Bellù, L.; Sims, R. The Future Challenges of Food and Agriculture: An Integrated Analysis of Trends and Solutions. *Sustainability (Switzerland)* **2019,** *11* (1), 222.

Caon, T.; Martelli, S. M.; Fakhouri, F. M. New Trends in the Food Industry: Application of Nanosensors in Food Packaging. In *Nanobiosensors*. Academic Press, 2017; pp. 773–804.

Dalmoro, V.; Zimnoch dos Santos, J. H.; Pires, M.; Simanke, A.; Baldino, G. B.; Oliveira, L. Encapsulation of Sensors for Intelligent Packaging. In *Food Packaging*. Academic Press, 2017; pp 111–145.

Fang, Z.; Zhao, Y.; Warner, R. D.; Johnson, S. K. Active and Intelligent Packaging in Meat Industry. *Trends Food Sci. Technol.* **2017,** *61*, 60–71.

Food Loss and Waste | FDA. (n.d.).

Food Waste. (n.d.).

Gregor-Svetec, D. Intelligent Packaging. In *Nanomaterials for Food Packaging: Materials, Processing Technologies, and Safety Issues*, 2018; pp 203–247. DOI: 10.1016/B978-0-323-51271-8.00008-5

Jia, W.; Liang, G.; Jiang, Z.; Wang, J. Advances in Electronic Nose Development for Application to Agricultural Products. *Food Anal. Methods* **2019,** *12* (10), 2226–2240.

Kuswandi, B.; Siddiqui, M. W., Eds.. Sensor-Based Quality Assessment Systems for Fruits and Vegetables. In *Sensor-Based Quality Assessment Systems for Fruits and Vegetables*, 1st ed.; CRC Press, 2020. DOI: 10.1201/9781003084174

Liu, W.; Zhang, M.; Bhandari, B. Nanotechnology—A Shelf Life Extension Strategy for Fruits and Vegetables. *Crit. Rev. Food Sci. Nutr.* **2020**. DOI: 10.1080/10408398.2019.1589415

Ma, L.; Zhang, M.; Bhandari, B.; Gao, Z. Recent Developments in Novel Shelf Life Extension Technologies of Fresh-Cut Fruits and Vegetables. *Trends Food Sci. Technol.* **2017,** DOI: 10.1016/j.tifs.2017.03.005

Mariah, M. A. A.; Vonnie, J. M.; Erna, K. H.; Nur'Aqilah, N. M.; Huda, N.; Wahab, R. A.; Rovina, K. The Emergence and Impact of Ethylene Scavengers Techniques in Delaying the Ripening of Fruits and Vegetables. *Membranes* **2022,** *12* (2), 117.

Mataragas, M.; Bikouli, V. C.; Korre, M.; Sterioti, A.; Skandamis, P. N. Development of a Microbial Time Temperature Indicator for Monitoring the Shelf Life of Meat. *Innov. Food Sci. Emer. Technol.* **2019,** *52,* 89–99.

Mohapatra, S.; Kar, R. K.; Biswal, P. K.; Bindhani, S. Approaches of 3D Printing in Current Drug Delivery. *Sensors Int.* **2022,** DOI: 10.1016/j.sintl.2021.100146

Müller, P.; Schmid, M. Intelligent Packaging in the Food Sector: A Brief Overview. *Foods* **2019,** *8* (1). DOI: 10.3390/foods8010016

Mustafa, F.; Andreescu, S. Chemical and Biological Sensors for Food-Quality Monitoring and Smart Packaging. *Foods* **2018,** *7* (10), 168.

Oyom, W.; Zhang, Z.; Bi, Y.; Tahergorabi, R. Application of Starch-Based Coatings Incorporated with Antimicrobial Agents for Preservation of Fruits and Vegetables: A Review. *Prog. Org. Coatings* **2022,** DOI: 10.1016/j.porgcoat.2022.106800

Pnuma, W. UNEP Food Waste Index Report, 2021.

Poyatos-Racionero, E.; Ros-Lis, J. V.; Vivancos, J. L.; Martínez-Máñez, R. Recent Advances on Intelligent Packaging as Tools to Reduce Food Waste. *J. Clean. Prod.* **2018,** *172*, 3398–3409.

Qu, P.; Zhang, M.; Fan, K.; Guo, Z. Microporous Modified Atmosphere Packaging to Extend Shelf Life of Fresh Foods: A Review. *Crit. Rev. Food Sci. Nutr.* Taylor Francis. **2022,** DOI: 10.1080/10408398.2020.1811635

Schaefer, D.; Cheung, W. M. Smart Packaging: Opportunities and Challenges. In *Procedia CIRP*,Vol. 72; Elsevier, 2018; pp 1022–1027.

Shao, P.; Liu, L.; Yu, J.; Lin, Y.; Gao, H.; Chen, H.; Sun, P. An Overview of Intelligent Freshness Indicator Packaging for Food Quality and Safety Monitoring. *Trends Food Sci. Technol.* **2021,** *118,* 285–296.

Siddiqui, J.; Taheri, M.; Alam, A. U.; Deen, M. J. Nanomaterials in Smart Packaging Applications: A Review. *Small* **2022,** *18* (1), 2101171.

Singh, A. K.; Ramakanth, D.; Kumar, A.; Lee, Y. S.; Gaikwad, K. K. Active Packaging Technologies for Clean Label Food Products: A Review. *J. Food Meas. Charact.* **2021,** 1–11.

Sohail, M.; Sun, D. W.; Zhu, Z. Recent Developments in Intelligent Packaging for Enhancing Food Quality and Safety. *Crit. Rev. Food Sci. Nutr.* **2018,** *58* (15), 2650–2662.

Srivastava, A. K.; Dev, A.; Karmakar, S. Nanosensors and Nanobiosensors in Food and Agriculture. *Environ. Chem. Lett.* 2017, *16* (1), 161–182.

Topuz, F.; Uyar, T. Antioxidant, Antibacterial and Antifungal Electrospun Nanofibers for Food Packaging Applications. *Food Res. Int.* **2020**. DOI: 10.1016/j.foodres.2019.108927

USDA. Food Waste FAQs | USDA. *Usda* (2014), **2017,** 1–5.

Xanthopoulos, G. T.; Templalexis, C. G.; Aleiferis, N. P.; Lentzou, D. I. The Contribution of Transpiration and Respiration in Water Loss of Perishable Agricultural Products: The Case of Pears. *Biosyst. Eng.* **2017,** *158,* 76–85.

Yang, Z.; Zhai, X.; Zou, X.; Shi, J.; Huang, X.; Li, Z.; ... Xiao, J. Bilayer pH-Sensitive Colorimetric Films with Light-Blocking Ability and Electrochemical Writing Property: Application in Monitoring Crucian Spoilage in Smart Packaging. *Food Chem.* **2021,** *336,* 127634.

Yong, H.; Liu, J. Recent Advances in the Preparation, Physical and Functional Properties, and Applications of Anthocyanins-Based Active and Intelligent Packaging Films. *Food Packag. Shelf Life* **2020**. DOI: 10.1016/j.fpsl.2020.100550

Yousefi, H.; Su, H. M.; Imani, S. M.; Alkhaldi, K.; Filipe, C. D.; Didar, T. F. Intelligent Food Packaging: A Review of Smart Sensing Technologies for Monitoring Food Quality. *ACS Sen.* **2019,** DOI: 10.1021/acssensors.9b00440

Yuan, H.; Li, N.; Fan, W.; Cai, H.; Zhao, D. Metal-Organic Framework Based Gas Sensors. *Adv. Sci.* **2022,** *9* (6), 2104374.

CHAPTER 9

ACTIVE EDIBLE COATINGS AND FILMS FOR THE PRESERVATION OF THE QUALITY OF AGRICULTURAL PRODUCE

LOKESH KUMAR[1], SUMAN SINGH[2], and KIRTIRAJ K. GAIKWAD[1]

[1]*Department of Paper Technology, Indian Institute of Technology Roorkee, Roorkee, Uttarakhand, India*

[2]*Department of Food Engineering, Institute of Food Science and Technology, VCSG Uttarakhand University of Horticulture and Forestry, Majri Grant, Dehradun, Uttarakhand, India*

ABSTRACT

Edible coatings and films are very popular preservation techniques for agricultural products due to their biodegradability and environment-friendly nature. Edible coatings and films are made from food-grade biopolymers such as polysaccharides, proteins, and lipids obtained from plants, marine life, and animals. These can be consumed along with the fresh produce, with or without peeling. Edible coatings and films protect vegetables and fruits from external contamination and enhance sensory attributes. Active agents such as antimicrobial, antioxidant, and antibrowning agents enhance shelf life. Edible packaging made from polysaccharides has high gas barrier properties. Edible films of protein show high mechanical strength. A composite film of polysaccharide and protein has better gas barrier and mechanical strength properties. Further, the addition of lipids increases the hydrophobicity of coatings and films. Due to the high moisture content and

Novel Packaging Systems for Fruits and Vegetables. Kirtiraj K. Gaikwad & Suman Singh (Eds.)

continuous respiration process, postharvest preservation techniques are adapted for fresh vegetables and fruits. Active edible coatings and films provide selective permeability to moisture and gases. This chapter aims to summarize various biomaterials for edible coatings and films, fabrication processes, nanotechnology, active agents, and their effects on vegetables and fruits. Moreover, the use of edible coatings and films for whole and cut vegetables and fruits is also discussed.

9.1 INTRODUCTION

Edible coatings and films are eco-friendly and a sustainable alternative to conventional synthetic polymer-based plastic. Edible film is standalone wrapping material, and coating is a thin layer of covering material directly applied to the vegetables and fruits. Edible coatings and films are fabricated from consumable, food-grade biopolymers such as polysaccharides (starch, pectin, and alginate), proteins (whey and zein), and waxes (carnauba and candelilla). These biopolymers are extracted from renewable natural resources like plants and animals. Edible films and coatings preserve the food from physical and chemical changes, reduce moisture and gas exchange between the food product and its surroundings, preserve the quality, and enhance the food product's shelf life. Food-grade additives like natural antioxidants, antibrowning, antimicrobial, flavoring, and coloring agents are also added to the biopolymer matrix. Antimicrobial agents retard microorganism growth on the food surface, and antioxidants and antibrowning agents subsequently reduce the rate of oxidation and browning tendency of perishable vegetable and fruit products. Coloring and flavoring agents increase consumer acceptance of vegetables and fruits (Gaikwad et al., 2016, 2019; Singh et al., 2018).

Edible films and coatings can be fabricated using various techniques. Edible film is mainly formed by solution casting and extrusion methods. Spraying, dipping, and panning are the most prominent coating methods for food products. Additives are blended with the film-forming solutions to modify and improve the desired properties. During the film-forming process, all the ingredients interact with each other and form new bonds such as covalent bonds, ionic bonds, and electrostatic hydrogen bonds. The mechanisms involved in the coating and film fabrication should be of the food process category, such as using food-grade solvents, adding food-grade plasticizers, and biopolymers, along with heating and pH modification. Controlling the fabrication process is crucial, as changing the processing conditions may alter the mechanism of the reaction (Kumar et al., 2022).

Edible polymers are biodegradable in nature, so compared to synthetic polymer-based packaging materials, they have low barrier and strength properties. These limitations can be overcome by adding certain food-grade additives. These additives can improve the moisture and gas (oxygen and ethylene) barrier properties along with their mechanical properties (Gaikwad et al., 2018; Upadhyay et al., 2022). Flexible films can change their shape under various kinds of loading conditions without failure. Film flexibility can be enhanced by adding food-grade plasticizers like glycerol and sorbitol. The main purpose of the packaging is to protect the food, but when certain active agents are incorporated in the films and coatings known as active edible coatings and films, they can enhance the life span of fresh vegetables and fruits. Essential oils, enzymes, and organic acids act as antimicrobial agents. When these are added in solution, the edible packaging can retard the growth of microorganisms and enhance the shelf life of fresh produce. Antioxidant agents are incorporated into the solution to retard the rate of oxidation of fresh produce. It maintains taste and nutrition and improves consumer acceptance of food. Antibrowning agents reduce the enzymatic activities in the fresh-cut fruits, so the color-browning tendency also decreases. When a single biopolymer material cannot meet all the desired packaging requirements, more than one biopolymer is used for packaging, known as composite edible packaging. These composite biopolymers have combined desirable properties of each biopolymer. These biocomposites have a high barrier to moisture and gases, along with better strength properties. Incorporating the nanoclay in the polymer matrix also enhances the overall moisture and gas barrier of the edible films (Deshmukh et al., 2022; Gaikwad et al., 2019).

Since the edible coatings and films are made from food-grade polymers, they are not only biodegradable but also sustainable for our environment (Sultana et al., 2022). When edible packaging is used for food packaging, it can be consumed along with the food product. Edible packaging is highly selective and depends on the type of food, temperature, relative humidity (RH), and mechanism of food quality deterioration. It is commercially used for packaging perishable fruits and vegetables. Since it is biodegradable, it can also be used for nonfood applications such as grocery bags, agriculture purposes, and coating papers. As for primary packaging, edible coatings and films are a potential alternative to synthetic polymer-based plastic. It cannot be used as secondary packaging since direct handling and transportation may affect its quality (Kumar et al., 2022).

9.2 COMPOSITIONS OF EDIBLE COATINGS AND FILMS

Edible coatings and films are made from polysaccharides, proteins, and lipids, which are obtained from natural sources. It creates a modified atmosphere around the fruits and vegetables and prevents moisture loss, gas exchange, and lipid transfer from fresh produce to the surrounding environment. Naturally available hydrocolloids such as polysaccharides, proteins, and lipids are able to fabricate continuous films. Edible coatings and films are fabricated from a combination of all these materials, incorporating active agents like antimicrobial, antioxidant, and flavoring agents. Films and coatings with active agents show additional functional properties, such as reducing microbial growth and reducing the rate of oxidation. Here we will discuss polysaccharide, protein, and lipid, and combinations of these known as composites, as primary packaging.

9.2.1 POLYSACCHARIDE-BASED EDIBLE COATINGS AND FILMS

Polysaccharides are a nontoxic, abundantly available replacement material for conventional synthetic plastic. Natural gums extracted from plants, that is, pectin, cellulose, starch, chitosan, carrageenan, and alginate, are highly used polysaccharides for edible coatings and films. Various monosaccharide, polysaccharide, and disaccharide units polymerize and form glycosidic bonds. Edible coatings and films fabricated from polysaccharides exhibit excellent gas barrier properties because polysaccharides form a tightly packed network structure. Gas barrier properties affect the ripening process and improve the life span of fresh vegetables and fruits. Polysaccharides are hydrophilic in nature, so they do not remain intact when moisture comes into contact with a film or coating. The low moisture barrier tendency of polysaccharides limits their use as primary packaging. Various methods, such as the addition of hydrophobic lipids and cross-linking with food-grade additives, have been employed to reduce the moisture transmissivity of films. Further, the incorporation of active agents provides functional properties to the coating and film. Antioxidant agents reduce the oxidation rate of vegetables and fruits; antimicrobial agents retard the growth of microbes; and nutrients provide nutritional additions to the vegetables and fruits (Mohamed et al., 2020).

9.2.2 PROTEIN-BASED EDIBLE COATINGS AND FILMS

Protein can be derived from milk, soybeans, corn, and gelatin and can form coatings and films for the packaging of fresh produce. Proteins are polymers of amino acids that bind together with peptide bonds. As with polysaccharides, proteins are also hydrocolloid in nature. They show a low water vapor barrier and tensile strength. Protein-derived films and coatings exhibit excellent gas barrier properties. Edible polymers are selected according to end barrier requirements. Hydrophilic protein is mixed with wax or lipid to improve the overall hydrophobicity of coatings and films. Here, too, active agents give functional properties to the coatings and films. Antimicrobial, antioxidant, and antibrowning agents preserve the quality of vegetables and fruits and improve the overall shelf life of agricultural fresh produce (Mohamed et al., 2020).

9.2.3 LIPID-BASED EDIBLE COATINGS AND FILMS

Lipids provide excellent moisture barrier properties as film or coating materials. Lipids are extracted from natural resources like plants, insects, and animals. The main functional groups associated with lipids are glycerides, phosphatides, phospholipids, fatty acids, and cerebrosides. Lipids mixed with hydrocolloids, such as polysaccharides and proteins, improve the overall water vapor barrier of coatings and films. But these coatings have more moisture permeability compared to pure lipid coatings. The low polarity of lipids makes them inert to react with water; thus, they provide excellent moisture barrier properties. The hydrophobicity of lipids makes films and coatings thicker and brittle. Solid fat and liquid oil, which have a major content of triglycerides, are also hydrophobic in nature. They can be extracted from plants and animals. Essential oils contain volatile compounds, which also fall under the category of lipids. Essential oils show antimicrobial and hydrophobic properties as well. Waxes such as carnauba, beeswax, and candelilla are edible in nature. They improve the gloss of the food surface and provide a moisture barrier, thus enhancing the shelf life of vegetables and fruits (Mohamed et al., 2020).

9.2.4 COMPOSITE-BASED EDIBLE COATINGS AND FILMS

Polysaccharides and proteins provide gas barrier properties and have film-forming properties, but they have poor moisture barrier properties. Lipids such as wax, essential oil, and resins are hydrophobic; they exhibit excellent moisture barrier properties. The ultimate aim of packaging is to protect and enhance the shelf life of vegetables and fruits. So, polysaccharides or proteins are combined with hydrophobic lipids so that final coating and film will provide superior properties. Composite edible films have high water vapor and gas barrier properties along with good mechanical strength. Such coatings and films are known as composite edible coatings and films. Sousa et al. (2021) reported that hydroxypropyl methylcellulose added with 20% beeswax coated palmer mangoes enhanced the shelf life up to 6 days. This coating also delays the respiration, ripening, and softening of mangoes. A composite edible coating of chitosan-gum arabic and *Cleistocalyx operculatus* extract was used to coat the banana. The coated bananas have a shelf life of 21 days at room temperature (Le et al., 2021).

9.3 EDIBLE COATING AND FILM FABRICATION

As primary packaging, the edible film is wrapped around the vegetables and fruits in the form of a pouch or bag, and edible coating is applied directly on the surface of fresh produce. The solubility of the biopolymer matrix along with functional additives is an important factor to be considered. Edible films can be formed by mainly two methods: solvent casting and the hot melt extrusion method. Edible coatings can be formed by spraying, dipping, and panning methods. In this section, all the coating and film-forming methods are discussed thoroughly. Figure 9.1 shows a schematic diagram for the fabrication of edible coatings and films.

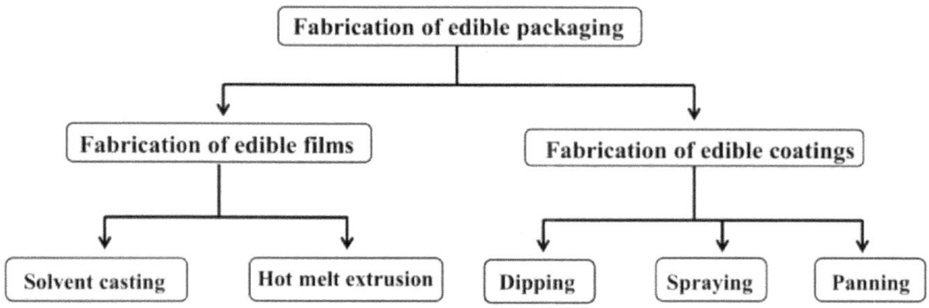

FIGURE 9.1 A schematic diagram of the fabrication methods for edible films and coatings for vegetable and fruit packaging applications.

9.3.1 FILM-FORMING PROCESS

Edible films, which enclose the vegetables and fruits, act as primary packaging. These films can be converted into pouches or sachets and consumed along with the food. Edible films are fabricated by two processes: solvent casting, known as the wet process, and melt extrusion, known as the dry process. In solvent casting (wet process), the solubility of the biopolymers along with all the functional additives is an important factor. In extrusion (dry process), glass transition and thermoplasticity should be acknowledged. Physical chemistry, the interaction among the molecules, determines the film's characteristics like mechanical strength, water and gas permeability, and structural chemistry, where the structure of molecules determines the film-forming process. Homogenous film structure and mechanical properties are affected by the cohesiveness among the edible film ingredients. Recently, active edible films have been in the trend. When active agents such as antioxidants, antimicrobials, nutraceuticals, sweeteners, and coloring agents are incorporated in the edible film-forming solution, they impart certain functional properties to the edible film. These active agents remain in the edible film and improve the value of vegetables and fruits (Kumar et al., 2022).

9.3.1.1 SOLVENT CASTING FILM FABRICATION PROCESS

Solvent casting (wet process) is the most effective and cost-efficient film-forming method at laboratory scale. The solvent casting method has three important steps: (i) solubilization, (ii) casting, and (iii) drying. In the first solubilization step, all the biopolymers, along with plasticizers and functional additives, are dissolved. In the second casting step, the film-forming solution is poured into the petri plate of suitable dimension and material. In the last drying step, the solvent evaporates and the film-forming solution changes into dried film. Since all the ingredients are food-grade, the solvents should also be food-grade. Water and ethyl alcohol are the most common solvents for casting edible films. The drying step involves the formation of intramolecular bonds between biopolymers to provide a suitable microstructure to the film. So forced air-dried films will have insufficient bonds, and films will show negative impacts on their physical and structural properties (Kumar et al., 2022). These films should be imperfection-free and consistent. Physical properties, mechanical properties, and barrier properties are greatly affected by film structure. For continuous film fabrication, the temperature of the air, flow rate, and adhesive nature of the surface upon which the film is cast are important parameters. The drying of the solvent by evaporation can be at ambient room temperature or at an elevated temperature. The drying temperature affects the physical properties, barrier properties, and morphology of the film. Hemicellulose laurate polymer-based film fabricated by solvent casting processes enhances the shelf life of green chilies by 15 days (Zhang et al., 2020).

9.3.1.2 HOT MELT EXTRUSION FILM FABRICATION PROCESS

The casting method is time-consuming since the evaporation of solvents (water or ethanol) takes time. It limits the use of the casting process on a large scale. The extrusion process reduces the time required to fabricate the edible films. This reduces the cost of film fabrication to a competitive range compared to synthetic polymers. Since extrusion uses no or very little amount of solvent, it is known as a dry process. In the extrusion process, high temperatures and shear phenomena melt the biopolymer. The extrusion process can be further categorized into three steps. First is the feeding zone, where biopolymers, plasticizers, and functional additives are added in the extruder. Next is the kneading zone, where the extruder screw mixes all the ingredients properly. The last zone is the heating zone, where the oven supplies heat to the surface of the extruder. This heat will melt the biopolymers and other ingredients. The extruder screw always rotates inside the extrusion chamber; it provides shear energy that combines with oven heat and helps in the melting of biopolymers. At last, one die is fixed at the end of the extrusion chamber. This die decides the size and thickness of the film. Shear energy and heat energy change the biopolymer structure and improve the film's properties. Biopolymers, which are sensitive to high shear and heat, cannot be extruded. Extrusion temperature will degrade the desired properties of films (Kumar et al., 2022).

Shear energy is produced by the rotation of the extruder screw, so it is a subclass of mechanical energy. Heat energy is produced by ovens, so it is part of thermal energy. The rotation of the screw (rpm) changes the shear energy produced, so it also changes the melt rate and properties of the extruded film. Along with these, barrels' temperature, moisture content present in the feed zone, and pressure at the die are important parameters for the final film. High-amylose corn starch-based edible films formed by the extrusion process at extrusion temperature of 100°C and 120 rpm have the best mechanical and barrier properties. These films also improve the quality of mango (*Mangifera indica L.*).

9.3.2 FABRICATION OF EDIBLE COATING

As primary packaging, edible coating encloses the outermost surface of the fruit and vegetables. This coating solution contains biopolymers, additives, and functional ingredients as well. Edible coating formation follows the same mechanism as the solvent casting method. First, the coating solution is dispersed over the fruit surface, and then adhesion between the fruit surface and the coating takes place. After evaporation of food-grade solvent, a thin layer of biopolymer coating forms on the surface of perishable vegetables and fruits. The coating prevents moisture and gas exchange between food and the surrounding atmosphere. If functional additives are also added, then the coating will preserve the vegetables and fruits while reducing the rate of deterioration. Dipping, spraying, and panning are the most useful methods for applying coating over the surface of perishable vegetables and fruits (Kumar et al., 2022).

9.3.2.1 DIPPING METHOD

The dipping method is a well-known coating method for vegetables and fruits. The viscosity, surface tension, dwelling, and density of the coating-forming solution decide the final coating thickness. This technique can form a thick coating on the fruit's surface. The dipping method can be classified into three steps. In the first step, thoroughly washed vegetables and fruits are immersed in the coating-forming solution. In the second step, the fruit remains in solution for some time, and the coating solution gets deposited on the fruit surface. This time is known as dwell time. This time period ensures the adhesion of the coating on the fruit surface. In the third step, the fruit is removed from the coating-forming solution and dried. In drying, the solvent from the wet coating evaporates and forms a thin coating over the surface of perishable vegetables and fruits. The drying can take place at room temperature or high temperatures by natural or forced convection methods with the help of a dryer (Kumar et al., 2022). When fresh-cut slices of apple were dipped in the whey protein nanofibrils-based edible coating, the rate of browning was reduced and the shelf life of the apple improved. Chitosan and oxidized starch-based composite edible coatings have antimicrobial properties and enhance the shelf life of coated *Carica papaya* L. Alginate-lactate-based edible coatings preserve the sensorial qualities and nutritional value of the strawberries. A sweet cherry coated with guar gum biopolymer with ginseng extract has better quality than an uncoated one.

9.3.2.2 SPRAYING METHOD

A coating solution with low viscosity can be directly applied to vegetables and fruits by spraying. The spraying method, with the help of a spray nozzle, breaks the coating solution into tiny droplets and disperses them over the surface of the perishable vegetables and fruits. For the fixed volume of coating-forming solution, tiny droplets will cover more area of the fruit surface. Edible coating can be sprayed by two methods: (i) pressure-assisted atomization and (ii) air spray-assisted atomization. In the pressure-assisted atomization method, pressurized air is used to form the droplets. This method is also well-known as the airless atomization method. With the help of a small nozzle, high-pressure energy is supplied to the coating solution, and this energy breaks the coating solution into small droplets. These droplets cover the surface area of vegetables and fruits, and after drying, form a thin layer of edible coating. In the air spray method, coating droplets are formed by high-speed forced air. High-speed air comes into contact with coating-forming solutions flowing at low speed. Friction between fluid and air disrupts the flow of coating solution, and the solution starts to form small droplets. These droplets come into contact with high-speed air and start to flow with it. Then these droplets come out of the nozzle at high speed and form a thin layer on the fruit's surface. This method is highly suitable for vegetables and fruits with a high surface area. The main disadvantage of the spraying method is the inability to spray highly viscous solutions. For solutions with high viscosity, the dipping process (immersion method) is always recommended. Edible coating of guar gum and xanthan gum by spraying method on cassava improves the postharvest life of cassava. Pea starch and guar gum biocomposite edible coatings sprayed on Valencia oranges maintain their overall freshness and flavor after 4 weeks when stored at 5°C. An edible coating made from chitosan and gelatin reduces water loss, reduces microbial spoilage, and enhances the firmness of red bell peppers.

9.3.2.3 PANNING METHOD

The panning method is most useful for the coating of large amounts of oval and round-shaped confectionary and food items in a single batch. Pan, which is a big bowl, is continuously rotated, and confectionary food items are placed inside the bowl. The coating solution is ladled or sprayed over the product. Continuous rotation of the pan helps the coating-forming solution spread to the surface of confectionery and food. Along with this, continuous air at room temperature or an elevated temperature is circulated. It evaporates the solvent and forms a coating layer on the surface of the food product. The amount of coating solution decides the final thickness of the coating (Kumar et al., 2022).

9.4 FUNCTIONS OF ACTIVE EDIBLE COATINGS AND FILMS

Edible films are fabricated from food-grade biopolymers, so they are biodegradable and nontoxic in nature. It can be consumed along with vegetables and fruits; if not eaten, even then edible packaging will degrade easily. After going to the landfill, it will not pollute the earth's soil or ocean. It is sustainable in nature, and if used as an alternative to synthetic plastic, it will reduce the plastic waste generated from food packaging. Edible packaging protects vegetables and fruits from moisture and gases by creating a protective film or coating layer. It can protect highly perishable fruits from ultraviolet light. It prevents aroma, color, and flavor alteration and provides a shiny smooth surface to the vegetables and fruits. This increases the probability of consumer acceptance. Active agents such as antioxidants, antibrowning agents, and antimicrobials are also incorporated into the edible film matrix. This packaging releases the active agents in a controlled manner and enhances the life span of the vegetables and fruits. Food with a high shelf life will have more demand compared to normal food. In this section, we will discuss the functional properties of active packaging in detail (Cerqueira et al., 2017; Kumar et al., 2022). Figure 9.2 shows the various functions of edible coatings and films for vegetables and fruits.

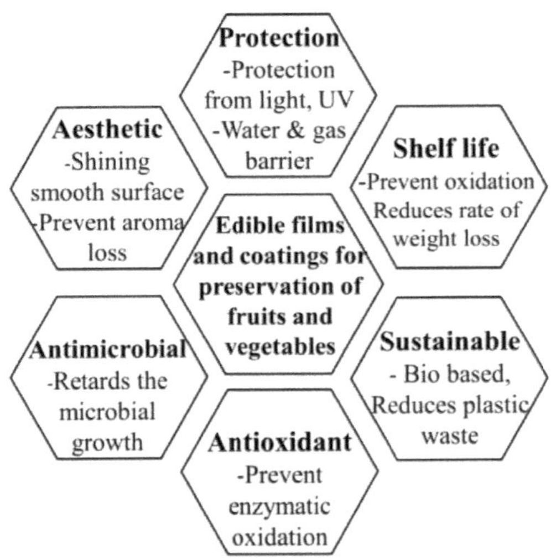

FIGURE 9.2 Various functions of edible coatings and films for the preservation of vegetables and fruits.

9.4.1 MOISTURE AND GAS BARRIERS OF EDIBLE COATINGS AND FILMS

The quality of fresh-cut and whole fruits deteriorates due to moisture transfer, oxygen transfer, and unwanted odor absorption. Moisture transfer from the fresh-cut fruits to the surrounding atmosphere will reduce the total weight of the fruit. If oxygen comes into contact with fresh-cut fruits, it will oxidize them. Oxidation of fruit will change its natural color and taste (Gupta et al., 2021). Edible coatings and films maintain the quality of vegetables and fruits, such as weight loss and oxidation, thereby improving the life span and consumer acceptability of vegetables and fruits. Edible film is a separate layer of biopolymer that wraps around the food, and edible coating is a thin layer of biopolymer that directly enrobes the food. Edible coatings and films create hindrances for the moisture and gas exchange between food and the surroundings. Moisture loss is directly related to weight loss, and gas transfer is related to the respiration and ripening of vegetables and fruits. When rice starch and sucrose ester as additive-based edible coatings are applied to the Cavendish banana stored at 20 ± 2°C, it reduces the weight loss, reduces the respiratory rate, and delays the ethylene synthesis. Uncoated banana ripens after 7 days but edible coated banana has a prolonged life span of 12 days.

Moisture and gas permeation is a bulk-volume phenomenon, so the rate of moisture and gas transfer depends on various factors. The most important factor is cohesiveness among all the film-forming ingredients. If the film-forming components have a nonbinding tendency with each other, this can lead to phase separation of the film. Separation of phases will make the film nonhomogeneous, and then moisture and gas permeability will increase. During the drying of the film, bubbles and pinhole porosity can be formed. Bubbles are basically entrapped air inside the film, and pinholes are created when gas leaves the film during drying. Considering all the precautions, edible coatings and films will reduce the rate of weight loss, so they will improve consumer acceptance of vegetables and fruits (Kumar et al., 2022).

9.4.2 ANTIOXIDANTS AND ANTIBROWNING AGENTS IN EDIBLE COATINGS AND FILMS

Antioxidants maintain the nutrition values and prevent discoloration and degradation of fresh vegetables and fruits. The fresh-cut fruits are highly reactive to oxygen gas. This oxygen can be available through the surrounding atmosphere or through the food surface itself. During the cellular metabolism reaction, oxygen radicals are released as byproducts. These oxygen radicals are called reactive oxygen species (ROS). They are very reactive and unstable oxygen molecules that interact with the food surface and change the organoleptic properties. Antioxidant agents are incorporated into the biopolymer matrix. These agents are released in a controlled manner and react with oxygen species. They break the chain reaction and protect the vegetables and fruits from oxidative spoilage. Many spices, herbs, and seeds are well-known food-grade antioxidant agents. Flavonoids, phenolic compounds, terpenoids, and tannins can be easily extracted from antioxidant agents without harming the plant species. Further, antibrowning agents are also added to the coating and film-making solution. They solve the color-browning problem of fresh-cut fruits. The main factor behind the browning of fruits is the presence of polyphenol oxidase (PPO) enzymes. These enzymes oxidize the phenolic compounds present in the fruits and generate dark color pigment. Oxidation also alters the natural taste, flavor, and nutritional value of vegetables and fruits. Citric acid and ascorbic acid are food-grade antibrowning agents. These agents reduce the rate of browning of fresh, perishable cut vegetables and fruits such as apples and potatoes. They are also incorporated into edible films and coating-forming solutions. These agents reduce the reactivity of polyphenol enzymes by lowering the pH of the film-forming solution or by binding with the active sites of the PPO enzymes. These antioxidants and antibrowning agents are added during the preparation of film to maintain the organoleptic properties for a longer period of time and improve the life span of fresh perishable cut vegetables and fruits (Kumar et al., 2021; Tanwar et al., 2021). Studies have reported that incorporating antioxidant and antibrowning agents in edible coatings and films enhances the life span and consumer acceptance of vegetables and fruits. Researchers have reported the antibrowning effects of citric acid and ascorbic acid. When these agents are incorporated in the alginate-based edible coating, sensory attributes such as the appearance and texture of fresh-cut apple are also enhanced. Thyme leaf extract incorporated in the chitosan-starch biopolymer matrix releases the flavonoids and terpenoids as the main active compounds. Table 9.1 shows various active antioxidant agents incorporated in the edible film matrix for vegetable and fruit packaging.

9.4.3 ANTIMICROBIAL AGENTS IN EDIBLE COATINGS AND FILMS

Microorganism growth on the food surface is one of the major reasons for the contamination and deterioration of perishable vegetables and fruits. Food experts have reported that about 25% of food spoilage is due to microorganism contamination. Loss of quality, color, texture, and nutrition

TABLE 9.1 Active Antioxidant Agents Incorporated in Edible Film Matrix for Vegetable and Fruit Packaging.

Category of active material	Active ingredient	Polymer matrix	Application	Result	Reference
Antioxidant	Ascorbic acid and moringa leaf extract	Papaya puree and pectin	Coating on minimally processed pears, stored at 8 ±1°C	Edible coating has antioxidant effect and improves the shelf life and sensory acceptance	(Rodríguez et al., 2020)
	α-dl tocopherol acetate	Sodium alginate biopolymer	Coated the sliced carrot by dipping method and stored at 10°C and 65% RH	Antioxidant film maintains the weight, pH, TSS, and control the growth of mold and yeast	(Keshari et al., 2022)
	Bacterial cellulose nanofiber (0.6% and 0.8%)	Chia seed mucilage-based polysaccharides matrix	Dipped coated strawberries were stored at 4°C and 75% RH	Coating preserves the bioactive compound of strawberries, retards the enzymatic browning, and enhances the shelf life	(Mousavi et al., 2021)
	Ascorbic acid	Carboxyl methyl chitosan and gelatin matrix	Sweet cherry dip-coated and stored at 0 ±0.5°C and 85–90% RH for 30 days	Antioxidant coating reduces the weight loss, decay ratio of fruit, maintains the firmness, and retards the browning	(Zhang et al., 2021)
	Gallic acid	Chitosan	White button mushroom in film stored at 90% RH and 4 ±1°C for 15 days	Edible film reduces respiration and browning rate of mushroom	(Liu et al., 2019)
	Loquat leaf extract	Sodium alginate	Coated nanfeng tangerines stored at 5 ±1°C for 60 days	Coating maintains the quality, reduces decay and growth of *Penicillium digitatum*	(Zhang et al., 2022)

values comes under the food spoilage category. The growth of bacteria, molds, and yeasts during storage and distribution causes the vegetables and fruits to spoil. When these microbes grow above a permissible limit, they deteriorate the fruit and make it nonconsumable. Microbial growth relies on various factors such as the pH of food, the availability of water and oxygen, and temperature. Earlier, the solution containing antimicrobial chemicals was sprayed on the fruits, or the fruits were dipped in the aqueous solution of antimicrobial chemicals. This method was not effective as antimicrobial agents present on the food surface neutralized very fast or diffused into the food itself, reducing the quantity of antimicrobial agents on the food surface (Kumar et al., 2022; Kumar and Kirtiraj, 2022).

Antimicrobial agents mixed in the edible biopolymer matrix have increased inhibitory action compared to the dipping and spraying methods. The incorporation of antimicrobial agents inside the edible matrix maintains the effective concentration of antimicrobial agents over the surface of food products. The controlled release of antimicrobial agents improves the effectiveness of active antimicrobial edible packaging. Antimicrobial agents can be divided into various categories, such as organic acids (acetic acid, benzoic acid, and ascorbic acid), bacteriocins (nisin and subtilin), and essential oils (thymus, oregano, mint, and cinnamon). These food-grade antimicrobial agents damage the living cells of the microorganisms. It further retards the growth of microorganisms and improves the life span of perishable vegetables and fruits. When lemon and the essential oil of bergamot were incorporated as active antimicrobial agents in the whey protein-based edible film matrix, they were effective against *S. aureus* and *E. coli*. These films also show improved water and oxygen barrier properties. Table 9.2 shows various active antimicrobial agents incorporated in the edible film matrix for vegetable and fruit packaging.

9.4.4 NUTRACEUTICALS

Consumer demands for natural products with functional properties are increasing (Gaikwad et al., 2017). Edible coatings and films that can hold active ingredients that enhance the nutritional value of vegetables and fruits are known as nutraceuticals. They are consumable, food-grade additives that possess health benefits and can also prevent diseases. They are extracted from natural sources and are generally available in powder form. The amount of nutraceuticals added to the edible packaging matrix must be

TABLE 9.2 Active Antimicrobial Agents Incorporated in Edible Film Matrix for Vegetable and Fruit Packaging.

Category of active material	Active ingredient	Polymer matrix	Application	Result	Reference
Antimicrobial	Lemon extract	Soy protein isolate	Coating on fresh-cut melons stored at 4°C	Edible coating controls the growth of mold and yeast and maintains sensory attributes	(Yousuf et al., 2020)
	Organic acid	Chitosan	Fresh-cut dip-coated apples, stored at 4°C for 35 days	Antimicrobial coating retards the microbial spoilage of cut apples	(Jin et al., 2020)
	Cumin essential oil	Modified potato starch	Dipped coated early crisp pears were stored at 25°C	Coating suppressed the respiration and weight loss; antifungal coating enhances the shelf life	(Oyom et al., 2022)
	Thyme essential oil	Starch- gellan	Edible coating on the apple stored at 25°C and 65% RH for 14 days	Antifungal coating reduces the gray mold (*Botrytis cinerea*) count	(Sapper et al., 2019)
	Carvacrol and thymol essential oils	Corn starch with high amylose content	Coated papaya and mango stored at 20°C for 18 days	Coating is antifungal against the growth of *Colletotrichum gloeosporioides*	(Ochoa-Velasco et al., 2021)
	Lactic acid bacteria, sodium caseinate, and oleic acid	Potato starch	Edible coating on the grapes stored at 20°C and 85% RH for 7 days	Edible coating controls the fungal growth of *Botrytis* and *L. plantarum*	(Marín et al., 2019)

studied carefully, as they also affect the mechanical properties as well as the moisture and gas exchange tendency of edible packaging. Nutraceuticals are incorporated in the edible coating-forming solution to enhance the overall nutrition value of the coated product. Nutraceuticals like ascorbic acid, an antioxidant, and pectin support cardiovascular activities and have benefits for human health. Protein is an essential element for human growth. Milk protein, a nutraceutical-based edible coating and film, enhances the overall nutritional value of fresh vegetables and fruits. Active edible coatings and films made from whey and casein protein provide moisture and oxygen barriers, better tensile strength, and protein as a nutrient.

9.5 NANOTECHNOLOGIES IN EDIBLE COATINGS AND FILMS

Nanotechnology is a branch of science involved in the study of nanoscale materials. Nanoscale materials have one of the characteristic dimensions in the nanorange (<100 nm). These materials are more reactive and show different properties compared to their bulk counterparts. The addition of nanomaterials in the form of nanoparticles, nanotubes, nanocomposites, or nanolaminates as active ingredients in the biopolymer matrix enhances the food packaging's functionality and performance.

9.5.1 EDIBLE NANOLAMINATE

Edible nanolaminates are multilayered structures made of food-grade biopolymers. Edible nanolaminates can be formed by layer-by-layer deposition of food-grade biopolymers. Here, the thickness of each layer is in the nanorange, and this thickness depends on the adsorption of biopolymer materials on the surface of fresh produce. The total thickness of laminate is the sum of all the thicknesses, so it depends on the number of nanolaminate layers. These layers can be chemically interlinked with hydrogen bonds, electrostatic bonds, and charge transfers. In the food packaging field, bonding with electrostatic bonds is highly explored. First, the fruit or vegetable, cut or whole, comes into contact with the laminate-forming solution. This solution is made from polysaccharides or proteins, which are food-grade biopolymers. Now this first laminate will distribute a certain charge all over the food surface. After that, this laminated fruit or vegetable is dipped into water for cleaning. The water treatment will remove unbound and excessive biopolymers. Now the laminate is completely bound to the food surface with a certain charge. After that, fruits or vegetables are dipped into the second laminate-forming solution. This laminate will have an opposite charge compared to the charge of the first laminate. Opposite charges will attract each other and form the second layer of nanolaminate. Again, excessive amounts of unbound material are washed with distilled water (Kumar et al., 2022). Ionic strength, pH, and temperature are some of the factors that affect the fabrication process and properties of the nanolaminates. Active compounds like antioxidant agents, antibrowning agents, antimicrobial agents, and flavoring agents can also be added to the biopolymer matrix. These active compounds released in a controlled manner from the polymer matrix will improve the shelf life and increase consumer acceptance of vegetables and fruits. Nanolaminate with the extract of *Flourensia cernua* helps to reduce weight loss and microorganism growth and enhances the shelf life of tomato fruit (Salas-Méndez et al., 2019).

9.5.2 BIONANOCOMPOSITE

Biocomposites are made of a biopolymer matrix along with reinforcement fillers. If these fillers have one of the dimensions in the nanorange (<100 nm), then they are known as bionanocomposites. Nanoparticles and nanotubes can act as nanofiller reinforcements. These reinforcement agents provide better mechanical and barrier properties compared to biopolymer materials. Montmorillonite (MMT), saponite, and hectorite clays are very popular for improving the moisture and gas barrier properties of biopolymers. Evenly distributed nanoclay creates a tortuous path for moisture and gas transmission through the composite (Cerqueira et al., 2017; Kumar et al., 2022). The edible films with reinforcement in nanocellulose improve the barrier and strength of the composite film (Gupta et al., 2021). Cellulose nanofibers as reinforcement agents enhance the strength and barrier properties of mango-puree-based nanocomposite films (Cerqueira et al., 2017).

9.6 APPLICATIONS OF EDIBLE COATINGS AND FILMS ON VEGETABLES AND FRUITS

Healthy eating habits among consumers have led to an increase in demand for fresh vegetables and fruits. They are an excellent source of nutrition because they contain vitamins, minerals, and fibers. Due to a busy modern lifestyle, consumers prefer easy and quick ways to consume vegetables and fruits. This has increased the demand for freshly cut vegetables and fruits. Most vegetables and fruits are perishable in nature because of their metabolism and respiration. Microorganism contamination of fresh vegetables and fruits, along with moisture loss, ripening, and respiration, make them perishable with a low shelf life. Edible coatings and films are an effective means of protecting and improving the life span of cut vegetables, fruits, and as a whole. All these natural activities affect the quality and are responsible for durability as they are made of living tissues. In the food supply chain, loss of moisture is directly related to weight loss, along with change in color, reduction in nutrients, and unpleasant flavor and odor due to respiratory processes that are undesirable for vegetables and fruits. Edible coatings made from polysaccharides and proteins can minimize all these effects on vegetables and fruits. Coating creates a modified atmosphere around the vegetables and fruits that reduces gas exchange, respiratory losses, enzymatic browning, texture deterioration, and off-flavor. Hydrophobic edible coatings, which can be made from lipids, resins, and waxes, can reduce water loss, leading to weight loss, and are beneficial for vegetables and fruits.

Various novel food packaging concepts have been developed to meet consumer demand while considering the safety and quality of fresh vegetables and fruits. Active packaging interacts with fresh produce. It intervenes in the metabolism process and retards deterioration and ripening. Active packaging affects lipid oxidation, browning, and microbial growth. Appropriate active agents are incorporated into the edible polymer matrix for their effective utilization. Active gas scavenging agents are added to the polymer matrix to reduce food spoilage from various gases available in the free headspace or generated due to different metabolic reactions. Active antioxidants, antibrowning agents, and antimicrobial agents are also introduced into the in-polymer matrix to retard the natural spoilage processes of vegetables and fruits. The supply chain of vegetables and fruits involves a lot of processes, from harvesting, handling, and packaging to transportation. Highly perishable fresh vegetables and fruits are susceptible to the growth of various microorganisms. Microorganisms from the category of molds and yeasts contaminate fresh vegetables and fruits, reduce their shelf life and consumer acceptance, and at the end, food turns into food waste. Active antimicrobial agents release in controlled manners when the microbial count increases a certain level and retard the growth of microorganisms on the vegetables and fruits. Antioxidant agents such as citric acid and ascorbic acid retard the rate of oxidation, and antibrowning agents reduce the rate of browning of perishable vegetables and fruits. All the active agents enhance the life span of fresh vegetables and fruits, reduce food waste, and enhance consumer acceptance (Kumar and Kirtiraj, 2022).

9.6.1 EDIBLE COATINGS AND FILMS FOR FRESH-CUT VEGETABLES AND FRUITS

Consumer demand for freshly cut vegetables and fruits that are natural and ready to eat, adapted to the busy modern lifestyle, has peaked. The main challenge for the cut vegetable and fruit industries is to preserve the freshness of the produce for a longer period of time. These can be obtained by following steps such as first cleaning fresh produce, then peeling, and at last cutting and storing. Fresh-cut vegetables and fruits have more surface area exposed to the surrounding environment, making them more susceptible to quality loss. The loss of firmness, color, moisture, and flavor affects the shelf life and increases consumer rejection. Modern advanced packaging technology has shown good results in enhancing the life-span of freshly cut vegetables and fruits (Kumar et al., 2022). The quality of cut fresh produce relies on internal and external factors as well. Physiology, morphology, biochemical defense, and genotype are internal factors, and chemical treatment, storage temperature, sharpness of the knife, and RH are external factors. Compared to whole fresh produce, cut produce is highly susceptible to deterioration because of the generation of wounds in tissues during the cutting process. Compared to vegetables, fruits have complicated mechanisms of metabolism and ethylene production, so it is more difficult to enhance the shelf life of fruits. The edible coating method has been explored to meet the consumer's requirements qualitatively and quantitatively. A modified atmosphere is created around the cut vegetables and fruits by the edible coating. It maintains the moisture in the tissue of freshly cut produce, reduces the rate of respiration, reduces ethylene gas production, etc. If active agents were also introduced in the coating, then the antimicrobial agents would retard the growth of microorganisms, and antioxidants would reduce the oxidation of fresh-cut produce.

A fresh slice of cucumber was dip-coated with edible chitosan, packed in a modified atmosphere of nitrogen gas, and stored at 5°C. Modified atmospheric packaging retards chlorophyll degradation, the respiration rate of tissue, physiological changes, and microbial safety, thus enhancing the shelf life of cut cucumber (Olawuyi et al., 2019). Fresh-cut carrots were dip-coated with edible pectin polymer and stored at 8°C. This coating enhances the shelf life of carrots up to 12 days. A bilayer edible coating with first layer of guar gum-crosslinked sesame protein and a second layer of calcium chloride-mango puree was used to coat the mango slice by layer-by-layer technique and stored at 5°C. This coating maintains the color and firmness, inhibits microbial growth, and improves the shelf life for 15 days (Sharma et al., 2019). Alginate, an edible coating containing sunflower oil and glycerol, was used for coating the fresh-cut cantaloupe slices. Slices were stored at 4±1°C for 40 days. This edible coating extends the life span of freshly cut melons by up to 24 days. Polysaccharides-based aloe vera gel-based edible coating containing calcium chloride was sprayed over the fresh-cut papaya and stored at 5°C and 90% RH. This coating reduces color change and ripening rate and maintains the firmness and sensory attributes of cut papaya (Farina et al., 2020).

9.6.2 EDIBLE COATING AND FILMS FOR WHOLE VEGETABLES AND FRUITS

Deterioration of vegetables and fruits can be caused by preharvest, harvest, and postharvest processes. Enzymes such as amylase, cellulase, oxidase, and pectinase are the main reason for the oxidative damage; they deteriorate sensory attributes and nutrition values. Vegetables and fruits soften due to the dissolution of plant tissues. Postharvest deterioration, such as changes in color, firmness, texture, flavor, and loss in nutritional values, occurs due to biochemical changes like the oxidation of anthocyanin, vitamins, and chlorophyll, along with the modification of starch to sugar. The growth of foodborne pathogens, fungi, and bacteria causes microbial spoilage of fresh produce, which compromises freshness and consumer acceptance. Moisture loss from fresh produce affects appearance, weight loss, and firmness. During harvesting, handling, and postharvest, damage to surface tissue or skin also leads to deterioration. Better technology, transport, communication, and coordination among marketers and producers have improved the demand and supply of fresh vegetables and fruits all around the year. Consumer demand for fresh vegetables and fruits that are minimally processed and free of chemicals is increasing. Minimally-processed fresh produce has natural flavor, freshness, and nutritional values. Edible coatings and films are food-grade semipermeable membranes around the fresh produce, preserving the flavor, freshness, and nutritional values. Edible coatings and films create barriers to moisture and gas transfer between fresh produce and their surroundings. Active agents are also added in the coating and film for targeted delivery of antioxidant, antibrowning, and antimicrobial properties and additional nutrients. These active agents retard the rate of oxidation, browning, and growth of microorganisms. Overall, edible coatings and films maintain weight, flavor, and firmness, enhances fresh produce's shelf life, and improves consumer acceptance.

Khatri et al. (2020) reported the effect of polysaccharides-based aloe vera gel and chitosan composite coating on the tomato. These tomatoes were stored at 4±1°C in a dark space for 6 weeks. A 1% aloe vera and 1% chitosan composition shows the best results, delays the ripening process, and extends the shelf life up to 42 days. Beeswax and coconut oil coated lemons were stored at 21 ± 2°C and 50 ± 5% RH under modified atmospheric packaging conditions. Uncoated lemons turned yellow after 6 days, but coated lemons were green after 18 days. Coating reduces respiration rate, weight loss, and ethylene production and retains the green color (Nasrin et al., 2020). Chitosan films with glycerol were cross-linked with citric acid. An active antioxidant film shows a better moisture barrier and enhances the shelf life of green chili. Rastegar et al. (2019) reported that sodium alginate-based edible coating maintains nutrition value and firmness, reduces weight loss, and enhances the shelf life when stored at 15 ± 1°C and 85 ±1% RH. A pullulan-based edible coating containing calcium chloride and lemon juice was coated on bananas. These coated bananas were stored at 25 ±1°C and 70% RH. Coating reduces weight loss, has a low browning index, improves color saturation, and improves the shelf life up to 20 days. A researcher has reported the effect of a resin gum arabic-based edible coating containing tulsi extract and sodium caseinate. Coated guavas were stored at 28 ± 2°C for 7 days. Coating delays ripening and increases consumer acceptance. It improves the shelf life to 7 days, compared to 4 days for uncoated guava.

9.7 CONCLUSIONS

Consumers' demand for minimally-processed vegetables and fruits has promoted the development of edible coatings and films that will enhance quality, nutritional value, and sensory attributes. All the ingredients, such as polymer matrix, plasticizers, and functional additives, used for the edible coatings and films are food-grade. Regulatory agencies such as the European Food Safety Authority and the Food and Drug Administration are responsible for the approval of food-grade materials. New edible polymers with enhanced barrier properties and strength are under investigation. Naturally available antimicrobial agents (essential oils, organic acids, and bacteriocins), antioxidant agents, and flavoring agents are also being researched. For primary packaging, edible coatings and films are excellent alternatives to synthetic polymer-based plastics. If not eaten with vegetables and fruits, it can be easily decomposed after disposal. Edible coatings and films enclose the vegetables and fruits, thus creating a modified atmosphere around them. This modified atmosphere can reduce water vapor loss, oxygen and ethylene interaction, and aroma transfer between agricultural produce and the environment. Nanotechnology combined with active edible packaging enhances the functionality of coatings and films. Nanoencapsulation of functional additives ensures a controlled release in the edible packaging. The processing of edible coatings and films is still in the infant stage. Due to the high production cost and low barrier and strength properties compared to already available synthetic plastics, mass production and commercialization are big challenges. Further research is required to modify the existing materials and explore new materials that can provide high barrier and strength properties with minimal cost.

KEYWORDS

- **edible films**
- **edible coating**
- **food preservation**
- **active packaging**
- **shelf life**
- **vegetables and fruits**

REFERENCES

Cerqueira, R.; Parente, M. Â.; Correia, R. N.; Ramos, S.; Leandro, Ó.; Teixeira, C.; ... Augusto, A. Edible Food Packaging. In *Edible Food Packaging*. 2017. DOI: 10.1201/b19468

Deshmukh, R. K.; Akhila, K.; Ramakanth, D.; Gaikwad, K. K. Guar Gum/Carboxymethyl Cellulose Based Antioxidant Film Incorporated with Halloysite Nanotubes and Litchi Shell Waste Extract for Active Packaging. *Int. J. Biol. Macromol.* **2022,** *201,* 1–13.

Farina, V.; Passafiume, R.; Tinebra, I.; Scuderi, D.; Saletta, F.; Gugliuzza, G.; ... Sortino, G. Postharvest Application of Aloe Vera Gel-Based Edible Coating to Improve the Quality and Storage Stability of Fresh-Cut Papaya. *J. Food Qual.* **2020.** DOI: 10.1155/2020/8303140

Gaikwad, K. K.; Lee, J. Y.; Lee, Y. S. Development of Polyvinyl Alcohol and Apple Pomace Bio-Composite Film with Antioxidant Properties for Active Food Packaging Application. *J. Food Sci. Technol.* **2016,** *53* (3), 1608–1619.

Gaikwad, K. K.; Lee, S. M.; Lee, J. S.; Lee, Y. S. Development of Antimicrobial Polyolefin Films Containing Lauroyl Arginate and their Use in the Packaging of Strawberries. *J. Food Meas. Charact.* **2017,** *11* (4), 1706–1716.

Gaikwad, K. K.; Singh, S.; Ajji, A. Moisture Absorbers for Food Packaging Applications. *Environ. Chem. Lett.* **2019,** *17* (2), 609–628.

Gaikwad, K. K.; Singh, S.; Lee, Y. S. Oxygen Scavenging Films in Food Packaging. *Environ. Chem. Lett.* **2018,** *16* (2), 523–538.

Gaikwad, K. K.; Singh, S.; Negi, Y. S. Ethylene Scavengers for Active Packaging of Fresh Food Produce. *Environ. Chem. Lett.* **2019,** *18* (2), 269–284.

Gupta, V.; Ramakanth, D.; Verma, C.; Maji, P. K.; Gaikwad, K. K. Isolation and Characterization of Cellulose Nanocrystals from Amla (Phyllanthus Emblica) Pomace. *Biomass Convers. Biorefin.* **2021,** 1–12.

Gupta, V.; Yadav, R.; Tanwar, R.; Gaikwad, K. K. κ-Carrageenan-Based Bio-Nanocomposite Film Reinforced with Cellulose Nanocrystals Derived from Amla Pomace for Food Packaging. *Biomass Convers. Biorefin.* **2021.**

Jin, T. Z.; Chen, W.; Gurtler, J. B.; Fan, X. Effectiveness of Edible Coatings to Inhibit Browning and Inactivate Foodborne Pathogens on Fresh-Cut Apples. *J. Food Saf.* **2020,** *40* (4), e12802.

Keshari, D.; Tripathi, A. D.; Agarwal, A.; Rai, S.; Srivastava, S. K.; Kumar, P. Effect of α-dl Tocopherol Acetate (Antioxidant) Enriched Edible Coating on the Physicochemical, Functional Properties and Shelf Life of Minimally Processed Carrots (Daucus Carota Subsp. Sativus). *Future Foods* **2022,** *5,* 100116.

Khatri, D.; Panigrahi, J.; Prajapati, A.; Bariya, H. Attributes of Aloe Vera Gel and Chitosan Treatments on the Quality and Biochemical Traits of Post-Harvest Tomatoes. *Sci. Hortic.* **2020,** *259,* 108837.

Kumar, L.; Ramakanth, D.; Akhila, K.; Gaikwad, K. K. Edible Films and Coatings for Food Packaging Applications: A Review. *Environ. Chem. Lett.* **2022,** *20* (1), 875–900.

Kumar, P.; Tanwar, R.; Gupta, V.; Upadhyay, A.; Kumar, A.; Gaikwad, K. K. Pineapple Peel Extract Incorporated poly(vinyl alcohol)-Corn Starch Film for Active Food Packaging: Preparation, Characterization and Antioxidant Activity. *Int. J. Biol. Macromol.* **2021,** *187,* 223–231.

Kumar, R.; Kirtiraj, D. Natural Antimicrobial and Antioxidant Compounds for Active Food Packaging Applications. *Biomass Convers. Biorefin.* **2022,** 0123456789.

Le, K. H.; La, D. D.; Nguyen, P. T. M.; Nguyen, M. D. B.; Vo, A. T. K.; Nguyen, M. T. H.; ... Duc Nguyen, D. Fabrication of Cleistocalyx Operculatus Extracts/Chitosan/Gum Arabic Composite as an Edible Coating for Preservation of Banana. *Prog. Org. Coat.* **2021,** *161,* 106550.

Liu, J.; Liu, S.; Zhang, X.; Kan, J.; Jin, C. Effect of Gallic Acid Grafted Chitosan Film Packaging on the Postharvest Quality of White Button Mushroom (Agaricus Bisporus). *Postharvest Biol. Technol.* **2019,** *147,* 39–47.

Marín, A.; Plotto, A.; Atarés, L.; Chiralt, A. Lactic Acid Bacteria Incorporated into Edible Coatings to Control Fungal Growth and Maintain Postharvest Quality of Grapes. *HortScience* **2019,** *54* (2), 337–343.

Mohamed, S. A. A.; El-Sakhawy, M.; El-Sakhawy, M. A. M. Polysaccharides, Protein and Lipid-Based Natural Edible Films in Food Packaging: A Review. *Carbohydr. Polym.* **2020,** *238,* 116178.

Mousavi, S. R.; Rahmati-Joneidabad, M.; Noshad, M. Effect of Chia Seed Mucilage/Bacterial Cellulose Edible Coating on Bioactive Compounds and Antioxidant Activity of Strawberries During Cold Storage. *Int. J. Biol. Macromol.* **2021,** *190,* 618–623.

Nasrin, T. A. A.; Rahman, M. A.; Arfin, M. S.; Islam, M. N.; Ullah, M. A. Effect of Novel Coconut Oil and Beeswax Edible Coating on Postharvest Quality of Lemon at Ambient Storage. *J. Agric. Food Res.* **2020,** *2* (December 2019), 100019.

Ochoa-Velasco, C. E.; Pérez-Pérez, J. C.; Varillas-Torres, J. M.; Navarro-Cruz, A. R.; Hernández-Carranza, P.; Munguía-Pérez, R.; ... Avila-Sosa, R. Starch Edible Films/Coatings Added with Carvacrol and Thymol: In Vitro and In Vivo Evaluation against Colletotrichum gloeosporioides. *Foods* **2021,** *10* (1), 175.

Olawuyi, I. F.; Park, J. J.; Lee, J. J.; Lee, W. Y. Combined Effect of Chitosan Coating and Modified Atmosphere Packaging on Fresh-Cut Cucumber. *Food Sci. Nutr.* **2019,** *7* (3), 1043–1052.

Oyom, W.; Xu, H.; Liu, Z.; Long, H.; Li, Y.; Zhang, Z.; ... Prusky, D. Effects of Modified Sweet Potato Starch Edible Coating Incorporated with Cumin Essential Oil on Storage Quality of 'Early Crisp.' *LWT* **2022,** *153,* 112475.

Rastegar, S.; Hassanzadeh Khankahdani, H.; Rahimzadeh, M. Effectiveness of Alginate Coating on Antioxidant Enzymes and Biochemical Changes during Storage of Mango Fruit. *J. Food Biochem.* **2019,** *43* (11), e12990.

Rodríguez, G. M.; Sibaja, J. C.; Espitia, P. J. P.; Otoni, C. G. Antioxidant Active Packaging based on Papaya Edible Films Incorporated with Moringa Oleifera and Ascorbic Acid for Food Preservation. *Food Hydrocoll.* **2020,** *103,* 105630.

Salas-Méndez, E. de J.; Vicente, A.; Pinheiro, A. C.; Ballesteros, L. F.; Silva, P.; Rodríguez-García, R.; ... Jasso de Rodríguez, D. Application of Edible Nanolaminate Coatings with Antimicrobial Extract of Flourensia Cernua to Extend the Shelf-Life of Tomato (Solanum lycopersicum L.) Fruit. *Postharvest Biol.and Technol.* **2019,** *150,* 19–27.

Sapper, M.; Palou, L.; Pérez-Gago, M. B.; Chiralt, A. Antifungal Starch–Gellan Edible Coatings with Thyme Essential Oil for the Postharvest Preservation of Apple and Persimmon. *Coatings* **2019,** *9* (5), 333.

Sharma, L.; Saini, C. S.; Sharma, H. K.; Sandhu, K. S. Biocomposite Edible Coatings Based on Cross Linked-Sesame Protein and Mango Puree for the Shelf Life Stability of Fresh-Cut Mango Fruit. *J. Food Process Eng.* **2019,** *42* (1), e12938.

Singh, S.; Gaikwad, K. K.; Lee, Y. S. Antimicrobial and Antioxidant Properties of Polyvinyl Alcohol Bio Composite Films Containing Seaweed Extracted Cellulose Nano-Crystal and Basil Leaves Extract. *Int. J. Biol. Macromol.* **2018,** *107,* 1879–1887.

Sousa, F. F.; Pinsetta Junior, J. S.; Oliveira, K. T. E. F.; Rodrigues, E. C. N.; Andrade, J. P.; Mattiuz, B. H. Conservation of 'Palmer' Mango with an Edible Coating of Hydroxypropyl Methylcellulose and Beeswax. *Food Chem.* **2021,** *346* (December 2020).

Sultana, A.; Kathuria, A.; Gaikwad, K. K. Metal–Organic Frameworks for Active Food Packaging. A Review. *Environ. Chem. Lett.* **2022,** *20* (2), 1479–1495.

Tanwar, R.; Gupta, V.; Kumar, P.; Kumar, A.; Singh, S.; Gaikwad, K. K. Development and Characterization of PVA-Starch Incorporated with Coconut Shell Extract and Sepiolite Clay as an Antioxidant Film for Active Food Packaging Applications. *Int. J. Biol. Macromol.* **2021,** *185*, 451–461.

Upadhyay, A.; Kumar, P.; Kardam, S. K.; Gaikwad, K. K. Ethylene Scavenging Film Based on Corn Starch-Gum Acacia Impregnated with Sepiolite Clay and Its Effect on Quality of Fresh Broccoli Florets. *Food Biosci.* **2022,** *46,* 101556.

Yousuf, B.; Srivastava, A. K.; Ahmad, S. Application of Natural Fruit Extract and Hydrocolloid-Based Coating to Retain Quality of Fresh-Cut Melon. *J. Food Sci. Technol.* **2020,** *57* (10), 3647–3658.

Zhang, X.; Xiao, N.; Chen, M.; Wei, Y.; Liu, C. Functional Packaging Films Originating from Hemicelluloses Laurate by Direct Transesterification in Ionic Liquid. *Carbohydr. Polym.* **2020,** *229,* 115336.

Zhang, Y. L.; Cui, Q. L.; Wang, Y.; Shi, F.; Liu, Y. P.; Liu, J. L.; Nie, G. W. Effect of Carboxymethyl Chitosan-Gelatin-Based Edible Coatings on the Quality and Antioxidant Properties of Sweet Cherry During Postharvest Storage. *Sci. Hortic.* **2021,** *289,* 110462.

Zhang, Y.; Shen, Y.; Chen, C.; Gan, Z.; Chen, J.; Wan, C. (Craig). Loquat Leaf Extract and Alginate Based Green Composite Edible Coating for Preserving the Postharvest Quality of Nanfeng Tangerines. *Sustain. Chem. Pharm.* **2022,** *27,* 100674.

CHAPTER 10

PROTECTIVE PACKAGING FOR DISTRIBUTION AND STORAGE OF FRUITS AND VEGETABLES

AADITYA PANDEY[1], ABHIJIT MAITI[1], and KIRTIRAJ K. GAIKWAD[2]

[1]*Department of Polymer and Process Engineering, Indian Institute of Technology Roorkee, Saharanpur, Uttarakhand, India*

[2]*Department of Paper Technology, Indian Institute of Technology Roorkee, Roorkee, Uttarakhand, India*

ABSTRACT

The global fresh produce packaging business is rapidly expanding, with a total market value of $120 billion predicted by 2025. The main driving forces in this market are customer expectations and low-cost packaging options. Protective packaging for fruits and vegetables is meant to match consumer quality requirements while keeping the product cost-effective. Vibrations and shocks are frequent during the transportation of horticulture products from the farmer's farm to markets throughout the world. These products are protected while packaging in order to withstand mechanical damage during delivery. Natural or manufactured materials might be used to cushion against these damages. Natural cushions include leaves, coconut fiber, rice hulls, etc., while varieties of materials made from paper and polymer are also used for this purpose. Products are packaged in corrugated fiber board (CFB) boxes to provide better handling and shipment due to the weight and environmentally favorable nature of paper-based packaging. Prepackaging treatments are gaining popularity in this industry these days. Pretreating the goods for enhanced shelf life throughout transportation

Novel Packaging Systems for Fruits and Vegetables. Kirtiraj K. Gaikwad & Suman Singh (Eds.)

includes precooling, acid treatments, heat treatments, and other methods. Packaging will be critical in meeting the growing demand in this industry, but there will be hurdles. This market contains perishable goods; therefore, price fluctuations are excessively variable across countries and seasons. In this industry, innovation necessitates sustainable and cost-effective solutions.

10.1 INTRODUCTION

Fruits and vegetables are agricultural produce and the source of the best available nutrition. The majority of the population is now focusing more and more on fresh produce. But it's not possible to produce all types of fruits and vegetables at every place; that's why packaging comes into the picture. Freshly produced fruits and vegetables are transported from one place to another within the same country, or sometimes to other countries too. Apples from the northern hill provinces of India are being enjoyed by the rest of India and actually by the world, and without protective packaging, it's not possible.

There are certain challenges associated with transportation. Protective packaging is designed to face these challenges. These challenges include mechanical shocks, vibrations, handling, and storage. Packaging cost is the main limiting factor in fruit and vegetable cases while ensuring quality at the same time. Higher-priced products can be packed easily to ensure smooth transportation with protective packaging, but due to the low cost associated with the majority of fruits and vegetables, protective packaging design has always been a tough challenge for packaging designers. Apart from cost-effectiveness, protective packaging for fruits and vegetables also has to ensure smooth handling and storage capabilities. Workers engaged in the fruit and vegetable market have too much work; that's why proper handling is not possible, which in turn becomes a challenge for packaging designers. In terms of materials used, protective packaging has to be made of materials that are low-weight and low-cost while ensuring enough strength for a particular fruit or vegetable. Due to this factor, paper-based packages are capturing this segment of fruit and vegetable packaging. Low weight, low cost, and forming into different shapes are the major reasons behind it. In the market, different types of paper-based packages can be seen, ranging from corrugated boards, shredded paper, honeycomb partitions, and pulp trays.

This chapter deals with the different types of packages that are being used as protective packaging for the transportation of fruits and vegetables. When we talk about transportation, we have to consider the standards of different countries. A special focus has been emphasized on this particular topic in this chapter. To give better insight to package designers of protective packaging for transportation (Qiu et al., 2020), numerical approaches have also been discussed. In the end, current technologies are being discussed, focusing on the challenges associated with them, and then new research has been discussed to get a sense of what could be the future in this field.

During transportation, fruits and vegetables undergo different types of stresses such as mechanical damage, vibration, compression, handling, stacking, etc. (Elik et al., 2019). The main aim of any packaging designer is to save fresh fruits and vegetables from these damages. These protective approaches can be categorized in two ways: one is cushioning materials, and the second is secondary or tertiary packaging. These are discussed in detail in an upcoming section.

10.2 CUSHIONING MATERIAL

Cushioning can be defined as the material that protects the product from the dynamic hazards of the transportation system. For fruits and vegetables, different types of cushioning materials are used to protect them from transportation damages, which are impossible to avoid in a distribution environment. These dynamic damages include vibration, compression, stacking, handling, etc. Dynamic damages are dependent on the absorbing capabilities of cushioning materials (Lin et al., 2020; Caleb et al., 2018). Also, the maturity stage of fruits or vegetables has to be investigated before using any cushion material; more ripened products require more cushioning compared to less ripened products (Lin et al., 2020).

Figure 10.1 shows the different types of properties required by cushioning materials and their classifications as per use and origin.

Different sorts of curves can be employed in the cushioning development process to gain a better understanding of the material's capabilities for specific applications in fruit and vegetable packing (Ge, 2019). Impact energy, bruising volume, stress energy, and maximum acceleration are used to create these cushion curves. A packaging designer can construct cushions for protective packing of fruits and vegetables by understanding these curves:

A. Dynamic cushion curve (ASTM D 1596)
B. Cushion curve predicted by stress–strain curve
C. Cushion factor curve (c-e curve).

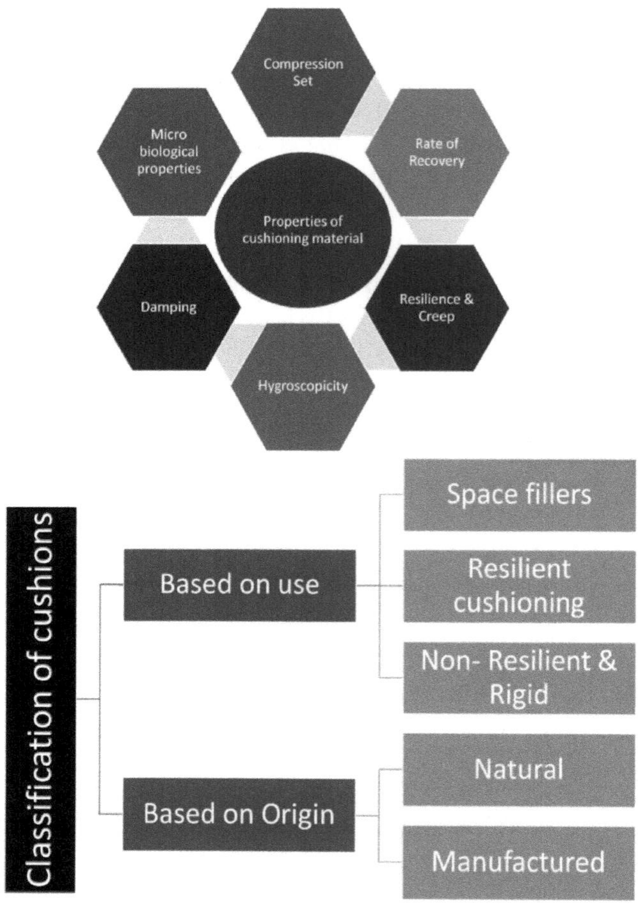

FIGURE 10.1 (a) Properties required for cushioning material and (b) classification of cushioning materials.

Understanding these curves is necessary to select the perfect material as per our requirements. These materials may be naturally derived from agricultural waste or may be fabricated from plants. Upcoming sections are focused on this categorization of cushioning materials.

10.2.1 NATURAL CUSHIONS

They include materials obtained from used packaging or agricultural waste, such as paper shreds, coconut husk, plant leaves, rice hulls, paddy straw, and similar agricultural waste commodities. These materials are conventional cushions used by farmers to transport their fruits and vegetables from their farms to the markets. These types of materials have lower costs or sometimes no cost, as they can be easily found in farms. Apart from cost benefits, these materials also have damping characteristics and shock absorption capacity. Rice hull shows outstanding impact absorption capacity when compared with bubble wrap and crumb rubber (Alnashwan et al., 2014). Also, rice hulls and coconut fibers have very good thermal insulation characteristics. There are some natural cushioning materials that also have antifungal and antimicrobial properties. The advantages associated with natural cushioning materials, along with their environmentally friendly characteristics, are the major reasons behind the extensive use of these materials in the fruit and vegetable supply chain during transportation, even in today's packaging market. There are also some disadvantages to these natural cushioning materials, including losing their cushioning properties in moist conditions. Also, poor resilience and higher compression set values are some of them. Due to these disadvantages in the fruit and vegetable packaging segments, new manufactured cushioning materials are used; these materials are discussed in the coming section.

10.2.2 MANUFACTURED MATERIALS

Every fruit and vegetable has different packaging requirements, and these requirements push packaging engineers to design dedicated cushioning materials for a particular fruit or vegetable. These manufactured cushioning materials have more elegant looks, due to which high-end horticulture products are being packaged with these cushioning materials only. Pulp tray, foam net, crate liner, bubble films, synthetic foams, thermoformed tray, etc. are widely used manufactured cushioning materials in the fruit and vegetable packaging market. These materials offer better shock absorption capacity, better moisture resistance, higher resilience, and resistance against infection growth among fruits and vegetables (Caleb et al., 2018). From the packaging market's point of view, these materials seem to be a perfect fit for industrial growth, but higher cost and availability at farms are major limiting factors for these materials. Also, when we compared these materials with natural cushioning materials, they are far behind on environmental grounds.

Apart from this classification, it's mandatory to discuss the basic test procedure to test the effectiveness of cushioning materials. The general practice to test this effectiveness is to use an impact test. Fruit is placed with cushioning material on the sample holder, and a pendulum of known mass is released from the particular angle to hit the sample. This impact energy is recorded as per the mathematical relation as follows:

$$E_I = mgl(1 - \cos\theta) \tag{10.1}$$

where *Ei* is impact energy (j), *m* is the mass of the pendulum (kg), *g* is gravitational acceleration (m/s^2), *l* is the length of the rope (m), and θ is the angle of release.

After impact, fruits and vegetables are stored at room temperature to observe the browning.

The bruise volume can be calculated as follows:

$$V_B = \left(\frac{\pi}{8}\right) * w^2 * d \tag{10.2}$$

where V_B is the bruise volume, *w* is the width of the bruise (mm), and *d* is the depth of the bruise (mm).

Once the impacts with particular impact energy and bruise volumes are recorded with certain repetitions, the probability of bruising can be found as follows:

$$P(Bruising) = \frac{Number\ of\ non-zero\ bruises}{Total\ number\ of\ repetitions} \tag{10.3}$$

The deflection of cushioning material is one of the most important factors to consider when designing and selecting a material for cushioning. When a package is impacted, the cushioning material deflects, and this deflection is the working length of the cushion. Working length is a function of impact height and acceleration. When constructing a new protective package for fruits and vegetables, the equations listed in Table 10.1 must be employed. The main factors in these equations are deflection, impact energy, and static and dynamic stress, and aside from physical testing with instruments, researchers must be able to estimate these factors in advance for optimal selection throughout the design phase.

TABLE 10.1 Different Equations and Mathematical Relations Related to Cushioning Designing and Material Selection (Ge, 2019; Goodwin and Young, 2011).

Equation	Basic science	Application for packaging designers
$D = \frac{2h}{G}$	Deflection	Estimation of working length for cushions
D = % deflection ∗ t	Transmission of impact from one end to other	Estimation of thickness of cushioning material
$S_t = \frac{weight}{load\ bearing\ area}$	Stress due to concentrated load	Drawing cushion curves for different heights and estimating peak accelerations
$E = \frac{S_t * h}{t}$	Energy storing capability of material, that is, area under stress–strain curve	Estimation of material's characteristics required for particular application and drawing cushion curve
$C = \frac{G_m}{(\frac{h}{t})}$	Ratio of force to energy	Drawing of a c-e curve to select cushioning designs from different materials

Where D is the deflection of cushion material in mm, h is the height of drop in mm, G is the impact acceleration (mm/s^2), S_t is the static stress in kg/m^2, G_m is the maximum acceleration during impact, t is the thickness of cushion, and E is the impact energy.

Cushions are the primary contact of fruits and vegetables during transportation; these cushioned products are packed in secondary packages. In the next section, secondary and tertiary packaging will be discussed in detail.

10.3 CORRUGATED FIBER BOARD (CFB)

Once the fruits and vegetables are packed with cushioning materials, to ensure their handling and transportation, they are packed in a secondary package; in most cases, it's a CFB box. Earlier, wooden boxes were used for this purpose, but due to their weight, they have been replaced with CFB boxes, which are lighter in weight and easy to handle. For transportation of fruits and vegetables, nowadays CFB boxes are the most commonly used packaging material for harvest items, even for international trade. Because CFB boxes are made of paper, their use in packing is both cost-effective and environmentally friendly.

Commonly used CFB boxes are 3-ply boxes, but in some cases, 2-ply boards have also been used as cushioning material for fruits and vegetables. A 2-ply box is nothing but a paper layer with a glued flute layer on it. Due to the operation process through which they are made, 2-ply boards are generally called single-face boards as well. In a 3-ply box, this single-face board is glued to an outer liner of paper, and these three layers of paper together are called a single-wall board or 3-ply board. The number of walls can be increased as per the requirement by gluing more than one single-face board together; that is, a double-wall board can be made with two single-face boards and one outer liner. Similarly, a triple-wall box can be made with three single-face boards and one outer liner. In a CFB box, the fluting layer provides strength to the structure, and different types of flutes can be used as per our requirements.

A box with the same paper combinations but different types of flutes has variations in their strength. This implies the importance of selecting of flute type for particular applications. In most cases for fruits and vegetables, a transportation B type of flute is used, but packaging designers are nowadays playing with smaller flute options so that the weight of the box can be minimized. Equation 10.4 should be used to calculate the grammage of corrugated board.

$$GSM_{CFB} = \sum_{i=0}^{n} (GSM_{liner})_i + \sum_{i=0}^{n} (CF)_i * (GSM_{Flute})_i \tag{10.4}$$

where GSM_{CFB} is the grammage of CFB box in g/m^2, GSM_{liner}, and GSM_{flute} are the grammages of paper used as liner and fluting medium, respectively in g/m^2, CF is the corrugating factor (dimensionless), which is different for each flute type and signifies how much paper is used to create 1 m of flute.

For a 3-ply box, eq 10.4 can be reduced to as follows:

$$GSM_{CFB} = GSM_{outer\ liner} + GSM_{inner\ liner} + CF * GSM_{flute} \tag{10.5}$$

With eq 10.5, one can easily find the total weight of the box by multiplying it with the board area.

The weight of the CFB box is an important factor in the entire supply chain of fruits and vegetables, as handling, storage, and transportation costs are all dependent on this weight. Thus, for packaging designers, this should be the primary focus while designing packages for fruits and vegetables.

There are several other factors that should be considered during the design process. Designers must be aware of the stack height of CFB boxes on pallets as well as during the entire transportation process. Understanding stack height helps in choosing the box with the correct box strength to withstand this much load. The entire designing process can be visualized with the following example:

For example, if a CFB box filled with Himachal apples is to be designed, and for international trades, pallet height standards allow 10 boxes in a stack, and each box carries 30–35 apples (200 g ± 20 g), then we first need to calculate the total load carrying capacity of the box, which is the bursting strength of the board, and then the stacking strength of the box should be calculated in order to select the paper combination for the CFB box. Each box in this example should be able to carry a load of 10 kg and have a stacking strength of 10*9 kg, or 90 kg. This computation is based on the assumption that any box may arrive at the bottom of the stack and that it will require nine boxes to reach the stack height of 10.

Now to calculate the stacking strength, box compressive strength (BCT) should be tested, or during the design process, it should be calculated with probable paper combinations. The following equations must be used to calculate the BCT of the 3-ply box (de la Fuente et al., 2018).

$$BS = (BF_1 * GSM_{outer\ liner} + BF_2 * GSM_{inner\ liner} + \frac{x}{100} * BF_{flute} * GSM_{flute}) * \frac{1}{1000} \tag{10.6}$$

where *BS* is the bursting strength of CFB in kg/m^2, *BF* is the burst factor (dimensionless) of the corresponding suffix, *GSM* is the grammage of the corresponding suffix in g/m^2, and x is the contributing factor of the used flute (in %), and it varies for each flute.

$$BCT = 5.87 * ECT * \sqrt{h * z} \tag{10.7}$$

where *BCT* is the box compression strength in kg, *ECT* is edge crush strength of the board in kg/m, h is the caliper of board in m, and z is the perimeter of box in m. *ECT* value depends on the type of paper used and its RCT (ring crush strength) value.

The abovementioned relationships are sufficient to design a package for fruit and vegetable transportation. Fruits and vegetables are transported from farms to different markets, even in supermarkets, in different countries and provinces. Thus, it's a good approach to have an understanding of the entire supply chain before using these equations because there are various handling, storage, and distribution mechanisms associated with a package.

Factors like the number of shiftings, types of handling mechanisms (forklifts, etc.), stacking pattern during storage at each point, and mode of transportation are also equally important while designing or choosing a box for fruit and vegetable transportation. Keeping in mind these factors, different standards and test methods are discussed in detail so that readers will have a good understanding of them.

10.4 TESTS AND STANDARDS FOR PROTECTIVE PACKAGING

In this section, different types of regulations are discussed that are applicable to protective packaging of fruits and vegetables during transportation. Internationally, there are various organizations engaged in regulating food contact materials (FCM). These organizations set up frameworks for countries, and every imported fruit and vegetable in these countries must comply with these regulations. In Europe, this work is done by the European Commission. Different reference laboratories have been set up by the EU in order to ensure that every fruit and vegetable entering the EU countries must possess the standards set up by EURL-FCM. Trade packages and cartons of fresh fruits or vegetables must mention the name and address of the packer, including country of origin, along with the lot number to trace, and postharvest technology applied during packaging must also be printed on the container/carton. In the USA, the Code of Federal Regulations (CFR) is followed for the transportation of fruits and vegetables. CFR corresponds to the packaging of fruits and vegetables. In India, the Food Safety and Standard Authority of India regulates the regulations for fruit and vegetable packaging for transportation purposes. All of the abovementioned authorities aim at maintaining the quality of harvested products and standardizing the processes so that consumers are aware of what they are eating. Apart from these authorities, there are some universal standards that need to be followed in the transportation of fruits and vegetables. These standards include different test methods standardized by ASTM for shipping containers, CFB boxes, and cushioning materials. Commonly used test methods are listed in Table 10.2 along with their applications, that is, where we should use them. Countries and importers strictly follow the certification for these tests so that imported quality to their country must be maintained no matter from where fruits and vegetables are being imported.

The majority of fruits and vegetables are shipped by air. Pallet loads are used to store or carry the items. Pallet loads should be formed as quickly as possible in the logistics chain and kept unbroken for as long as possible to maximize distribution cost savings. For international transportation purposes, ISO has standardized the pallet sizes as follows:

1. 1200*800 mm
2. 1200*1000 mm

These pallet sizes are helpful in making transport easier and reducing transport costs. For sea transports, 20 and 40 ft containers are used, and packaging designers for fruits and vegetables must keep in mind the standard sizes for these containers so that maximum space can be utilized during shipping.

TABLE 10.2 Different Tests and Standards for Cushions, Pallets, and Containers (Trevisani et al., 2014; AGMPM, 2010).

Test/dataset	Standard	Application
Impact test	ASTM D880	Shipping containers
Dynamic shock cushioning characteristics	ASTM D1596	Packaging materials
Drop test	ASTM D5276	Loaded containers by free fall
Concentrated impact test	ASTM D 6344	Transport packages
Mechanical shock test	ASTM D 3332	Shock fragility of products
Compression test	ASTM D 7030	Short-term creep performance of CFB containers
Shock test	ASTM D 6537	Instrumented package shock testing for package performance
Vibration test	ASTM D3580	Vertical linear motion vibration test
Vibration test	ASTM D 4728	Random vibration testing of shipping containers
Pressure test	ASTM D 7660	Guide for conducting internal pressure test on UN packagings
Vocabulary	ISO 445:2013	Pallets for material handling
Quality parameters	ISO 18333	Acceptable quality of wooden pallets
Principal dimensions	ISO 6780	Flat pallets for intercontinental materials handling
Repairment protocols	ISO 18613	Maximum number of defects and damages allowed before a pallet repairment
General test methods	UNI EN 13626	Pallet containers

10.5 CURRENT TRENDS

Horticulture products have variations in their shelf lives, and their ripening mechanisms are also different. As technology advances, fruits and vegetables are also experiencing this change. Protective packaging for transportations is nowadays combined with postharvest techniques and protective packaging materials. These postharvest techniques include precooling, postharvest treatments (such as chlorination, wax coatings, acid treatments, and ethylene inhibitors), and thermal treatments (Mahajan and Kapoor, 2021).

Precooling helps remove the excess heat from the fruits and vegetables in order to reduce the ripening rate. It helps in minimizing respiration rate, ethylene production, microbial activity, and metabolic processes, and thus quality is preserved and shelf life is prolonged during transportation. Depending upon the horticulture produce demand, different types of precooling can be applied, such as room cooling, hydrocooling, and vacuum cooling. Postharvest treatments are also gaining attention nowadays as they provide sufficient treatment to fruits and vegetables to face the atmospheric conditions. Room cooling is applied through the natural circulation of cold air in packages to reduce the unfavorable excess heat from the shipments. It is quite an inexpensive practice, and cold-room packaging is very useful in maintaining the quality of fruits and vegetables during transportation. Package icing and hydrocooling are other methods of precooling; these methods are relatively more expensive than room cooling methods.

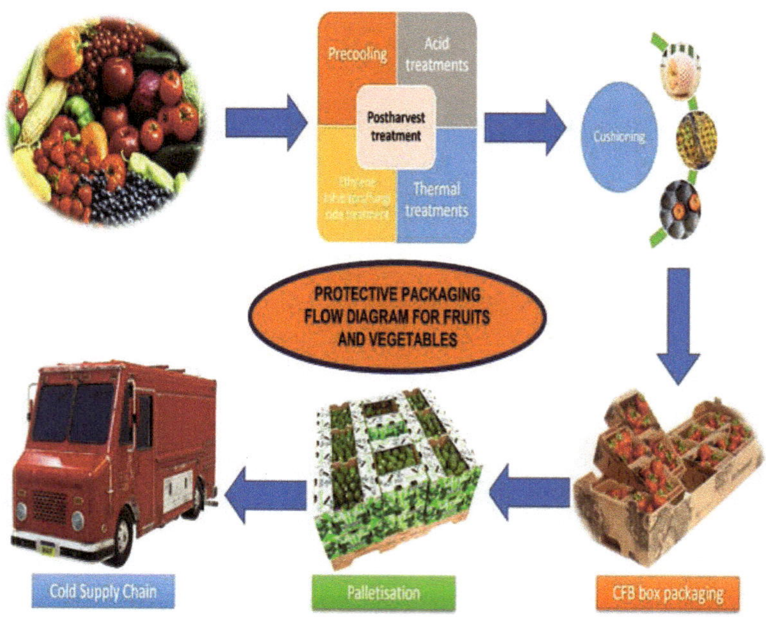

FIGURE 10.2 A flow diagram for harvest product packaging.

In hydrocooling, fruits and vegetables are immersed in cold water before packaging, and in package icing, water-tolerant packages are used to pack the fruits and vegetables along with flaked or crushed ice. In all of these methods of precooling, the primary aim is to remove the excess heat, which is favorable for microbial growth and respiration mechanisms. Once the heat is removed from the fruits and vegetables, microbial growth is delayed during transportation.

Chlorination, wax treatment, acid treatments, and ethylene inhibitors are the commonly applied postharvest treatments. Chlorination utilizes the disinfecting characteristics of chlorine to disinfect the surface microbes from the fruits and vegetables, reducing the chances of microbial spoilage during transportation. Several chlorine salts are being used in the washing water to wash the fresh fruits and vegetables, but optimum concentrations need to be decided as pH plays an important role in this method. Other than salts, chlorine gas is also being utilized in water streams to decontaminate the surface and help inactivate the pathogenic microflora (Mishra et al., 2018). One has to keep in mind that if the pH is less than 5, it leads to the formation of toxic gases, and if the pH is more than 8, chlorine loses its antimicrobial properties. Thus, one has to ensure and maintain the pH in the range of 6–7 for better results with the chlorination method. Also, after chlorination, fruits and vegetables need to be washed with clean water. In another method of postharvest treatments, different types of waxes are used as edible coatings on harvest products' surfaces, which help regulate the metabolic process and control moisture and oxidations in fruits. An edible coating on the surface works as a partial barrier against gas and physical impacts. Shellac and beeswax are commonly used coatings for fruits and vegetables. Edible coatings are nowadays facing a sustainability issue, and to address this issue, natural-based edible coatings are gaining attention. Different types of essential oils are used for coating purposes, and these oils include essential oils from *Thymus vulgaris*, *Ocimum canum*, *Zingiber officinale*, etc. To retard the browning process of fruits and vegetables, acid treatments are also done, but using an acid with harvested products must comply with the regulations of the authorities. Ethylene, the main cause in the ripening process, can also be controlled by ethylene inhibitors, and that's why most fruits and vegetables are now being packaged after treatment with ethylene inhibitors (Navarro-Martínez et al., 2021), benzothiadiazole, 1-methyl cyclopropane (1-MCP), cycloheximide, silver thiosulfate, silver nitrate, benzo(1,2,3)thiadiazole-7-carbothioic acid S-methyl ester (BTH), etc.

Hot water treatment and vacuum heat treatment are the most commonly used thermal treatments nowadays as postharvest tools to save the fruits and vegetables during transportation. Hot water treatment helps in the uniform redistribution of coated wax to improve the physical barrier properties on fruits and vegetables' surfaces, and at the same time, helps in notable decreases in dermal cracks (Malik et al., 2021). On the other hand, vacuum thermal treatment helps reduce the chances of infection in horticulture products due to fruit flies and other insects through the application of steam. In this method, corrugated boxes are stacked in humid and heated rooms, and saturated water vapors are then applied to the surroundings of these boxes to kill larvae and insect eggs before shipment.

Other than these post-harvest treatments, nowadays new types of packages are also being used for transportation purposes, but their use is limited to high-value products (Caleb et al., 2018). Clamshell packages for kiwis and dates are the most popular nowadays. Apart from new packages, the cold supply chain is the major advancement in the fruit and vegetable transportation system. Temperature and humidity conditions are the most important factors during the supply chain of harvest products (Zheng et al., 2022). This supply chain is very expensive, and that's why it's limited to expensive horticulture products only. Developed countries are nowadays focusing more and more on these supply chains, as food wastage is the major challenge currently. As per the data published by the World Forum Organization, the USA accounts for food wastage equivalent to \$218 billion per year, followed by \$143 billion from the European countries. Raut et al. (2019) reviewed the different types of models identified to reduce the food wastage associated with the supply chain of fruits and vegetables. In their work, the authors have found that the majority of food loss in the supply chain is due to improper handling and a lack of cold logistics infrastructure. They predicted that with proper cold logistics, 20–30% of food wastage can be reduced in fresh harvest products. It can be concluded that temperature and humidity conditions are the main driving forces in fruit and vegetable protective packaging.

In the next segment, different challenges faced by the current transportation system for fruits and vegetables are discussed.

10.6 CHALLENGES IN HARVEST PRODUCTS' PACKAGING FOR TRANSPORTATION

Currently, the fruit and vegetable packaging market has a variety of products, and each product's requirements for packaging are different. These requirements include the product's natural deterioration factors such as temperature, humidity (Zheng et al., 2022), ethylene concentration (Morya and Sharma, 2019), oxygen, and CO_2 concentrations, as well as external factors such as vibration and mechanical damages during transportation and handling. For example, coconuts are strong enough to absorb mechanical shocks, but mangoes are not. Thus, the packaging of mangoes should be able to withstand these mechanical damages. Currently, protective packaging for fruits and vegetables sees variation in the supply chains of various countries. When an alphonso mango from India reaches different parts of the world, it has to be maintained in quality, but roads are not similar everywhere, and the cold supply chain is not developed in most of the developing countries. This is the most difficult aspect of creating protective packaging for a packaging engineer because the package must be able to withstand all of the above situations. Other than the supply chain and the product's characteristics, the cost of packaging is the major challenge for packaging engineers for fruits and vegetables' protective packaging. The cost of the products varies drastically from one product to another and also from one season to the next. Consider a situation when mangoes are coming to your market in elegant display packaging. Will you buy them at a higher price? Thus, designing an optimum cost-protective sustainable package for a particular product itself is a challenging task (Foodware365, 2021). We can see in our nearest local markets that few fruits have elegant packaging, while others have very rough packaging. This is due to the value associated with those products. Thus, providing a cost-effective, sustainable protective package for fruits and vegetables is a challenge in this rapidly growing market.

10.7 CONCLUSIONS

Demand for fresh fruits and vegetables is growing rapidly worldwide, and to meet this demand, these products must be supplied with protective packaging so that the quality is maintained at the consumer end. Currently, fruits and vegetables are packed after post-harvest treatments with cushioned material and tertiary or secondary packaging in the form of a CFB box. Temperature-based supply chains are nowadays getting popular in this segment, which ensures good quality at the consumer end, but there are limitations with these supply chains, which include the price rise of products and a lack of infrastructure in most of the countries. This chapter gives a detailed discussion about different aspects of protective packaging designs for cushions and secondary packaging options. Different prepackaging treatments are also described in detail. These pretreatments are different for different fruits and vegetables. Also, protective packages are mostly designed to fulfil the requirements of international trade, that's why significant discussion has been done on important test methods and standards for fruit and vegetable packaging. These standards actually allow exporters and importers to meet consumer expectations and legal liabilities. In the end, the challenges of this market are briefly explained to spark the readers about new innovation in this field.

KEYWORDS

- **protective packaging**
- **cushions**
- **CFB**
- **pretreatments**
- **sustainability**

REFERENCES

AGMPM. List of Fatigue Standards and Fracture Standards Developed by ASTM, 2010; pp 9–11.

Alnashwan, W.; Aloumi, B.; Malasri, S.; Kist, M.; Othmani, A.; Fotso, R.; … Sanchez-Luna, Y. Shock Absorption of Crumb Rubber and Coconut Fiber. *Int. J. Adv. Packag. Technol.* **2014,** *2*, 119–128. https://doi.org/10.23953/cloud.ijapt.14

Caleb, O. J.; Martin Geyer, A.; Mahajan, P. V. Packaging of Fresh-Cut Fruits and Vegetables. In *Innovative Packaging of Fruits and Vegetables: Strategies for Safety and Quality Maintenance*; Siddiqui, M. W., Rahman, M. S., Wani, A. A., Eds.; Apple Academic Press, 2018; pp 199–208.

de la Fuente, J.; Kisch, R.; Singh, J.; Roy, S.; Peleman, D. Investigation of Compression Strength of Bliss Style Corrugated Fiberboard Boxes. *J. Appl. Packag. Res.* **2018,** *10* (4), 45–64.

Elik, A.; Yanik, D. K.; Istanbullu, Y.; Guzelsoy, N. A.; Yavuz, A.; Gogus, F. Strategies to Reduce Post-Harvest Losses for Fruits and Vegetables. *Int. J. Sci. Technol. Res.* **2019,** *5* (3), 29–39. https://doi.org/10.7176/jstr/5-3-04

Foodware365. Tackling the Ongoing Challenge of Fresh Produce Packaging, 2021.

Ge, C. Theory and Practice of Cushion Curve: A Supplementary Discussion. *Packag. Technol. Sci.* **2019,** *32* (4), 185–197. https://doi.org/10.1002/pts.2427

Goodwin, D.; Young, D. Cushioning. In *Protective Packaging for Distribution: Design and Development*; DEStech, 2011; pp 81–102.

Lin, M.; Chen, J.; Chen, F.; Zhu, C.; Wu, D.; Wang, J.; Chen, K. Effects of Cushioning Materials and Temperature on Quality Damage of Ripe Peaches According to the Vibration Test. *Food Packag. Shelf Life* **2020,** *25*. https://doi.org/10.1016/j.fpsl.2020.100518

Mahajan, B. V. C.; Kapoor, S. Postharvest Handling and Diseases of Horticultural Produces. In *Postharvest Handling and Diseases of Horticultural Produce*; Singh, D., Sharma, R. R., Devappa, V., Kamil, D., Eds.; CRC Press, 2021; pp 33–42.

Malik, A. U.; Hasan, M. U.; Hassan, W. U.; Khan, A. S.; Shah, M. S.; Rajwana, I. A.; ... Anwar, R. Postharvest Quarantine Vapour Heat Treatment Attenuates Disease Incidence, Maintains Eating Quality and Improves Bioactive Compounds of 'Gola' and 'Surahi' Guava Fruits. *J. Food Meas. Charact.* **2021,** *15* (2), 1666–1679. https://doi.org/10.1007/s11694-020-00763-z

Mishra, V.; Abrol, G. S.; Dubey, N. Chapter 14 - Sodium and Calcium Hypochlorite as Postharvest Disinfectants for Fruits and Vegetables. In *Postharvest Disinfection of Fruits and Vegetables*; Siddiqui, M. W., Ed.; Academic Press, 2018; pp 253–272. https://doi.org/https://doi.org/10.1016/B978-0-12-812698-1.00014-5

Morya, S.; Sharma, A. Packaging of Fruits and Vegetables. In*Advances in Horticultural Crop Management and Value Addition*; Singh, D. S. K., Kaur, D. S., Eds.; 2019; pp 363–374.

Navarro-Martínez, A.; López-Gómez, A.; Martínez-Hernández, G. B. Potential of Essential Oils from Active Packaging to Highly Reduce Ethylene Biosynthesis in Broccoli and Apples. *ACS Food Sci. Technol.* **2021,** *1* (6), 1050–1058. https://doi.org/10.1021/acsfoodscitech.1c00071

Qiu, F.; Hu, Q.; Xu, B. Fresh Agricultural Products Supply Chain Coordination and Volume Loss Reduction Based on Strategic Consumer. *Int. J. Environ. Res. Public Health* **2020,** *17* (21). https://doi.org/10.3390/ijerph17217915

Raut, R. D.; Gardas, B. B.; Narwane, V. S.; Narkhede, B. E. Improvement in the Food Losses in Fruits and Vegetable Supply Chain - A Perspective of Cold Third-Party Logistics Approach. *Oper. Res. Perspect.* **2019,** *6* (C).

Trevisani, A.; Iaccheri, E.; Fabbri, A.; Guarnieri, A. Pallet Standards in Agri-Food Sector: A Brief Survey. *J. Agric. Eng.* **2014,** *45* (2), 90–95. https://doi.org/10.4081/jae.2014.220

Zheng, D.; Chen, J.; Lin, M.; Wang, D.; Lin, Q.; Cao, J.; ... Chen, K. Packaging Design to Protect Hongmeiren Orange Fruit from Mechanical Damage during Simulated and Road Transportation. *Horticulturae* **2022,** *8* (3), 258. https://doi.org/10.3390/horticulturae8030258

CHAPTER 11

PACKAGE DESIGNING ASPECTS FOR MARKETING OF FRUITS AND VEGETABLES

PRAGYA SRIVASTAVA[1], DAKURI RAMAKANTH[2], LOKMAN HAKIM[1], and KIRTIRAJ K. GAIKWAD[1]

[1]*Department of Paper Technology, Indian Institute of Technology Roorkee, Roorkee, India*

[2]*Department of Polymer and Process Engineering, Indian Institute of Technology Roorkee, Roorkee, India*

ABSTRACT

Fruits and vegetables have been a powerhouse of nutrients to the global community since ancient times, and the demand has increased rapidly due to urbanization and globalization. Among all, fruits and vegetables are the highest waste contributors across the globe due to inadequate packaging. Both necessity and demand are driving the scientific community to shift their focus onto the advanced packaging design and technologies to safeguard and extend the shelf life of fruits and vegetables. Package design plays a crucial role in selecting a product and protecting it. This chapter introduces basic terminology related to fruits and vegetables packaging, followed by an overview of fruits and vegetables and their characteristics. Later, the packaging design aspects and the supply chain of fruits and vegetable packaging are discussed. Additionally, insight into the traditional packaging design and trends in fruits and vegetable packaging has been provided.

Novel Packaging Systems for Fruits and Vegetables. Kirtiraj K. Gaikwad & Suman Singh (Eds.)

11.1 INTRODUCTION

Fruits and vegetables supply a wealth of nutrients to the human body, enhance immune systems, and reduce the risk of various diseases. The overall fruit and vegetable supply chain in India is riddled with issues and challenges. Many people around the world consume less than the amount produced due to a variety of issues such as availability, price, lack of knowledge and awareness, losses and wastage of fresh produce in the supply chain, improper transportation infrastructure, and lousy market analysis; thus providing severe hurdles to India's agricultural development. Fruit and vegetable packaging problems are caused mainly by their post-harvest metabolism, which is respiration, ripening, and senescence, and it can only be slowed, not stopped. When enzymes like pectinase and amylase break down starches and pectin, the fruit softens and becomes sweeter. Ethylene, a naturally occurring gas that begins and supports the ripening process, is another important element in fruit ripening. Supplying the world's rapidly growing population while adhering to food safety rules is challenging for the food industry. Longer shelf-life and higher quality stability for foods are becoming increasingly important. Packaging technologies are developing fresh produce packaging in response to market demands. Food spoiling or overripening may occur due to shipment delays or temperature abuse during transportation, resulting in financial loss. Many factors influence the rate of fruit and vegetable respiration, including variety, temperature, environmental component, and humidity. The natural respiration of fruits and vegetables causes an increase in carbon dioxide and a decrease in oxygen in the atmosphere inside the packing, making it difficult to maintain product quality. Thus, the packaging is an integral part of the food production process that can enhance food quality and provide safety, shelf life, convenience, and economic feasibility. Packaging provides a barrier between foods and the outside world that protects them from contamination and biological risks.

Fresh cut fruits and vegetables, on the other hand, have increased respiration rates due to mechanical cutting stress. As a result, fruit and vegetable packaging design should focus on creating an appropriate gas atmosphere in the package's headspace, such as optimal oxygen and carbon dioxide gas. Reduced oxygen levels in packing have been shown to successfully inhibit the ripening of products. However, too little oxygen can cause anaerobic respiration, which causes off-flavors and product degradation. Increased carbon dioxide concentrations can also slow down respiration, though carbon dioxide tolerance levels vary by food type. Tomatoes and grapes, for example, have a carbon dioxide tolerance of around 2%, whereas melons and berries have a tolerance of up to 15% (Siddiqui et al., 2018). To maintain quality, gas permeable or perforated packaging materials are required for fresh produce such as fruits and vegetables. To extend the shelf life of fresh produce, the in-packing environment must collaborate with packaging materials of varying permeability to achieve the best gas atmosphere for packaging. Another key element that might alter the shelf-life of produce is microbial activity. If fresh produce is not stored or handled properly, it may quickly decay. Preventing additional contamination of fresh produce at the packaging level is as important as decontamination before packaging.

11.1.1 BASIC TERMINOLOGY

Shelf-life—A food product's shelf life can be defined as the time span during which it is still safe for human consumption. ***Fresh produce***—Produce is a broad term that refers to a variety of farm-produced crops and goods, such as fruits and vegetables. The phrase "produce" generally denotes that the products in concern are fresh and of the same quality as when they were picked. India produces a diverse range of horticulture crops, including fruits and vegetables, and is the world's second-largest fruit and vegetable grower, accounting for more than 90% of total horticulture production, earning it the nickname "the fruit and vegetable basket." ***Climacteric fruits/non-climacteric***—Fruits are classified as climacteric or non-climacteric based on their respiratory characteristics. Climacteric fruits are those that enter the climacteric phase after harvest, meaning they continue to mature after being picked. Such fruits ripen after picking and release much more ethylene than non-climacteric fruits. Examples are pears, peaches, apples, mango, banana, etc. Ethylene production is minimal in non-climacteric fruits. They can't ripen further once harvested. Examples are lemon, pineapple, melon, grapes, berries, etc.

11.2 FOOD WASTE

Food waste refers to food that is abandoned or lost uneaten. Not only does food go to waste, but so do the resources necessary to produce it, from labor to land and water. Wastage of food is one of humanity's most complex problems. Around one-third of the world's food is thrown away at various points throughout the supply chain (Waste, 2019). According to a research, food is either thrown out along the supply chain or at the end of the preparation and consuming process (Al-Obadi et al., 2022). In developed countries, food loss is mainly owing to rigorous adherence to the expiry dates of food goods as specified on the package. It's also because food retailers must ensure that even perishable food items fit esthetic requirements in terms of color, shape, and size. However, over shopping leads to food waste at the consumer level (Lipinski et al., 2016). Waste occurs as a result of labeling misunderstanding between the manufacturing and expiry dates (Hussain and Tanu, 2021). Low-income societies have lower food-quality expectations, and as a result, food waste generation per person is low. However, due to the impact of rising populations and economic constraints, it is anticipated that overall food waste in emerging nations such as India is comparable to that in affluent ones.

Food waste has been an issue for decades, and it is only getting worse. People only started paying attention when the COVID-19 pandemic struck in 2020. Things that affluent Indians took for granted, like getting food or worrying about how long their fruits and vegetable supplies would last, became unexpectedly inconvenient. While it is commonly known that the globe wastes one-third of the food it produces, data show that fruits and vegetables are the most wasted food items. Every year, at least half of the world's fruits and vegetables are lost or wasted owing to factors such as drought, pests, storage, transportation, and retail issues (DTE Staff, 2018). The primary purpose of a package is to protect its contents. Because 30% of all food produced worldwide is wasted throughout the supply chain, optimal packaging might be one of the answers to reduce this huge number. Losses in the supply chain before the product reaches the customer are a problem in developing nations. Adequate packaging may help safeguard fruits and vegetables while also extending their life span so that they reach consumers safely. Food is wasted more at the household level in industrialized nations owing to wasteful behavior. One of the issues leading to packaging waste might be inappropriate packing sizes and difficult-to-empty containers. When it comes to package sustainability, the protective function is sometimes overlooked, with discussions focusing solely on the kind and quantity of material utilized in its manufacture.

Food waste accounts for around 30% of overall food output. Using suitable packing might be one way to reduce this enormous proportion. Understanding the proper packaging and the settings in which it must function will lead to package design optimization and the development of practical, cost-effective packaging. According to the Food and Agriculture Organization (FAO), lack of packaging is one of the leading causes of food waste, particularly in developing countries (Wohner et al., 2019). Packaging's contribution to food loss in industrialized countries is less clear. Food packaging is an integral part of food safety and can aid in the reduction of food waste. Inadequate packing, such as overly big package sizes, unsuitable material, contaminated packaging, and technical problems in the packaging process, causes fresh produce to decay. Food loss from the food service sector is increasing as more food is consumed outside the home. Packaging designers should think about how their design decisions affect food loss prevention. To avoid food loss due to packaging, primary, secondary, and tertiary packaging must be designed with interrelationships in mind. To increase their product's capacity to prevent food waste, package design must consider the requirements of packaging across the entire supply chain. Consumer behavior research and education and collaboration between manufacturers, retailers, and producers should be used to accomplish this.

11.3 OVERVIEW OF AGRICULTURAL PRODUCE (FRUITS AND VEGETABLES) AND THEIR CHARACTERISTICS

Following product harvest, many biological processes such as respiration and ethylene generation continue. Thus, 25–30% of horticultural produce is wasted due to inefficient marketing and inadequate post-harvest infrastructure (Singh and Kaur, 2019). Water content in fruits and vegetables ranges from 70 to 95%. In any typical atmospheric situation, they dry quickly. As a result, the plant loses its rigidity, and the cells shrink, resulting in wilting and shriveling. Appropriate packaging extends the shelf life of fresh produce by preventing moisture loss and thereby wilting. The most significant disadvantage of purchasing ready-to-eat fresh-cut fruits is their rapid decomposition and degeneration, which reduces their shelf life and results in an unpleasant appearance and palatability. Fruits and vegetables are living goods that ripen, and plant tissue breaks down as the aging process progresses. Following product harvest, many biological processes such as respiration and ethylene generation continue. Minimizing damage is essential to get the utmost shelf life out of fresh food. Microorganisms such as molds, yeasts, and bacteria are the leading causes of fruit and vegetable deterioration. As a result, proper handling and packing of fruits and vegetables are critical for maintaining their freshness and quality. Fruit and vegetable packing is essential for extending their shelf life. A package's purposes are to contain, protect/preserve, aid in safe and efficient transportation and distribution, and, ultimately, inform and motivate a consumer's purchase decision.

11.3.1 CHARACTERISTICS OF PRODUCTS

Horticultural products are highly perishable and are easily influenced by weather conditions, distribution hazards, and microbial deterioration. It is necessary to understand the biology of horticultural produce in order to build a suitable package. This poses a unique challenge in terms of fresh product storage and packaging. The following product features must be considered while developing a packaging system.

11.3.1.1 RESPIRATION

Fresh produce continues to breathe even after harvesting. The oxidation of sugars produces carbon dioxide, water, and heat during respiration. Oxygen is utilized, and CO_2 is emitted during the respiration process. The high rate of respiration has a direct relationship with the rate of product deterioration. The respiratory activity of fresh produce affects its storage life. Respiration is reduced, and senescence is delayed when produce is stored at a low temperature, therefore extending storage life. Controlling the quantities of oxygen and carbon dioxide around fresh produce can also help reduce respiration rate. The ripening process is aided by ethylene, a natural plant hormone generated during respiration. The ability of fruits to produce ethylene varies widely. As shown in Table 11.1, an excess of ethylene produced by fruits with moderate-to-high production rates causes senescence in fruits that demonstrate a ripening phase rise in respiration.

TABLE 11.1 Classification Based on the Amount of Ethylene Produced by Fruits and Vegetables (Kader, 1980).

Fruits and vegetables	Ethylene generation rate (μl/kg-h at 20°C)
Pomegranate, strawberry, citrus, cherry grape	0.01–0.1
Blueberry, kiwifruit, pineapple, raspberry, peppers, persimmon	0. 1–1.0
Plum, passion fruit, tomato, banana, honeydew melons, mango, fig, orange, lemon	1.0–10.0
Apple, apricot, papaya, peach, pear, avocado, nectarine	10.0–100.0
Sapote, passion fruit, cherimoya, mamey apple	> 100.0
Green pepper, sapodilla	Production rate not determined

11.3.1.2 TRANSPIRATION

The process by which fresh produce lose moisture is known as transpiration. During postharvest processing and storage, fresh fruits, and vegetables lose moisture through their skins (Becker et al., 2014). Fruits and vegetables have a high water content, ranging from 70 to 95%. Rapid drying of the product owing to significant loss of moisture causes wilting, shriveling, and loss of rigidity in the environment. Fresh produce must be stored in a low temperature, high humidity environment to minimize transpiration losses and maximize market quality and shelf life. Various skin coatings and moisture-proof films can be employed during produce packing to significantly reduce transpiration and lengthen storage life.

11.4 PACKAGING DESIGN ASPECTS OF FRUITS AND VEGETABLES

Fruits and vegetables are living products that ripen, and plant tissue breaks down as the aging process progresses. Molds, yeasts, and bacteria are the most prevalent microorganisms that cause fruit and vegetable deterioration. These organisms can have a negative impact on the quality of fruit and vegetable, reducing their shelf life. Moreover, fruits and vegetables are often seasonal and regional. As a result, proper packaging design for fresh produce is essential for minimizing their degradation and ensuring their availability throughout the year and across the country. Produce departments around the country are continuing to favor packaged goods over bulk fruits and vegetables. This is owing to packaging's significant supply chain benefits, including ease of shipping, decreased shrink, increased food safety, fewer labor at retail, cleaner displays, and consumer convenience.

The process of creating and producing a container to hold, protect, convey, dispense, store, identify, and advertise a product is known as packaging design (Srivastava et al., 2022). The container can be built of various materials, including plastic, paper, glass, or metal, depending on the nature of the product. Consumers like to purchase products that arrive in a portable package, perform as expected, and deliver all the information they require about the product (Singh et al., 2021). Fruit and vegetable packaging is a farming process that involves transporting fresh produce from the field to the consumer. It's a great challenge to design a package that satisfies the specific issues and requirements of fresh fruits and vegetables due to anaerobic degradation in fruits and vegetables, which ruins them in a matter of hours. Selecting the appropriate fruit and vegetable package can be critical to a farmer's market survival. As consumers, we interact with fruits and vegetables regularly, but we often overlook the importance of packaging. The prime functions of the package include containing, protecting, and identifying. The package should be able to preserve the contents, protect them from the environment, and keep them safe during transportation and handling. Another important function of packaging is identification, which provides consumers information about the product while simultaneously attempting to advertise it by drawing attention to it. As the number of products on the market grows, companies are seeking for more efficient ways to compete, and packaging and packaging design has undoubtedly become a key instrument for differentiation. With the rising self-service concept and changing consumer preferences, the package is becoming more popular as a sales marketing tool and a stimulator of impulsive purchase behavior. Packaging design has become a company's most important asset due to its ability to attract and retain consumers.

The following are some of the essential general fruit and vegetable packaging standards:

- The package should have good barrier properties that protect against microbe contamination, a barrier to moisture loss and oxygen ingress, and protection against odors or environmental toxicants.

- Packaging material should not contain any type of toxic chemicals.
- The package must meet handling and marketing standards in terms of size, shape, strength, and weight.
- The package must be stackable and inter-lockable.
- Packages should be re-usable and recyclable for convenient disposal.
- The package should be transparent and UV-protected.
- The package should be tamper-resistant or tamper-evident, meaning any interference with the contents should be visible.
- Packaging should be cost-effective while also preserving the integrity of the product.
- Packaging should have an appealing appearance, that is, it should evoke positive emotions in consumers, and it should be printable.
- The package should be compatible with fresh produce.

11.4.1 PACKAGING CLASSIFICATION

11.4.1.1 DEPENDING ON THE LEVEL OF PACKAGING

The level of packaging means how a product is packed and distributed. A product can be damaged in several ways, including during transit, handling, retail stocking, and even when consumers carry the product home. To ensure that the product is appropriately protected, so the wastage is kept to a minimum, it's vital to create a balance between the levels of packaging. Primary, secondary, and tertiary packaging are the three layers of packaging, as shown in Figure 11.1.

a) Primary packaging

Primary packaging, often known as a consumer or retail packaging, is the package in which the product comes into immediate contact (Figure 11.1). It holds a product for sale. The purpose of primary packaging is to protect the goods while informing and attracting customers. Examples of primary packaging for fresh produce include plastic bags, trays with flow pack wrappers, and stretch films.

b) Secondary packaging

It is the external packaging of the primary packaging that groups multiple packets and further protects or labels the product. Examples of secondary packaging for fresh produce include cardboard cartons, cardboard/plastic crates, trays, shrink-wrapped bundles, etc. (Figure 11.1).

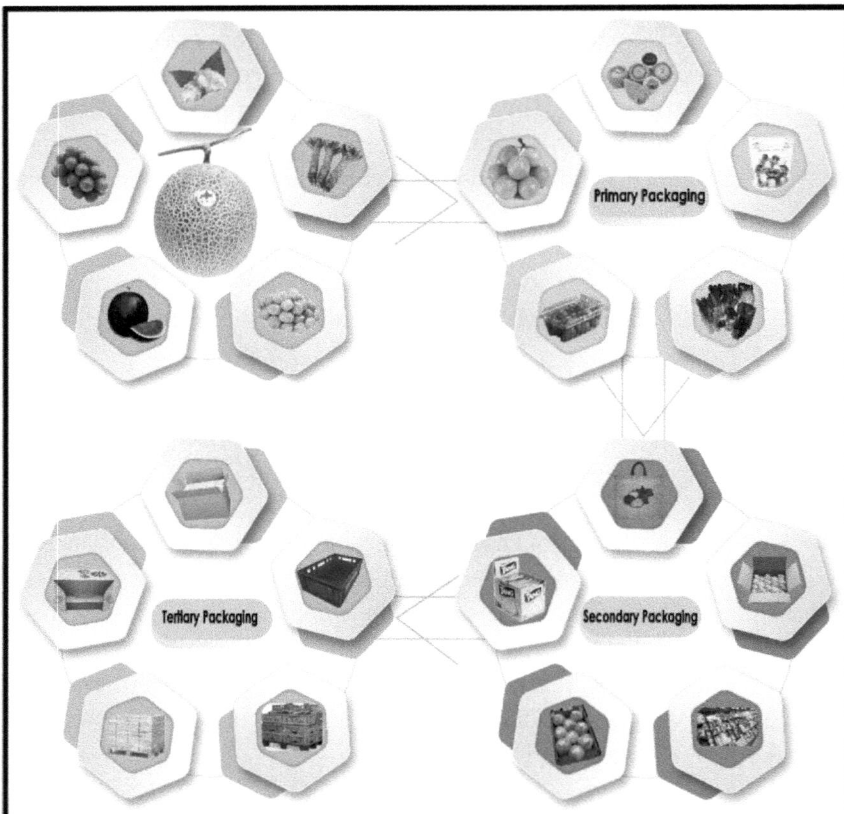

FIGURE 11.1 Different levels involved in fruits and vegetable packaging, including examples of primary, secondary, and tertiary packaging levels.

c) Tertiary packaging

Tertiary packaging is used for bulk product handling, shipment, and storage purposes. It protects both the product and its secondary and primary packing. Examples of tertiary packaging for fresh produce include pallets used to transport bulk shipments and stretch wrap used to secure stacks of cartons and wooden crates, as shown in Figure 11.1.

11.4.1.2 DEPENDING ON FRESH PRODUCE SUPPLY

On the basis of the purpose of supplying fresh fruit and vegetables, packages are divided into two categories: retail and shipping. Retail packaging is designed for consumers, and they are generally smaller in size, with a capacity of 1/2 to 1 dozen fruits and vegetables. The marketing aspects of the product determine the type of consumer package. Shipping or transport packages do not have a marketing function, and they are designed for long-distance transit and come in various sizes. These bags must have good mechanical properties and should be able to withstand impacts, vibration, and compression while being transported. Fresh produce is transported in a variety of packaging materials.

11.4.1.3 DEPENDING ON ITS RIGIDITY

Fruit and vegetable packaging can be classified into three varieties based on their physical properties: flexible, semi-rigid, and rigid materials. Any package or part of a package whose shape may be easily modified is defined as flexible packaging. In comparison to rigid packing, they take up less space. Flexible plastic films are inexpensive and have good moisture and gas barrier qualities. Cellulose, polyvinylidene chloride (PVDC), polyethylene (PE), polyethylene terephthalate (PET), polypropylene (PP), and ethylene-vinyl acetate are the most popular polymers used in flexible packaging (EA) (Ritsko, 2003). Rigid containers protect the contents against physical damage during transportation and distribution. They are generally made of wood, plastic, metal, or cardboard and are available in the shape of crates, barrels, drums, and sacks for final shipment. Semi-rigid packing materials include paper and board, whereas rigid packaging materials include crates, glass, metal, cans, and others.

11.5 THE SUPPLY CHAIN OF FRUITS AND VEGETABLE PACKAGING

Fresh produce is the most perishable and essential component of human nutrition. They are seasonal in nature, and during the surplus time, price falls, and production becomes unprofitable because of forced sales. As a result, a rise in fruit and vegetable output will be of little use if the product is not adequately handled, processed, or consumed. The costs of preventing losses are always less expensive than the costs of producing (Shiksha, 2013). As a result, fruit and vegetable packaging has a vital role in extending their shelf life, hence reducing losses. To meet the country's demand for fruits and vegetables, proper management of fruits and vegetables is required at all stages of the supply chain. Fresh fruits and vegetables face various physical and environmental challenges before reaching our refrigerator. The supply chain for fruits and vegetables is shown in Figure 11.2.

FIGURE 11.2 A schema of the supply chain of fresh produce such as fruits and vegetables. (1) Collection of fresh produce from the farm. (2) Transportation of fresh produce to the warehouse. (3) Storage of fruits and vegetables in the warehouse. (4) Distribution to different locations.

The term "harvest" refers to the act of gathering a crop. It is a crucial procedure that determines the quality and storage of the product. Harvested products are metabolically active and go through ripening and senescence. Inadequate process management can lead to considerable nutritional and quality losses and financial losses for all players in the supply chain, from growers to consumers. So, the fresh produce undergoes optimal post-harvest treatments to slow down natural senescence and maturation and lower the chances of microbial growth and contamination (Mahajan et al., 2014). Grading fruits and vegetables is an essential part of post-harvest management after they have been harvested. Grading is the process of classifying vegetables and fruits into different grades based on their physical characteristics such as size, shape, color, weight, and volume in order to maximize market value. The physical attributes are affected by agroclimatic conditions. Packaging fresh produce is an important element to be focused on so that fresh produce can readily follow its long and complex path from grower to consumer. Because of inefficient marketing and improper post-harvest infrastructure, 25–30% of fruits and vegetables go to waste (Singh and Kaur, 2019). Proper handling and packaging are essential for preserving the freshness and quality of fruits and vegetables during the supply chain. Throughout their whole life cycle, from development to final consumption, the produce should not be changed in any way, including its basic structure or usage. So, selecting appropriate packaging for commercial-scale marketing necessitates significant consideration.

11.6 TRADITIONAL PACKAGE DESIGN OF FRUITS AND VEGETABLES

The usage of a fiberboard carton is the most prevalent method of packaging in this industry. However, to avoid abrasion damage, most products require additional internal packaging, such as tissue paper wraps, trays, or pads. Furthermore, bags, crates, baskets, bulk bins, and palletized containers have been used for handling, transporting, and marketing fresh produce. In the packaging of fresh fruits and vegetables, polyethylene (PE) and polypropylene (PP) bags are commonly utilized. Because PE and PP have low water vapor permeability, both films and bags are sometimes perforated to allow the product to breathe. When it comes to fresh fruit shipping and handling, delicate fruits are packed in smaller packs with fewer layers of fruit to avoid compression damage. Moulded trays were used to physically separate each piece of vegetable. Shock and vibration cause damage to produce cells, resulting in an increase in respiration and the release of enzymes that initiate browning reactions. Therefore, individual fruits are wrapped in tissue or waxed paper, which increases physical protection while also reducing disease microorganisms spread within a pack. Owing to perishability, most fresh produce are still offered unpackaged in perforated polyethylene (PE) bags or nets at retail or wholesale (Ščetar et al., 2010), which poses a difficulty in long-term transportation due to brushing and subsequent decay of the loose packaged product. When long-term storage is desired, the packing film should have a high gas permeability and anti-fogging properties. Permeation is the key to increasing the shelf life of fresh fruit.

11.6.1 PACKAGING MATERIAL USED IN PACKAGING OF FRESH FRUITS AND VEGETABLES

The primary purpose of a fresh produce packaging material is to store fruits and vegetables in such a way that their quality is preserved and storage life is extended. Thus, the necessity for products to have longer shelf lives necessitates the development of superior barrier qualities in packaging materials. Food and perishables may be securely shipped across town or across the country with the correct materials and planning. Fruits and vegetables are packed with a variety of materials.

11.6.1.1 PLASTIC BAGS/POLYETHYLENE FILM

Plastic bags are the most popular type of consumer packaging for fruits and vegetables. The materials are incredibly inexpensive, and automated bagging equipment minimizes packing costs even more. Film bags are transparent, allowing for easy observation of the contents. Plastic films come in a variety of thicknesses and grades. In the production of plastic bags for food packaging, low-density polyethylene (LDPE) resins are commonly utilized.

11.6.1.2 SHRINKWRAP

Shrink wrapping is one of the most recent developments in produce packaging. It is a simple, user-friendly method and easily adopted by small-scale farmers and companies. Potatoes, sweet potatoes, apples, sweet corn, cucumbers, onions, and a variety of tropical fruit have been successfully packaged using shrink wrapping (Boyette et al., 2020). Cryovac, polyolefin, and LDPE are some of the heat-shrinkable films used. These films are the most effective at extending both shelf-life and storage life, as well as preserving fruit quality compared to unwrapped fruits (Technologies, 2009). This technique involves overwrapping the produce with heat-shrinkable sheets of the desired thickness through the use of a machine. The procedure prevents the fruits from deteriorating physiologically. Fruit is individually shrink-wrapped to preserve its quality throughout transit and storage by maintaining appropriate gas and humidity levels. As a result, under suitable storage conditions, it doubles or even triples the fruit's storage life. Also, these unit packs help protect the product from abrasion, dust, and filth.

11.6.1.3 BROWN PAPER BAGS

In the food industry, paper bags are an excellent packaging option. Paper bags are not only more flexible and durable than plastic bags, but they are also less expensive to produce. These bags are biodegradable and can last a long time if used properly. Paper bags are a versatile packing option for a number of fresh foods. Also, paper bags can be customized to fit the packaging demands and branded to reflect the brand identity. When compared to plastic bags in the same situation, paper bags have a higher tear tolerance. Paper bags are made from kraft pulp and renewable wood. Furthermore, paper bags that are robust enough to be reused reduce the likelihood of them being thrown away. Despite being composed of paper, these bags have a weight capacity of roughly 25 pounds (Jersey, 2021).

11.6.1.4 RIGID PLASTIC PACKAGES/CLAMSHELLS

Clamshell packaging is a hinged thermoformed container traditionally used in the food industry. Clamshells are becoming increasingly popular because they are affordable, adaptable, provide excellent product protection, and present a visually appealing consumer package. They're most typically associated with high-value produce consumer packaging such as small fruit, berries, mushrooms, and other products. Also, for pre-cut produce, clamshells are frequently used. In terms of sizes and forms, the most typical clamshell volumes in regular grocery stores are 1-pound clamshell volumes for strawberries, 6-ounce clamshell volumes for raspberries and blackberries, and pints for blueberries (Bareuther, 2016). Clamshells have several advantages, including crystal-clear visibility of the contents, anti-fogging treatment, a durable lidding system that is easy for consumers to open and close, tamper-evident, and breathing functionality for the contents to ensure they breathe properly.

11.6.1.5 PLASTIC NETTING

Nets are typically used for packaging fresh fruits and vegetables. The polymer used for netting has an excellent elongation at break, high resistance, as well as strong pressure, and welding capabilities. Plastic food netting made from sustainable raw materials is also becoming more popular. Food that is wrapped in nets rather than plastic bags has a better ability to "breathe." This keeps moisture out of the container, preventing premature germination or even rotting. Packaging nets offer advantages over unpackaged food in terms of procurement and transportation, especially for large or standardized sales quantities. Typically, plastic film is used to secure products, which severely hinders appropriate ventilation. Plastic netting, on the other hand, secures the load and provides enough ventilation.

11.6.1.6 WOOD PACKAGING

For generations, wood has been used for food preparation, packing, storage, and transportation. There are a lot of advantages to using wood packing. Wood is a more robust, more substantial material that can endure much stress. The wood packing is designed to endure the severe mechanical stresses that are common during shipping and storage. Because wood is moisture and erosion resistant, it is rarely affected by moisture and keeps products dry, and wood containers can also be stacked, saving space. It's a budget-friendly option and doesn't corrode. When compared to plastic or other shipping materials, wood packing is less expensive. Choosing wood packaging over other materials helps the environment in some respects because it is recyclable. To meet the packaging and shipping demands, there are now over 1500 different sizes and kinds of wood product packaging material available. It includes wooden crates, wire-bound crates, wood pallets, pallets bin, and corrugated fibreboard. Wooden boxes are an excellent choice for shipping fruits and vegetables. Tomatoes are best transported in wooden trays, whereas watermelon and other forms of melon, as well as other large fruits and vegetables, are best transported in huge wooden crates. Wire-bound crates are most commonly employed for sweet corn, snap beans, and other produce that requires hydro cooling (essential for ripening). They're durable and stackable, so they save a lot of space. The fresh produce packaging market consumes 190 million of the 700 million pallets produced in the United States each year, and one pallet can hold anywhere from 20 to 100 individual packets of fresh produce, depending on their size (Skid and Pallets, 2019). Pallet bins are commonly used to transport goods from the field to the packing facility. The capacity varies from 12 to more than 50 bushels. They have a 5-year shelf life when stored outside. They can endure more than 10 years and be reused several times if properly stored.

11.6.1.7 CORRUGATED FIBERBOARD

Cardboard is frequently confused with corrugated fiberboard. It is available in a variety of weights and forms, and it is less expensive than other wood shipping materials. As a result, it is the most common alternative for exporting products. Corrugated fibreboard is made up of one or two flat linerboards and a fluted corrugated sheet. Paperboard is used for the corrugated medium and linerboard, which is a paper-like substance. The paper must be thicker than 0.008 inches to be considered paperboard (Ranganatham, 2012). It is used to make cardboard boxes and is created on "flute lamination machines" or "corrugators." The impact abrasion and compression strength of the corrugated board are excellent. Delamination of the corrugated material might occur if the storage humidity is too high.

11.7 TRENDS IN THE PACKAGING OF FRUITS AND VEGETABLES

There is a growing trend in the consumption of fruits and vegetables these days. Fruit and vegetable consumption per person increased from 281 grams per day in 2005–2006 to 397 grams per day in 2017–2018 (Maden et al., 2021). One of the critical reasons for the development of horticulture is a shift in food consumption habits, with consumers migrating away from cereals like wheat and rice and moving toward healthier options like fruits and vegetables.

For transport packaging of fresh fruits, several standard packaging materials have been modified, and new materials/technologies have been developed with the goal of providing rigidity, longer storage, stackability, ventilation, and cushioning. The main reasons for this include rising demands for product safety, shelf-life extension, cost-efficiency, environmental concerns, and consumer convenience. Innovative modified- and controlled-atmosphere packaging, edible coating techniques, and active and intelligent packaging technologies are being created, tested, and optimized in laboratories worldwide to better fulfill these diverse demands. The ready-to-eat, fresh-cut consumer products market has shown consistent development in recent years. However, cutting, on the other hand, accelerates the senescence process, resulting in very short shelf life for the products. Fresh-cut products have much higher respiration rates due to the cell stress caused by cutting. As a result, a variety of techniques are used to enhance shelf life. The most important characteristics of these materials are discussed here.

11.7.1 MODIFIED ATMOSPHERE PACKAGING (MAP)

Polymeric films have been used for packaging fresh foods, and they offer a number of benefits, including water loss management, skin abrasion prevention, and reduced contamination of the produce during handling. Fresh produce respiration results in decreased O_2 and greater CO_2 levels within the packaging. MAP is a food preservation technique that involves changing the gaseous atmosphere inside a pack and using packaging materials with an acceptable level of gas barrier to retain the modified atmosphere at a safe level (Romani et al., 2017). Within the pack, a modified atmosphere (MA) can be created in two methods. The active modification involves inflating the pack to create a small vacuum and replacing the atmosphere with a suitable gas mixture. CO_2, O_2, or ethylene absorbers (active packaging) may be added to the pack to manage the concentration of these gases. The atmosphere is created in passive modification systems by the fresh produce breathing within the pack. The final equilibrium environment will be determined by the fresh produce and packaging film's characteristics. The prevention and delay of fruit senescence (ripening) are one of the critical advantages of a controlled environment.

11.7.2 ACTIVE PACKAGING

Active packaging is defined as a system that tries to improve the function of packaging by interacting with food and/or the environment. Different active packaging methods in fruits and vegetables include antioxidants, ethylene scavengers, and moisture absorbers. By eliminating oxygen from the empty atmosphere in packaging and preventing the growth of aerobic microbes, oxygen absorbers improve the shelf life of fresh produce. Gas scavenging or gas flushing is used to provide an appropriate environment within the package. Oxygen absorbers can be used to reduce O_2 partial pressure inside the package as well as absorb O_2 that has entered from the outside environment (Deshmukh et al., 2022; Ramakanth et al., 2022). Ethylene scavengers are made up of potassium permanganate, activated carbon, and mineral substrates and can be used in films, sachets, paper bags, and cardboard boxes. Potassium permanganate can react with ethylene to produce acetate ethanol, allowing ethylene to be eliminated. In contrast, carbon-based ethylene scavengers capture ethylene in the environment and then break it down. Due to the high-water activity of fresh fruits and vegetables, packaging conditions of 85–95% relative humidity are recommended. Low humidity causes produce to wilt, whereas high humidity increases undesired bacteria growth. Humidity absorbers are designed to reduce humidity levels in packages to a safe level, allowing fresh fruits and vegetables to last longer. Many humidity absorbers are available, including sorbitol, sodium chloride, potassium chloride, and bentonite. Fast humidity absorbers include sodium chloride, potassium chloride, and sorbitol, while bentonite absorbs moisture more slowly than salts and sorbitol (Wilson, 2007).

11.7.3 ANTIMICROBIAL PACKAGING

The growth of unwanted bacteria on fresh produce is one of the most important variables influencing produce shelf life, which can lead to spoiling and present food safety risks. Antimicrobial packaging can help to decrease microbial contamination and hence improve fresh produce's shelf life. Antimicrobial packaging necessitates the employment of antimicrobial agents, which come in two varieties: volatile and non-volatile. Organic acids, enzymes, bacteriophages, and polysaccharides are examples of non-volatile substances that must come into continual contact with packaging material and product surfaces. Volatile substances, such as ethanol, essential oils, and sulfur dioxide, can quickly diffuse into the atmosphere without coming into contact with products. Antimicrobial packaging that is non-volatile is typically utilized in vacuum-packed products, whereas volatile antimicrobial packaging is appropriate for bag-packed products. Consumers currently prefer to employ ingredients and materials produced from natural sources in food processing due to health and safety concerns. Hence, essential oils are becoming more popular as antibacterial agents.

11.7.4 INTELLIGENT PACKAGING

Intelligent packaging is a term used to describe packaging technologies that can monitor the condition and quality of foods without affecting the in-package environment or shelf life. Intelligent packaging can connect with users and operate as a support tool by informing them of changes in the internal and external environment, as well as product conditions. Indicators, data carriers, and sensors are standard components of intelligent packaging. Indicators can provide consumers with quality information, such as notifying them if the produce has expired or if it has been exposed to any potential risks or harsh situations. Data carriers, such as radio frequency identification tags (RFID), are commonly employed for tracing and storage management. Sensors are made to quickly quantify corresponding analytes. The freshness indication is often used for fruits and vegetables. Freshness indicators for fruits and vegetables are based on headspace gas pH changes. Because total volatile acids are lost as fresh fruits, and vegetables spoil, pH rises with time. By just sensing a pH increase, the dye in the indicators will change color (Akhila et al., 2022).

11.7.5 EDIBLE COATINGS AND FILMS

One of the most ingenious methods for extending the commercial shelf life of fruits and vegetables is to coat them with edible coatings that act as a gas barrier. Edible materials can now be turned into edible films and coatings due to technological advancements. The research and deployment of bio-based packaging materials to extend the shelf life and improve fresh items' quality have received much attention in recent years. Food quality is maintained, shelf life is extended, and waste is reduced to some extent with edible packaging (Kumar et al., 2022). The type of food, such as fruit, dairy, or meat, as well as storage conditions influence the edible packaging material selection. Controlling the coating solutions' wettability, which affects the coating thickness, is also important for the efficiency of edible coatings safeguarding fruits and vegetables.

11.8 CONCLUSION

At this rate of increasing global food waste, there is a high chance that a severe and unrecoverable food shortage can be seen in near future. Appropriate packaging along with responsible production and consumption of produced food can deal with this problem. Especially for fresh produce like fruits and vegetables, a better packaging with improved safety is needed without compromising on aesthetics. Good packaging focuses on improving all primary, secondary, and tertiary layers, which can safeguard the product during storage and transportation. Types of packaging that are most commonly seen in the market for fruits and vegetables include wooden and plastic crates, corrugated boards, plastic netting, shrink wraps, paper, and poly bags. Both flexible and rigid packaging crucially improve the value of the product. New trends in fresh produce packaging like active and intelligent film design are briefed. In this chapter, the major packaging design aspects that need to be considered while proposing a design are discussed. It is concluded that packaging developed for post-harvest produce is designed by considering their respiration, transpiration, and ripening rates alongside the possible hazards that the packaging may undergo during storage, transportation, and marketing at retail stores.

KEYWORDS

- **package design**
- **fruits and vegetables**
- **agricultural produce**
- **sustainability**
- **marketing**

REFERENCES

Akhila, K.; Ramakanth, D.; Gaikwad, K. K. Development of Novel Gallic Acid- and Cellulose Acetate-Coated Paper as pH-Responsive Oxygen Indicator for Intelligent Food Packaging. *J. Coat. Technol. Res.* **2022**. https://doi.org/10.1007/s11998-022-00622-0

Al-Obadi, M.; Ayad, H.; Pokharel, S.; Ayari, M. A. Perspectives on Food Waste Management: Prevention and Social Innovations. *Sustain. Prod. Consum.* **2022,** *31*, 190–208. https://doi.org/10.1016/J.SPC.2022.02.012

Bareuther, C. M. Clamshell Packaging New Possibilities. *Produce Business Magazine*, Produce Business, 2016. https://www.producebusiness.com/clamshell-packaging-trends-open-new-possibilities-produce-market/

Becker, B. R.; Fricke, B. A. Transpiration and Respiration of Fruits and Vegetables, *Universidad of Missouri*, 2014, vol 18.

Boyette, M.; Sanders, D. C.; Rutledge, G. A. *Packaging Requirements for Fresh Fruits and Vegetables, NC State Extension Publications*, 2020.

Deshmukh, R. K.; Akhila, K.; Ramakanth, D.; Gaikwad, K. K. Guar Gum/Carboxymethyl Cellulose Based Antioxidant Film Incorporated with Halloysite Nanotubes and Litchi Shell Waste Extract for Active Packaging. *Int. J. Biol. Macromol.* **2022,** *201*, 1–13. https://doi.org/10.1016/j.ijbiomac.2021.12.198

DTE Staff. *Each Year, Half the Fruits, Vegetables World Produces are Wasted: FAO*; Down To Earth, 2018. https://www.downtoearth.org.in/news/health/each-year-half-the-fruits-vegetables-world-produces-are-wasted-fao-62085

Hussain, S.; Tanu M. G. *How Can India Plug the Food Wastage Hole at the Consumer and Retail Level?* The Wire, 2021. https://thewire.in/food/how-can-india-plug-the-food-wastage-hole-at-the-consumer-and-retail-level

Jersey, W. N. *7 Notable Features to Use Paper Bags in the Food Industry-Jetpaperbags*. JET Paper Bags, 2021. https://jetpaperbags.com/7-notable-features-to-use-paper-bags-in-the-food-industry/

Kader, A. A. Prevention of Ripening in Fruits by Use of Controlled Atmospheres. In *Food Technology*, 1980; vol *34* (3), pp 51–54.

Kumar, L.; Ramakanth, D.; Akhila, K.; Gaikwad, K. K. Edible Films and Coatings for Food Packaging Applications: A Review. *Environ. Chem. Lett.* **2022,** *20* (1), 875–900. https://doi.org/10.1007/s10311-021-01339-z

Lipinski, B.; Hanson, C.; Lomax, J.; Kitinoja, L.; Waite, R.; Searchinger, T. Toward a Sustainable Food System Reducing Food Loss and Waste. World Resource Institute, June 2016; pp 1–40.

Mahajan, P. V.; Caleb, O. J.; Singh, Z.; Watkins, C. B.; Geyer, M. Postharvest Treatments of Fresh Produce. *Philos. Trans. Math. Phys. Eng. Sci.* **2014,** *372* (2017), 20130309. https://doi.org/10.1098/rsta.2013.0309

Ramakanth, D.; Akhila, K.; Gaikwad, K. K.; Maji, P. K. UV-Activated Oxygen Scavenging System Based on Natural Rubber Latex from Hevea brasiliensis for Active Packaging Applications. *Ind. Crops Prod.* **2022,** *178*, 114658. https://doi.org/10.1016/j.indcrop.2022.114658

Ranganatham, M. DT-9: Lesson 4. Characteristics Of Paper, Corrugated Paper, Fiber Board And Wood, 2012. http://ecoursesonline.iasri.res.in/mod/page/view.php?id=6199

Ritsko, J. J. Packaging Technology. In *Digital Encyclopedia of Applied Physics*, 2003. https://doi.org/10.1002/3527600434.eap305

Romani, S.; Zurera, G.; Djekic, I.; Tarr, G. *Modified Atmosphere Packaging - An Overview. ScienceDirect Topics*, Science Direct, 2017. https://www.sciencedirect.com/topics/food-science/modified-atmosphere-packaging

Ščetar, M.; Kurek, M.; Galić, K. Trends in Fruit and Vegetable Packaging. *University of Zagreb, Faculty of Food Technology and Biotechnology, Pierottijeva 6, 10000 Zagreb, Croatia, June 2014*, 2010.

Shiksha, K. CPE: Lesson 8. Fruits and Vegetables Processing. In *Crop Process Engineering*, 2013. http://ecoursesonline.iasri.res.in/mod/page/view.php?id=807

Siddiqui, M. W.; Rahman, M. S.; Wani, A. A. In *Innovative Packaging of Fruits and Vegetables: Strategies for Safety*, 2018.

Singh, S. K.; Kaur, S. In *Advances in Horticultural Crop Management and Value Addition*, 2019.

Singh, A. K.; Ramakanth, D.; Kumar, A.; Lee, Y. S.; Gaikwad, K. K. Active Packaging Technologies for Clean Label Food Products: A Review. *J. Food Meas. Charact.* **2021**, *2021*, 1–11. https://doi.org/10.1007/S11694-021-01024-3

Skid, P; Pallets. *Why Wood Packaging is Best for Shipping Fruits and Vegetables*. Pallets Move The World, 2019. https://pasadenaskidandpallet.com/why-wood-packaging-is-best-for-shipping-fruits-and-vegetables/

Srivastava, P.; Ramakanth, D.; Akhila, K.; Gaikwad, K. K. Package Design as a Branding Tool in the Cosmetic Industry: Consumers' Perception vs. Reality. *SN Bus. Econ.* **2022,** *2* (6), 1–38. https://doi.org/10.1007/S43546-022-00222-5

Technologies, P. Fruits for Enhancing Shelf-life Indian Council of Agricultural Research. **2009,** *15* (3), 1–24.

van der Maden, E.; Ringo, E.; Eunice, L. Scoping Study on Fruits and Vegetables; Results from Tanzania. *Wageningen Economic Research*, 2021.

Waste, T. M. A. F. *Food Waste. A Global Problem, Too Good To Go*, 2019. https://toogoodtogo.com/en-us/movement/education/food-waste-a-global-problem.

Wilson, C. L. Intelligent and Active Packaging for Fruits and Vegetables. In *Intelligent and Active Packaging for Fruits and Vegetables*, 2007. https://doi.org/10.1201/9781420008678

Wohner, B.; Pauer, E.; Heinrich, V.; Tacker, M. Packaging-Related Food Losses and Waste: An Overview of Drivers and Issues. *Sustainability* **2019,** *11*, 264. https://doi.org/ 10.3390/SU11010264

CHAPTER 12

ADVANCE PACKAGING FOR SAFETY AND SECURITY OF AGRICULTURAL PRODUCE IN SUPPLY CHAIN

SHIVA SINGH[1], KAUSHIK GHOSH[2], and PRADIP K. MAJI[1]

[1]*Department of Polymer and Process Engineering, Indian Institute of Technology Roorkee Saharanpur Campus, Saharanpur, UP, India*

[2]*Department of Chemistry, Indian Institute of Technology Roorkee, Roorkee, UP, India*

ABSTRACT

The methodology of land cultivation, harvesting vegetables, and growing livestock is known as agriculture. It produces most of the world's foodstuff and textiles. Agricultural products involve cotton, leather, wool, and so on. Wood for paper, goods, and building are also offered by agriculture. Agricultural products are an integral part of our daily diet. The security of these products is essential for the health, well-being, and livelihoods of billions of people worldwide. While going through a supply process, its susceptibility to contamination increases. Packaging plays a vital role in this purpose. Incorporating advanced packaging techniques in the food supply chain can lead to safe, secure, and healthy agricultural production. In these terms, this chapter deals with processes involving food safety, the importance of a secure supply chain, the requirement of packaging, and the applications of advanced packaging systems involved in the safety and security of agricultural products.

Novel Packaging Systems for Fruits and Vegetables. Kirtiraj K. Gaikwad & Suman Singh (Eds.)

12.1 INTRODUCTION

In recent years, an increment in food awareness has been noticed by most consumers. This is due to drastic changes in people's eating patterns and the tremendous growth of the agricultural and food businesses. Customers nowadays are more worried about the quality and safety of their food consumables. Food security is defined as food that is devoid of contaminants and ingredients that could lead to the development of microorganisms that are hazardous to people's health and safety. Mishandling of food has been exposed, resulting in the spread of diseases. There have been other reports of similar incidents (Ashley, 2016).

Foodborne illness is described by the World Health Organization (WHO) as illnesses induced by the infiltration of external bodies into food, resulting in food poisoning in many situations. Foodborne infections are on the rise and have an impact on people's health all around the world. For example, in the United States, one out of every six persons is sickened by an unhealthy diet, causing 128,000 admissions to hospitals and approximately 3000 fatalities each year. Foodborne infections are unlikely to be entirely eradicated due to financial concerns, but they can be decreased via regular supervision of the supply chain. Customers want administrations to supervise food goods to preserve it is freshness and protect them from duplicitous food companies (Haji et al., 2020).

Agricultural product supply chains are complicated and vibrant and are determined in several phases that impact the cost, value, and transport period of a product. These phases can take place at any point along the supply chain, from the agriculture phase to when a product reaches the final consumers. Several elements play a crucial role in effective food supply chain supervision. For example, Pang et al. (2015) pointed out an adequate agricultural product supply chain which may assure appropriate delivery and consumer service while also lowering total expenses of the supply chain. Moreover, effective management of the supply chain can boost turnover and currency flow in organizations involved in the process as well as those that participate in it. As a result, it's vital first to comprehend what makes a good supply chain. Essentially, a well-functioning supply chain should be able to deliver the appropriate quantity of food while adhering to safety regulations, resulting in high-quality food that is both accessible and inexpensive. Many know-how have been incorporated into the supply system to limit hazards, increase efficacy, and cope with complications (Paam et al., 2016).

Furthermore, current implements have been considerably more user-friendly. They are now utilized for various responsibilities along the supply chain, planting, harvesting, preparation, processing, wrapping, loading, shipment, and accounting observation. Expertise is generally employed to speed up processes, satisfy rising worldwide food requirements at the correct moment, and trail food sources to evaluate food conditions and safety. The increasing usage of machinery in the supply system has improved and maintained the quality and safety of perishable foods. New technology has additionally given manufacturers and producers the ability to supervise the various stages of the food supply chain more cost effectively, according to Khan et al. (2018).

The term "food supply chain" indicates a system that defines the conversion of agricultural goods in a manner that can be placed on a plate and eaten. Perishable foods are frequently found in the outputs of a food supply chain. Perishable food is a type of commodity that is susceptible to quality loss during the manufacture, storage, shipping, or handling stages. The perishable food supply includes fresh goods such as fruits, bakery items, dairy products, vegetables, and meats. Fresh foods and perishable products, in particular, are more difficult to manage than other supply chain imports because of their small shelf lives. The essential of this is dependent on examining temperature and time parameters at every phase of the supply chain due to their substantial influence on quality of food (Aung and Chang, 2014).

Fundamentally, the longer the food is transported after it has been manufactured or processed, the more likely the food quality would deteriorate. Perishable food materials are also temperature-sensitive, making them particularly vulnerable to destruction through production, packaging, handling, and transit. Temperature can additionally affect the storage, production, distribution, and shelf-life of the fresh product, according to a study conducted by Smith and Sparks. As a result, several foods should be stowed at various temperatures to ensure their healthfulness and safety. Interferences to provide controlled ecological requirements and safeguard the quality of food, reduction of trash and profits should be implemented (Markina, 2018; Balaji and Arshinder, 2016).

The measures implicated in the delicate food supply chain consist of a variety of intermediate steps in every industry that crops or raw materials go through prior to being vended. The initial phase entails the various processes engaged in the agricultural divisions' production and processing.

Seeding, harvesting, shipping, processing, and distribution to the food processing factory are all crucial operations in this phase. Materials are delivered to processing factories, sorted, chopped, filtered, or chemically processed, cleansed, packed, and sent to suppliers. Concerning retailing, consumable foods necessitate additional conservation features crucial to delivering an excellent quality product to consumers at the termination of the supply chain, which is opposite to conventional wholesale to vendors following primary processing, as cereals and grains do (Büyüközkan and Göçer, 2018; Bosona and Gebresenbet, 2013). The processes involved in the food supply chain are depicted in Figure 12.1.

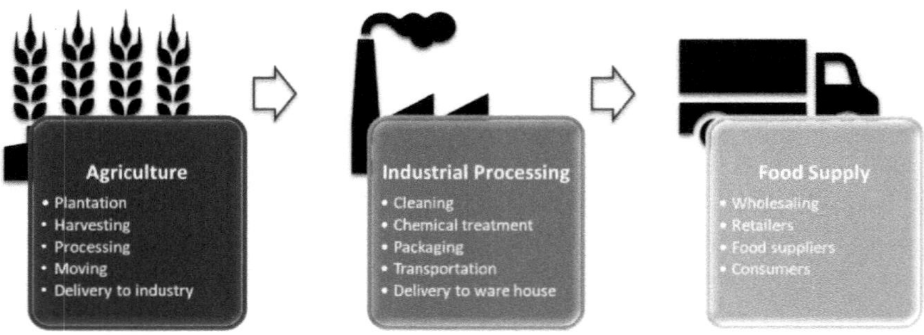

FIGURE 12.1 The process involved in the agro produce supply chain.

12.2 THE CONCEPT OF FOOD SAFETY AND SECURITY

Food security is a term that is inherently unobservable. The phrase has eluded a definite and operational but contextual meaning for the most part. Latest descriptions of food security have managed to place a greater emphasis on persons having access to both quantitatively and nutritionally satisfactory food. The notion also has numerous different aspects, each of which has its own set of implications. For example, "food security" is typically characterized as a state in which all parties have constant commercial and physical contact with appropriate food. Food safety exists when everyone's nutritional requirements for a valuable and healthy lifestyle are met (Vågsholm et al., 2020). The three elements of food security implicit in this concept are permanence, convenience, and access. Food convenience refers to the fact that, on average, accessible food sources are sufficient in amount and quality to fulfill intake requirements. Permanence implies reducing the risk of food intake falling lower than those needs due to seasonal changes or additional unforeseen factors. Lastly, access emphasizes the reality that many people go hungry even when there is a plentiful supply of food because they lack the supplies to generate or acquire the food they require. Food safekeeping is unquestionably a societal concern. Food safety is a critical aspect of a nation's ability to feed itself and maintain its independence (Pérez-Escamilla, 2017). On the contrary, food security is an extremely dynamic subject, as persistent hunger, regional food lacks, and unexpected rises in the costs of food may significantly impact public opinion and response. Although people's perception of food security may be hazy, it remains a sensitively potent idea in terms of what the public expects in this area. As a result, it is the responsibility of every government globally to ensure that food structures can offer all individuals a dependable, on time, and appropriate supply of healthy food on a long-term basis (Clapp et al., 2022; Cole et al., 2018).

12.3 STATE OF THE ART IN POLICY MAKING AND SECURITY FOR FOOD SECURITY

Alongside the rising realization that conventional farming policies are failing to meet their aims to some extent, there are additionally developing thoughts that globalization and the varying nature of food usage have driven the demand for policy making and its improvement. The primary cause driving the formation of these concepts is the need for the food system to be efficient in a more competitive world. According to Maxwell, all stages in the food chain, from generation to ultimate consumption, should be productive from a social welfare viewpoint. Construction strategies must include vibrant relative advantage; selling margins must give no more than ordinary profits over time, and consumer pricing should represent genuine scarcity values (Boratyńska and Huseynov, 2017; Thow et al., 2018).

While developing nations have attempted to change their agricultural policies to some extent, many have encountered significant challenges, and progress toward efficient, competitive, and effective private markets has been gradual. The motives for this narrow accomplishment contain: (1) domestic markets' official insufficiencies and organizational faintness; (2) national or local government's lack of commitment and strong skepticism; (3) flaws in the development and execution of restructuring agendas; and (4) private sector distrust of policy-makers' commitment to market reform. As a result, these limits are frequently invisible throughout the reform process and difficult to manage. Perhaps the most difficult challenge is balancing private division contribution and the government's balancing roles in lowering transaction expenses, encouraging active race, and improving the changeover for low-income manufacturers, particularly farmers, and clients (Bhunnoo, 2019). These are signals for the government to develop improved approaches and policies to improve the food system's (industry's) competitive edge. Conventionally, particularly with staple foods, most nations have favored the food self-sufficiency tactic, which encourages a nation to encounter all or a significant portion of its food demands from internal manufacture (as India's Make in India), rather than the riskier external-reliance method, which promotes a nation to pursue an outwardly oriented trade administration in the hopes of producing sufficient money from exports of more modest harvests to backing its food requirements (Mozaffarian et al., 2018). Underneath this self-capability strategy, the administration seeks to defend its essential food sector by keeping local production's competitiveness using regulatory tools for example input subsidies, cost controls, importation tariffs, and quotas (Lencucha et al., 2020). However, such a strategy is gradually questioned since it is said that it does not represent the countries' "true" relative (and competitive) lead. Such policies have the potential to skew the market whereas in the essence of market liberalization, nations are urged to eliminate their usage of unfair business policies and procedures while strengthening their genuine competitive benefit, which indicates reasonable contest and results in a more effective financial system (Gunaratne et al., 2021). Policy changes are required in this area. For a successful food safety policy, each national leadership must shift its policy focus from just safeguarding its food sector to establishing a "genuine competitive advantage," with resource allocation reflecting optimal results in real-time (Zerbian and de Luis Romero, 2021). Competitive advantage tactics are necessary to achieve supply efficiency—both amount and quality—so that production can compete successfully with imports. As a result, it facilitates sustainability and boosts local production, resulting in increased self-sufficiency (Simelane and Worth, 2020). More crucially, increased reasonable benefit in the food business may lead to longer-term food security by increasing chain associates' pay packet, which leads to increased buying power and, in turn, higher food quality.

12.4 CORRELATION BETWEEN FOOD SECURITY POLICY AND SUPPLY CHAIN MANAGEMENT

An assessment of the works on food safety and security yielded the following assumptions: (1) The idea of food safety measures has evolved through period of time. Nevertheless, the core characteristics of food safety measures are stability and accessibility, both of these are linked to food supply and reach. (2) Food safety measure thought has progressed from an emphasis on cumulative food accessibility to a wider framework of particular behavior that includes a living viewpoint. (3) In principle, a government must have an integrated strategy to guarantee that all people have substantial and financial reach to enough food to satisfy their nutritive requirements. Price stabilization is a popular tool used by developing nations in East and Southeast Asia to keep consumers' buying power and farm gate prices stable. However, it has been severely criticized because this type of policy is typically ineffectual and expensive to implement. (4) Globalization and the varying nature of the food system have driven the necessity for policy reforms and a rising acknowledgment of old policy ineffectiveness. One of the primary causes driving the birth of these concepts is the need for a more efficient and competitive food system in a more competitive economy. What might this conceptual framework gain from a literature survey in supply chain management? (1) Supply chain management is a comprehensive strategy for generating value and gaining a competitive edge. Supply chain managing is advantageous to all chain members by attempting to break the barriers between every of the chain's components to improve productivity and competitiveness. The idea of supply chain administration has expanded the idea of alliances to include multifirm cooperation to improve and regulate the full flow of supplies from the dealer to the end consumer. (2) In terms of food policy, government (public) policies, may possibly have a key influence on the establishment of joint supply chains. Conversely, supply chain supervision may aid in the management of long-term relationships and collaboration between small and big businesses, as well as between the private and public sectors. Supply chain management, via cooperation and partnership, may also assist businesses in responding dynamically to varying customer expectations, both in regional and worldwide marketplaces. (3) Supply chain collaboration may boost business efficiency across the board by improving logistics, data management, and affiliation management. More significantly, supply chain management enables equal value distribution among supply chain partners. (4) As many companies participating in supply chain association grows, so does the industry's national competitiveness. This is where food policy should be established in the future. The usage of the supply chain management idea to advance a country's food security is possibly viable because (1) supply chain administration can improve the competition of private companies in the interior of the agroindustry arrangement, such as the rice factories, thereby enhancing the rice logistic network as a system's countrywide competitive benefit; (2) the basic idea of supply chain administration (i.e., refining productivity by decreasing costs, sharing dangers, and managing affairs) can be applied to any industry; and (3) On a more appropriate level of abstraction, the required results of food safekeeping policy (i.e., convenience, permanence, and access) may be accomplished by implementing the logistic or supply chain administration idea, which promotes enterprises to collaborate and share value more equitably. This condition will benefit not just major commercial companies but also smallholders, especially small farmers because their involvement will be a contributing factor of a more viable supply system. Analytically, the potential of applying the logistic management idea to advance food safety has been seen in Indonesia, where indication exists that the values of logistic management have been partially applied, particularly in the current rice supply chain, where produce diversity is used and thus accessibility and propose the quality of the produce are protruding (Ssennoga et al., 2019; Abideen et al., 2021). Moreover, the rising significance and influence of agri-food SCM, as well as its interaction with agribiotechnology breakthroughs and applications, has been proven to have beneficial consequences for Malaysian food security and growth (Dharmalingam et al., 2021; Zhong et al., 2017).

12.5 PROCESSE FOR SAFE AGRI PRODUCE IN SUPPLY CHAIN: TECHNOLOGY INCORPORATION

This section examines current and future prospects of perishable food products industry by employment of different technologies to advance the efficiency of the food supply chain. It also inspects the advantages and disadvantages of employing various technological advancements in the food supply chain, which includes production, manufacturing, packaging, and transportation. It also elucidates the requirement for traceability in the supply chain and the methods for achieving traceability through the use of technology. Figure 12.2 displays a process diagram of vegetables and fruits and their whole supply chain movement. The perishable food supply chain includes agricultural raw materials, processing, packing, transportation, and retailing. Vegetables and fruits, for example, are the crops that lie within the category of agricultural raw materials. The fruits and vegetables are transferred to specific storage within processing plants after being harvested, threshed, and dried. They are then sorted, cleaned, chopped, or chemically treated. The fruits and vegetables are then subjected to secondary processing before being transferred to the quality control department to verify their authenticity against particular requirements. Then, depending on the transit length, shelf life, and storing temperature, they are bundled with the appropriate form of packaging. The vegetables and fruit are then shipped to their end destination and sold in retail locations in accordance with customer demand.

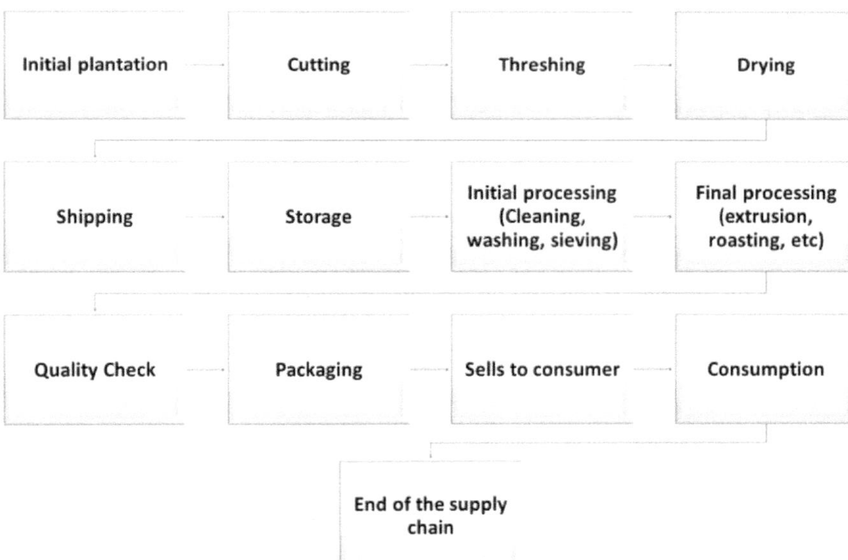

FIGURE 12.2 Process involved in the perishable agriculture produce supply chain.

12.6 SUPPLY CHAIN FOR AGRICULTURAL PRODUCES

The agricultural sector is important since it represents the beginning step of food production. According to Figure 12.1, agriculture is mainly anxious with the sowing, planting, and reaping of crops in the consumable food supply chain. As the food business implements developing technology breakthroughs, agricultural investments have increased year after year. In reality, contemporary digital technology in agriculture may help farmers improve their operations and make better strategic decisions. According to Kamilaris et al. (2019), soil and water sensors are the most extensively utilized technology in agriculture. The production and processing and delivery of perishable foods need advance systems which can control the ambient conditions according to Sufiyan et al. (2019). Along the supply chain these environmental control systems are key variables helps in evaluating the amount of consumed food and the exploited one. To simplify the process of crop imaging and assessment, satellites to monitor weather and imaging are used by the agriculture industry. Technologies with advanced automation has highly popularized, like using autonomous vehicles to identify and sort between seedlings and weed, to spray pesticides and insecticides on the crop, to harvest the fully developed crop, and do other tasks, have the potential to cut operator labor costs. Unmanned ground vehicles (UGVs) are also used to monitor and analyze the plant's performance (Ray, 2017).

Aside from that, autonomous intelligent cars are being employed in poultry farming for jobs such as collecting and sorting eggs, milking and feeding dairy cows, and cleaning, among other things. The chief executive officer (CEO) of Freelance Robotics, William Pagnon, acknowledged in an interview with Reuters that the business is involved in the development of several kinds of independent and intelligent machinery for agricultural applications, such as irrigation systems and tractors. Drones in farming practices have long been utilized in crop monitoring and pesticide application. They may also be used to deliver fertilizer and insecticides to crops. Unmanned aircrafts are widely used in agriculture fields has resulted in an increase in the effectiveness of pesticide spraying operations. In addition to monitoring and inspecting livestock, pilotless airplanes have been shown to be beneficial. Although agricultural operations benefit from the use of technology, the biggest problem in adopting such technological breakthroughs in the agriculture sector is a shortage of funds available for investment. For their part, farmers are more likely than nonfarmers to want simple and easily coordinated procedures in their agricultural operations to optimize their resource utilization (Kros et al., 2019). Cost-effective and easy solutions with few challenges may boost the possibility of acceptance by farmers, predominantly those having little formal education, to better encourage the usage of such innovations among these farmers.

12.7 ROLE OF SAFETY AND SECURITY IN AGRI PRODUCE SUPPLY CHAIN

The goal of the food store and restaurant industries is to receive premium-quality food at a low cost from the supplier while still making a profit and offering competitive prices to customers in the marketplace. Brand loyalty, consumer satisfaction, supply chain efficiency, and overall pleasure are all improved as a result of this. To reach these objectives, the grocery and restaurant industries must closely monitor each stage of the supply chain. When food is lost or discarded at any point within the supply chain process, it is a cause for worry. Unfortunately, according to studies, over 30% of all food produced is thrown away or otherwise discarded. This has a negative influence on food safety and security, the economy, and the environment, among other things. It is possible that food waste and loss may diminish the quantity of fresh produce available on the market, resulting in higher prices, and fewer options for low-income individuals (Naik and Suresh, 2018).

Furthermore, if food quality degrades to an extent that food must be offered at a lower price or thrown away, it has the potential to negatively impact the well-being and livelihood of farmers and producers. Using food planning to coordinate and regulate all areas of the supply chain may help to prevent food waste and loss in the supply network. Food management requires keeping an eye on the supply chain to ensure that all of the meals available are of high quality, taste well, and are safe to consume. The purpose is to verify that any commodities sold to retail establishments meet the standards established by health officials and government officials, among other things. Food inspectors are in responsible of preventing and identifying contamination, which may result in food loss, higher cost, and food insecurity if not done properly or quickly. If efficient food management procedures are not put in place, the whole supply chain may suffer as a result (Nchanji and Lutomia, 2021).

12.8 WHERE DOES PACKAGING FIT?

Primary packaging, or packaging that comes into straight contact with the product, is perilous to the accomplishment of any food processing operation, regardless of the industry. Any method will be rendered ineffective unless it is complemented by adequate package protection measures.

In thermal processing of food products, the use of hermetic packaging is very crucial to ensure food safety (Joshi et al., 2020). This is mostly employed in canning (filling of product in an unsterile glass, metal, laminate or polymer container with an unsterile produce, by heat processing and sealing it for viable sterility) or aseptic packing (hosting a sterile product, usually undergone for a short processing time at high- temperatures), or both. Packages for dry foods, fermented foods, natural foods, and processed foods must all be properly enclosed and maintained to be effective (Esteso et al., 2021).

Food safety is usually described in terms of disease-causing food risks. The disease caused by nutritional deficiency, or, more importantly, famine, is not usually classified as a dietary danger. It is fair to incorporate adequacy in the concept of food safety (Wicaksono and Illés, 2022). A crucial component of food and nutrition security is the preservation of nutrients via packaging, which also enables food to be carried to distant regions. Packaging that enhances shelf life also helps to maintain nutrients while allowing food to be delivered to remote locations (Ganeshkumar et al., 2017). We can extend the hazard list as summarized in Table 12.1.

TABLE 12.1 Hazard Involve in the Supply Chains which can be Reduced via Packaging Systems.

Sr. No.	Hazard	Outcome
1	Contamination by chemicals	Most feared by costumers
2	Physically damaged or abused produce	This is most seen hazard by consumers
3	Microbiological contamination	Most dangerous to consumers
4	Biological attacks by insects of rodents	Not consumable produce
5	Biological actions like sprouting	Consumption can lead to severe diseases

12.9 IMPORTANCE OF PACKAGING IN THE FOOD SUPPLY OF AGRI PRODUCE

In food processing the most critical aspect, while dealing with the perishable food products is packing. Research done by Verghese in his study suggests that in a year, that from the overall food produced for human consumption, approximately 40% of it has been thrown away or discarded as waste. This discarded food can equalize to around 10% energy intake for a week's worth of food for one person (Verghese et al., 2015). As a result, the role of packaging in this regard is crucial, as packaging restricts and preserves the food that humans will consume. Furthermore, because the safety and quality of provided food are intensely dependent on the packaging stage, packaging quality circumlocutorily effects customer satisfaction. It can be seen from Figure 12.1, that the packing of the products lies in the second part of the supply network. The employment of digital and robotic technology in the packaging of food ingredients is prevalent, with filling, handling, and shipping all being performed digitally. As Khan et al. (2018) have pointed out, that recently in the food industry these advance technologies are emerging and have begun to dominate, and they are set to become much more prominent in the near future. Food processing, handling, assembling, packaging, and delivering are just a few of the applications for which robots have been developed. It is possible that simple tasks can be automated in the supply network of food and perishable goods, such as packaging to facilitate the creation of new value-added services with better efficiency of the system. To maintain the quality of food items as high as possible many packaging industries have created several types of intelligent and smart packaging. Two such methods, moisture-absorbing packaging and multilayer barrier, are used to eradicate moisture from food and are described in further detail below. Another innovation is modified atmosphere packaging (MAP), which involves injecting gases into fresh commodities such as fruit, meat, and vegetables to safeguard their in-house environment. Additionally, oxygen scavengers, which are powdered particles that are placed to a box to reduce oxygen, are becoming increasingly often used in the packaging of products. Many firms have lately used ecologically friendly packaging so as to involve in the protection of the environment. Nanotechnology is a new method of preserving freshness, reducing waste, ensuring food safety, and improving quality in perishable food packaging (Pal, 2017). Advanced packaging includes active packaging (AP), which is a sort of advanced packaging. To maintain the quality of quality and cleanliness of the food, the packaging material tend to interact with the food, also resulting in improvement of shelf life and safety of food (Almasi et al., 2020). In smart and sophisticated packaging, sensors are employed in conjunction with neural networks and thresholds (Chen et al., 2020). It has been possible to combine food processors with other pieces of equipment to produce a range of items in one packaging line. In the field of advanced packaging machinery and automation systems, some of the major machine types include cartooning, labelling, shrink-sealing, case-and-tray formation, caping (including robotic systems), palletizing, feeding, placing and picking (robotic systems), sterilizing, cleaning, and inspection/detection machines. Furthermore, automated machines are used in the food business in the area of assembly line production. Thanks to robotic equipment, meat and chocolates can now be packaged quickly and effectively on trays (Bedaka et al., 2021). Furthermore, numerous automated forms of technology have been deployed to automate the final product because of the requirement for higher food output. Food productivity, manufacturing, preservation, and handling have all increased thanks to contemporary control network technologies in the food industry. One of the latest key developments is smart packaging in the food systems, which assists in maintaining the quality of products all over the supply network. Smart packaging is becoming more popular. Smart labels and radio-frequency identification (RFID) have been touted as a simple way to incorporate RFID into conventional package formats (Wu et al., 2020). The outcome of this is that food safety labels have shown to be successful in protecting food from theft as it moves through the logistics chain (Chen et al., 2020). Furthermore, to keep track of the temperature and standard of the produced food, smart labels have been used. Customer product safety assurance has been provided by these labels, which enable consumers to check product details by reading codes and assure that the products has not been interfered with (Urbano et al., 2020). The use of smart labels has also been shown to be beneficial in preventing recalls, minimizing the repercussions of recalls, and recognizing any form of adulteration across the supply chain (Liegeard and Manning, 2019). Additional significant benefit of smart packing materials is that it absorbs oxygen and assist in increasing the shelf life without the use of preservers. As a result, this sort of advancement and technology satisfies customer demand for clean-label products while also increasing shelf life and product distribution to additional geographical places. Scholars are constantly investigating and updating packaging techniques to improve client happiness, resulting in further innovation in the realm of packaging. Given that packaging is a delicate issue that is intricately tied to customer input in the first place, more consideration should be given to this stage.

12.10 APPLICATION OF SMART PACKAGING IN THE SUPPLY CHAIN

When it comes to packaging systems, the terms "intelligent packaging" and "smart packaging" are commonly used synonymously or in a misunderstanding of one another. AP and intelligent packaging are both types of SP that can monitor and react to external and internal variations in a product (intelligence) by engaging with an external boundary (optical or electrical). When it comes to product shelf life and freshness, communicating quality data to customers, improving product security and safety and improving product traceability across the supply chain are the ultimate goals of adopting special packaging (SP). The primary purpose of AP, which is a substitute to conventional packaging, is to promote and uphold high quality while also prolonging the shelf life of food goods. AP is becoming more popular. Different components capable of absorbing/releasing matters from or into packed food to evade rotting may be included into the system to do this. Intelligent packaging, on the contrary, is widely used to track and monitor the conditions of packaged items, as well as to record and communicate information about the product's status throughout the storage and transportation phases. For this reason, hardware components such as gas detectors, RFID, freshness and ripening indicators, and time-temperature indicators (TTI) devices are often utilized in intelligent packaging systems. For data storage and delivery, other "data carriers" systems may be used to display the data once it has been received (Müller and Schmid, 2019). A brief classification of smart packaging is given Figure 12.3.

One of the most significant strategies of addressing food waste and security is via technical breakthroughs in the field of food sanitation. Unique packaging materials and designs, temperature-controlled and energy-efficient storage solutions, and sophisticated monitoring systems are just a few examples of what we can offer you. Various cutting-edge packaging strategies, including AP and SP, have been examined and demonstrated to be a vital tools in decreasing food waste while also assuring food security and exceeding customer anticipations, among other things (Vilariño et al., 2017). Although advances in SP electronics, such as electrochemical sensors—e-noses and e-tongues—have made significant progress, present devices are frequently sophisticated and costly, and are not yet prepared to be incorporated into practical packaging (Poyatos-Racionero et al., 2018). However, despite the fact that the terms "best before" and "sell by" or "use by" have become commonplace in the food industry, they do not provide information on the condition of the food within the packaging. As a result, "dynamic shelf-life systems" should be designed in such a way that it is straightforward to comprehend. The packaging material's impact on environment is a source of concern, that's why it is necessary to conduct research into mapping the environmental performance assessment of packing materials. This study will help package designers by providing them with recommendations (Heller et al., 2019).

New technologies are needed to secure food quality. In the pharmaceutical and food sectors, technological improvements in SP systems are developing and being implemented, not just to enhance traditional packaging operations, but as a technique or a remedy to increase the shelf-life of eatable goods, increase production efficiency, reduce food waste, eliminating the need of preserving chemicals, and, most notably, provide high value and variation to ensure consumer well-being and gratification. In the pharmaceutical and food sectors, advanced SP technologies are developing and being implemented not only to supplement traditional packaging activities but also as a tool or a substitute for conventional packaging functions. A key area of concern continues to be the development of improved food packaging methods that target particular chemical or biological indicators in food. Pre-existing information about the important microbiological agents, their presence under different environments in various types of food produce, and the liberation of reaction chemicals generated throughout the rotting phenomenon are used to determine which market to target for the product (Janjarasskul and Suppakul, 2017).

Food degradation is a multistep process that may be caused by a number of different physical, chemical, enzymatic, or microbiological events in the environment. Bacterial growth and metabolism may result in pH changes, off-odors, the generation of toxic chemicals, gas production, and the development of slime. Chemical routes such as lipolysis, irradiation, and oxidation, may produce unpleasant scents as well as adverse effects on the environment. In addition to inherent qualities, extrinsic factors (such as humidity, temperature, and pH) may have an influence on the chemical, physical, and biological degradation of food (physicochemical and structural). Among the technologies that may aid in the reduction of food spoilage are integrity indicators, ripeness indicators, microwave doneness indicators, rancidity indicators, food spoilage indicators, and RFIDs (Han et al., 2018).

Smart RFID tags make an effort to determine the contained product quality established on the incorporation of sensing components in the tag. A sensitive coating, a litmus paper, an optical label, and conductivity or pH electrodes are among the chemical components that allow these sensor-enabled RFID labels to sense variations in food quality such as dielectric constant (ε_r), conductivity, viscosity, pH, food volatilities, and gases. The application of RF structure onto volatile responsive films of different food results in the detection of changes in response measured as a change in color of specific dyes sensitive to the volatiles rejected from a particular food or due to alteration in pH. Another example is when the food spoils, its dielectric constant varies which can be monitored by a colorimetric sensing device (Bibi et al., 2017). As a result of the breakdown of food, volatile compounds are released. Sulfuric compounds, ethanol, and carbon are among the consequences of bacterial breakdown that can be used as various types of food spoilage indicators. Sulfuric compounds, dimethylamine, trimethylamine, histamine, ammonia, carbon, and ethanol are among the molecules that can be used as various types of food spoilage indicators. The volatile compounds in food have been discovered and measured using solid-phase microextraction integrated with gas chromatography-mass spectrometry, near-infrared spectroscopy, ultraviolet–visible spectroscopy, and, as well as other analytical techniques. While smart sensor-activated RFIDs, have been hailed as user-friendly technology and is cost effective, nonintrusive, and for food packaging, most of these solutions are complicated, costly, and time-consuming to implement (Badia-Melis et al., 2018). As a result, since these packaging solutions feature indicators that interact with foods and metabolites in the headspace, in addition to external environmental elements, they have the potential to become strong tools for decreasing food waste. This makes it possible to keep track of the state of the food produce in question.

Chromogenic chemo sensors are one more up-and-coming disposable technology, not only because they are less costly, adaptable, and easy to print on the packaging but also because the naked eye may easily recognize their color change over time through transparent films. The lack of specificity in these sensors is a drawback. TTIs visualize a product's whole temperature profile (or a portion of it) by capturing both time and temperature impacts. TTIs have been produced based on their functioning processes (chemical, enzymatic, microbiological, and mechanical). TTIs, on the contrary, are unable to produce the metabolic changes that take place in food (Poyatos-Racionero et al., 2018). In 2015, for instance, researchers developed a unique time-temperature integrator based on the Maillard process. This kind of integrator can detect and change the temperature of the meal between 4 and 32°C. As a function of temperature and time, the color change rate constant was measured, and the Arrhenius model was used to get the color change rate constant. They demonstrated the sensor's ability as a predictor of food microbial growth at a variety of different temperatures using the sensor (Rokugawa and Fujikawa, 2015). Kulchan et al. (2016) established a colorimetric indicator to evaluate the rancidity of oxygen-responsive instant milk powder. It was determined that the indicator tags included a mixture of two dyes which are pH-sensitive, methyl red and bromothymol blue, which were likely to react with compounds volatile in nature, created during the storage process as a result of oxidation. Bromothymol blue's pH declined from 7.6 to 5.8 with a color transformation from blue to yellow and a pH variation from 7.6 to 5.8, while methyl red's pH decreased from 6.2 to 4.5 with a color change from yellow to red and a pH change from 7.6 to 5.8. The shift in color from bright green to orange might be perceived as a warning or rejection signal (Kulchan et al., 2016).

According to a study published last year by the Boston Consulting Group (BCG), progress in reducing food waste throughout the whole supply chain has been sluggish owing to the absence of infrastructure, effort, law, and coordination across the entire supply chain. Increases in supply chain efficiency and infrastructure, alone, are likely to cut food waste by $270 billion, or $1.5 trillion by 2050, according to the World Food Programme. SP systems may contribute to this goal by minimizing wastage of food and allowing a more environmentally friendly supply chain to be established. Using data carriers as examples, supply chain efficiency will be improved by assuring traceability, automatization, theft prevention, and counterfeit protection across the supply chain to reduce costs and increase efficiency (Han et al., 2018). RFIDs will improve the entire supply chain by aiding with managing inventory and traceability, therefore improving the overall quality and safety of the food being distributed. TTIs will also assist in the surveillance of an acceptable temperature profile or the establishment of a cold chain over an entire distribution network. As a consequence of this, it was expected for SP to improve the efficiency of distribution of the supply chain by reducing the food waste and by properly recognizing food that is decaying. However, it will also alleviate worries about food safety. Moreover, not only labor and material price associated with analyzing packaged items are decreased but also cost savings associated with preventing food waste, when SP is used, are decreased. Developing novel bioactive substances is expected to be a trend in the future, with the potential to boost market prospects, increase consumer demand for products, and be adopted by new food enterprises. Furthermore, it is hoped that further advancements in food waste data collection will benefit policy making as well as the marketing of these package designs in the future. Advanced manufacturing techniques are also necessary for reducing both the cost of mass production and the difficulties of incorporating smart devices into current packaging lines. It is necessary to do more study into the safety issues, as well as the practicality and ability to incorporate the technology into a wide-ranging variety of applications (Han et al., 2018). In conclusion, consumers should be up-to-date about current packaging technologies, including the costs and benefits associated with them, as well as their readiness to make an investment.

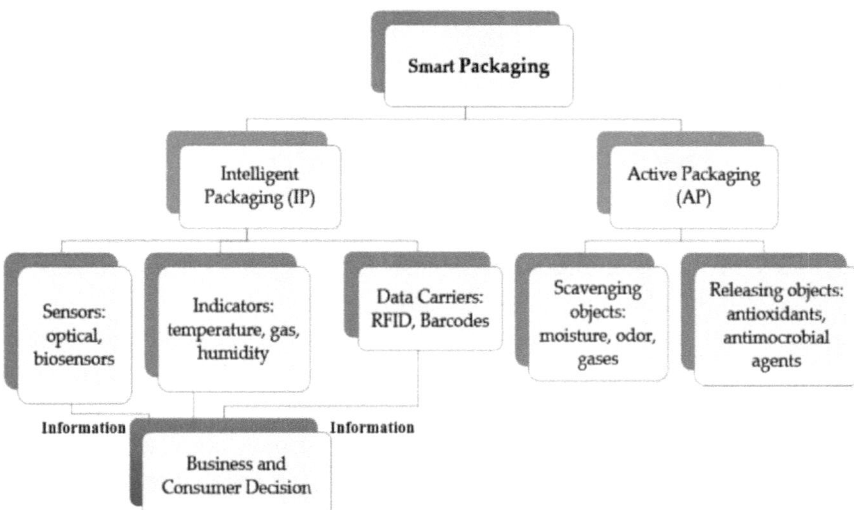

FIGURE 12.3 Classification of smart packaging systems

Source: Adapted from Beshai et al., 2020.

12.11 ROLE OF ANTICOUNTERFEITING TECHNIQUES IN THE AGRO PRODUCE SUPPLY CHAIN

Anticounterfeiting in the supply chain is a widespread concern that impacts practically every business. Contraband electrical components, automobile parts, consumer products, medications, and even counterfeit wines are all available. While counterfeit products cause billions of dollars in losses to manufacturers and distributors every year, the hazards to customers might be far greater than this. Defective counterfeit automobile components or customer items may swelter or catch fire, and counterfeit medications are responsible for the deaths of more than a million people each year in the United States. Despite the fact that manufacturers, distributors, shippers, and government organizations are actively attempting to eliminate counterfeit items from the supply chain, it is difficult to distinguish between legitimate and bogus products.

When counterfeit items enter the supply chain, it may be difficult, if not impossible, to distinguish them from legitimate products. To defeat counterfeiters, it is necessary to implement a reliable method of certifying items from the point of origin to the point of ultimate delivery. Fortunately, there is now cloud-based security technology available that can provide one-of-a-kind, unassailable digital identities for items, allowing them to be monitored at every stage of the supply chain (Lu et al., 2022).

12.11.1 WHY COUNTERFEIT PARTS ARE HARD TO IDENTIFY

Unfortunately, it is the responsibility of suppliers to ensure the timely delivery of products and components. Fake goods are dealt with by no centralized institution or system since government agencies and law enforcement are not up to the job. For example, despite recent crackdowns by government authorities, China remains one of the world's foremost manufacturers as well as the world's biggest provider of counterfeit products. Given the absence of any central police body to oversee the transportation of commodities, it is up to each individual nation to do all in its power to eradicate bogus items from both imports and exports.

Another issue to contend with is the overall quality of exporting products. Factory seconds are sometimes misrepresented as high-quality products. For example, in the Chinese wholesale market, Grade-A goods may be manufactured by factories that are authorized to manufacture authentic branded products, whereas lower-grade goods may be rerouted as factory seconds or produced in the same factories using fairly low-grade materials, depending on the situation. The fact that low-grade items are being passed off as high-quality products means that even people in charge of the supply chain are ignorant that the goods are bogus. The greatest method, since there is no way to police counterfeiters, would be to trace down the things that have been created. A safe, open ecosystem for product authentication that is simple to access and use has been established, and suppliers may now authenticate shipments from any merchants that are a member of the verification ecosystem by using the system (Chen et al., 2021).

12.11.2 FOOLPROOF IDENTIFIERS ELIMINATE COUNTERFEITS

For applications such as secure authentication, distributed ledger technology is especially suited, which makes it a particularly enticing option. Distributed ledgers can be said as digital systems that validate assets or communications by keeping data in several cloud sites at the same time. They are becoming more popular as a means of authenticating assets and transactions. The difference between a distributed ledger and standard databases is that a distributed ledger makes use of numerous nodes that must work together to authenticate each other.

The fact that distributed ledger technology is impenetrable makes it perfect for use in the authentication of goods shipments. The data is encrypted, and to get access to it, authorized users must be validated by each node, resulting in a consensus on the authenticity of the information. The benefit of adopting a distributed ledger system for commodities in the supply chain is that it allows for end-to-end validation, which means that items may be validated at any point along the supply chain (Xiong et al., 2020).

It is necessary to begin with a digital identification ecosystem built on a distributed ledger to authenticate and trace the movement of products and resources. Individual manufacturers may develop their own secure ecosystems, or a trade organization or other group might establish an open security consortium to which members can subscribe to protect products while in transit or at rest. As soon as the ecosystem is in place, each product may be assigned a unique identification that is protected by the distributed ledger system, which can then be used to track its progress. More crucially, anybody in the supply chain may authenticate items by scanning the identification with a quick-response code (QR) code reader, which is widely available.

Each unit is assigned a unique set of encrypted credentials. Authentication does not need knowledge of the contents of the box, allowing for the protection of the identity of high-value commodities while still ensuring that they are delivered correctly. The QR code tracks the movement of items throughout the distribution channel and may be viewed at any time using a free smartphone app or a QR code reader that can be downloaded. In addition, since the identities are generated via the use of a distributed ledger system, the codes cannot be faked or counterfeited. To monitor specific components, the digital security ecosystem is capable of creating an unlimited number of tracking numbers, which might be valuable for recalls and inventory management. For example, in the pharmaceutical industry, each batch of medications transported may be branded with a unique QR code that contains information such as the manufacturing source, production date, formulation, and other pertinent information about the drug. In the case of an issue or recall, the QR code may be used to identify every contaminated item that has been produced (Xie and Tan, 2021).

Distributed ledger technology and other security technologies provide suppliers with the tools they need to drive counterfeiters out of the supply chain by using distributed ledger technology and other security technologies. If you are unable to apprehend the criminals, you may at the very least reduce their earnings by tightening the supervision over products in transit. Cloud technology and breakthroughs such as distributed ledger technology are both cost effective and watertight, and the greater the number of producers that use new security solutions for product authentication, the less counterfeit goods will find their way into the distribution chain.

12.12 FSSAI REGULATIONS AND GUIDELINES FOR AGRO PRODUCE SUPPLY AND SAFETY: IMPORTANT ASPECTS

It is the goal of the Food Safety and Standards Authority of India to bring the quality and safety of unorganized fruits and vegetable retail markets up to the standards of established and organized fruits and vegetable retail shops. Using this effort, local vendors will be integrated with organized merchants, including ecommerce platforms, to improve the state of unorganized organic and nonorganica fruits and vegetable retail marketplaces throughout the country. It will also increase customer confidence in the quality, safety, and freshness of the food.

12.12.1 GENERAL REQUIREMENTS

According to the Food Safety and Standards Regulation 2011, merchants of fruits and vegetables are required to be registered. Food safety display boards with green color codes relevant to fruit and vegetable retail businesses, as well as the FSSAI registration number, must be placed on the shop's outside. A food safety team may be formed for the purpose of managing activities and communication linked to clean and the fresh fruit and vegetable marketplace. No one shall sell or offer for sale any fruits or vegetables that have been cultivated with carbide gas. The fruits may be artificially ripened using ethylene gas supplied from any safe source up to a concentration of 100 parts per million (ppm). It is forbidden for the source of ethylene gas to come into close contact with fruits. Anyone who keeps unapproved substances on the premises will be prosecuted. The use of stickers on fruits and vegetables should be avoided if they do not provide any useful information such as traceability, grade, price, or other relevant information. It is possible to prevent the direct application of stickers on items by using a functional variable. It is possible to package a few fruits in translucent thin films on which a sticker may be affixed. It is essential that if stickers are put directly on the skins of fruits or vegetables, the adhesives and ink used on the stickers are of high quality and do not have any bad effects on human health when consumed. All fruits and vegetables sold at the market must be devoid of any dangerous compounds such as coloring matter, mineral oils, or other toxic substances. Beeswax, carnauba wax, or shellac wax shall be used only for the wax coating of fruits at a concentration that does not exceed good manufacturing practices. Waxes other than those specified above are not approved for use in fruit coating. A label stating that the box of fresh fruit has been wax coated and the name of the wax should be included with the product (Agarwal, n.d.).

12.12.2 HANDLING OF AGRI PRODUCES

It is important to handle fruits and vegetables with care when displayed on display to minimize bruising and product damage. Produce such as fruits and vegetables should not be put directly on the floor or on the platforms. It should be placed on display units, boards, boxes, and other similar items. Organic fruits and vegetables should not be overstacked in display units to avoid scorching and bruising of the fruit and vegetables. A visual inspection of the fruits and vegetables on display should be performed frequently to look for evidence of damage, breakage, physical contaminations, and spillages. Any product that is of low quality should be withdrawn from the market. Any fruit or vegetable that has been chopped for display purposes must be thrown after it has been used (Agarwal, n.d.).

12.12.3 SOURCE OF FRUITS AND VEGETABLES AND THEIR TRACEABILITY

It may be necessary to prevent the procurement of fruits and vegetables from areas or fields with high levels of heavy metals in the soil and the use of drainage water for irrigation of fruits and vegetables. Fruits and vegetables must be purchased from legitimate producers and suppliers. Records of their information must be kept to prevent the introduction of unapproved or adulterated goods into the supply chain. All organic fruits and vegetable providers must have the proper certification and adhere to the Food Safety and Standards (Organic Foods) Regulations, which came into effect in 2017 (Agarwal, n.d.).

12.12.4 TRANSPORTATION OF FRUITS AND VEGETABLES

To ensure safety, security, cleanliness, and no damage to the commodity, all vehicles or modes of transportation shall be thoroughly inspected by experienced personnel, this can reduce the promotion of contamination, deterioration, and loss to the product. The vehicles shall be free from pests, holes, produce debris, objectionable order, and visible molds. It should be clean and hygienic. Transportation vehicles shall be adequately covered to prevent contagion or loss due to weather, rodents, insects, or other pests. All vehicles carrying temperature-sensitive food items or goods should be fitted with proper refrigerating equipment to maintain suitable temperatures which can be recorded by devices for proper documentation (Agarwal, n.d.).

12.12.5 *DIGITAL INTEGRATION*

To facilitate the proper food supply and fulfill the requirements of the produce, digital platforms can be used extensively. FSSAI is creating a digital application as an aggregator platform for collating the information of organic or non-organic fruits and vegetable vendors associated with clean and fresh fruits and vegetables available in the market. This application will help the e-commerce business or other online platform retailers to track a range of fruits and vegetables available in the market. The technology and e-commerce partners may be brought on board to develop or implantation of technology tools and create the application for propelling this initiative. Further, convergence with various government schemes will be explored for incorporation into the platform (Agarwal, n.d.).

12.13 CONCLUSIONS

This chapter includes a brief introduction to the supply chain for agricultural produce. It also covers the importance and day-to-day role of food safety and security. The importance of policy making for agricultural produce has been discussed briefly. Furthermore, a brief relation of government policies with supply chain management has been done to clarify the concept of the food supply chain and the importance of policy making. The importance of technology incorporation was summarized for the safe agricultural products supply chains. The importance of the safety and security of this agricultural produce in day-to-day life was also discussed. This chapter also summarized how the packaging fits in the supply chains and its significance during food supply. As the product's packaging plays a crucial role in its freshness, smart packaging is vital in protecting it from various hazards such as chemical, physical, or biological contaminations. Seeing the importance and regulation of food production, the counterfeiting of such products is always a challenge; different anticounterfeiting approaches have been discussed to address this issue. This chapter will surely clear the critical doubts of the readers related to agricultural production supply chain management, safety, security, and anticounterfeiting.

KEYWORDS

- **food safety**
- **supply chain**
- **anticounterfeiting**
- **packaging**
- **smart packaging**
- **agricultural product security**

REFERENCES

Abideen, A. Z.; Sundram, V. P. K.; Pyeman, J.; Othman, A. K.; Sorooshian, S. Food Supply Chain Transformation through Technology and Future Research Directions: A Systematic Review. *Logistics* **2021,** *5* (4), 83. https://doi.org/10.3390/LOGISTICS5040083

Agarwal, P. Clean and Fresh Fruit and Vegetable Market Guidance Document, n.d.

Almasi, H.; Jahanbakhsh Oskouie, M.; Saleh, A. A Review on Techniques Utilized for Design of Controlled Release Food Active Packaging. **2020,** *61* (15), 2601–2621. https://doi.org/10.1080/10408398.2020.1783199

Ashley, J. M. Introduction. In *Food Security in the Developing World*, 2016; pp 1–18. https://doi.org/10.1016/B978-0-12-801594-0.00001-4

Aung, M. M.; Chang, Y. S. Temperature Management for the Quality Assurance of a Perishable Food Supply Chain. *Food Control* **2014,** *40* (1), 198–207. https://doi.org/10.1016/J.FOODCONT.2013.11.016

Badia-Melis, R.; Mc Carthy, U.; Ruiz-Garcia, L.; Garcia-Hierro, J.; Robla Villalba, J. I. New Trends in Cold Chain Monitoring Applications-A Review. *Food Control* **2018,** *86*, 170–182. https://doi.org/10.1016/J.FOODCONT.2017.11.022

Balaji, M.; Arshinder, K. Modeling the Causes of Food Wastage in Indian Perishable Food Supply Chain. *Resour. Conserv. Recycl.* **2016,** *114*, 153–167. https://doi.org/10.1016/J.RESCONREC.2016.07.016

Bedaka, A. K.; Lee, S. C.; Mahmoud, A. M.; Cheng, Y. S.; Lin, C. Y. A Camera-Based Position Correction System for Autonomous Production Line Inspection. *Sensors* **2021,** *21* (12). https://doi.org/10.3390/S21124071

Beshai, H.; Sarabha, G. K.; Rathi, P.; Alam, A. U.; Jamal Deen, M. Freshness Monitoring of Packaged Vegetables. *Appl. Sci.* **2020,** *10* (21), 7937. https://doi.org/10.3390/APP10217937

Bhunnoo, R. The Need for a Food-Systems Approach to Policy Making. *Lancet* **2019,** *393* (10176), 1097–1098. https://doi.org/10.1016/S0140-6736(18)32754-5

Bibi, F.; Guillaume, C.; Gontard, N.; Sorli, B. A Review: RFID Technology Having Sensing Aptitudes for Food Industry and their Contribution to Tracking and Monitoring of Food Products. *Trends Food Sci. Technol.* **2017,** *62*, 91–103. https://doi.org/10.1016/J.TIFS.2017.01.013

Boratyńska, K.; Huseynov, R. T. An Innovative Approach to Food Security Policy in Developing Countries. *J. Innov. Knowl.* **2017,** *2* (1), 39–44. https://doi.org/10.1016/J.JIK.2016.01.007

Bosona, T.; Gebresenbet, G. Food Traceability as an Integral Part of Logistics Management in Food and Agricultural Supply Chain. *Food Control* **2013,** *33* (1), 32–48. https://doi.org/10.1016/J.FOODCONT.2013.02.004

Büyüközkan, G.; Göçer, F. Digital Supply Chain: Literature Review and a Proposed Framework for Future Research. *Comput. Ind.* **2018,** *97*, 157–177. https://doi.org/10.1016/J.COMPIND.2018.02.010

Chen, S.; Brahma, S.; Mackay, J.; Cao, C.; Aliakbarian, B. The Role of Smart Packaging System in Food Supply Chain. *J. Food Sci.* **2020,** *85* (3), 517–525. https://doi.org/10.1111/1750-3841.15046

Chen, C. L.; Shang, X.; Tsaur, W. J.; Weng, W.; Deng, Y. Y.; Wu, C. M.; Cui, J. An Anti-Counterfeit and Traceable Management System for Brand Clothing with Hyperledger Fabric Framework. *Symmetry* **2021,** *13* (11), 2048. https://doi.org/10.3390/SYM13112048

Clapp, J.; Moseley, W. G.; Burlingame, B.; Termine, P. Viewpoint: The case for a Six-dimensional Food Security Framework. *Food Policy* **2022,** *106*, 102164. https://doi.org/10.1016/J.FOODPOL.2021.102164

Cole, M. B.; Augustin, M. A.; Robertson, M. J.; Manners, J. M. The Science of Food Security. *Npj Sci. Food* **2018,** *2* (1), 1–8. https://doi.org/10.1038/s41538-018-0021-9

Dharmalingam, B.; Giri Nandagopal, M. S.; Thulasiraman, V.; Kothakota, A.; Rajkumar. Short food supply chains to resolve food scarcity during COVID-19 pandemic—An Indian model. *Adv. Food Sec. Sustain.* **2021,** *6*, 35. https://doi.org/10.1016/BS.AF2S.2021.08.001

Esteso, A.; Alemany, M. M. E.; Ortiz, Á. Impact of Product Perishability on Agri-food Supply Chains Design. *Appl. Math. Model.* **2021,** *96*, 20–38. https://doi.org/10.1016/J.APM.2021.02.027

Ganeshkumar, C.; Pachayappan, M.; Madanmohan, G.; Ganeshkumar, C.; Pachayappan, M.; Madanmohan, G. Agri-food Supply Chain Management: Literature Review. *Intell. Inf. Manag.* **2017,** *9* (2), 68–96. https://doi.org/10.4236/IIM.2017.92004

Gunaratne, M. S.; Radin Firdaus, R. B.; Rathnasooriya, S. I. Climate Change and Food Security in Sri Lanka: Towards Food Sovereignty. *Humanit. Soc. Sci. Commun.* **2021,** *8* (1), 1–14. https://doi.org/10.1057/s41599-021-00917-4

Haji, M.; Kerbache, L.; Muhammad, M.; Al-Ansari, T. Roles of Technology in Improving Perishable Food Supply Chains. *Logistics* **2020,** *4* (4), 33. https://doi.org/10.3390/LOGISTICS4040033

Han, J. W.; Ruiz-Garcia, L.; Qian, J. P.; Yang, X. T. Food Packaging: A Comprehensive Review and Future Trends. *Compr. Rev. Food Sci Food Saf.* **2018,** *17* (4), 860–877. https://doi.org/10.1111/1541-4337.12343

Heller, M. C.; Selke, S. E. M.; Keoleian, G. A. Mapping the Influence of Food Waste in Food Packaging Environmental Performance Assessments. *J. Ind. Ecol.* **2019,** *23* (2), 480–495. https://doi.org/10.1111/JIEC.12743

Janjarasskul, T.; Suppakul, P. Active and Intelligent Packaging: The Indication of Quality and Safety. **2017,** *58* (5), 808–831. https://doi.org/10.1080/10408398.2016.1225278

Joshi, S.; Singh, R. K.; Sharma, M. Sustainable Agri-food Supply Chain Practices: Few Empirical Evidences from a Developing Economy **2020,** *24* (3). *Https://Doi.Org/10.1177/0972150920907014*. https://doi.org/10.1177/0972150920907014

Kamilaris, A.; Fonts, A.; Prenafeta-Boldú, F. X. The Rise of Blockchain Technology in Agriculture and Food Supply Chains. *Trends Food Sci. Technol.* **2019,** *91*, 640–652. https://doi.org/10.1016/J.TIFS.2019.07.034

Khan, Z. H.; Khalid, A.; Iqbal, J. Towards Realizing Robotic Potential in Future Intelligent Food Manufacturing Systems. *Innov. Food Sci. Emerg. Technol.* **2018,** *48*, 11–24. https://doi.org/10.1016/J.IFSET.2018.05.011

Kros, J. F.; Liao, Y.; Kirchoff, J. F.; Zemanek Jr., J. E. Traceability in the Supply Chain. *Int. J. Appl. Logist.* **2019,** *9* (1), 1–22. https://doi.org/10.4018/IJAL.2019010101

Kulchan, R.; Boonsupthip, W.; Jinkarn, T.; Suppakul, P. Developing a Novel Colorimetric Indicator for Monitoring Rancidity Reaction and Estimating the Accelerated Shelf Life of Oxygen-Sensitive Dairy Products. *Int. Food Res. J.* **2016,** *23* (3), 1092–1099.

Lencucha, R.; Pal, N. E.; Appau, A.; Thow, A. M.; Drope, J. Government Policy and Agricultural Production: A Scoping Review to Inform Research and Policy on Healthy Agricultural Commodities. *Glob. Health* **2020,** *16* (1), 1–15. https://doi.org/10.1186/S12992-020-0542-2/TABLES/5

Liegeard, J.; Manning, L. Use of Intelligent Applications to Reduce Household Food Waste. *Crit. Rev. Food Sci. Nutr.* **2019,** *60* (6), 1048–1061. https://doi.org/10.1080/10408398.2018.1556580

Lu, Y.; Li, P.; Xu, H. A Food Anti-counterfeiting Traceability System Based on Blockchain and Internet of Things. *Procedia Comput. Sci.* **2022,** *199*, 629–636. https://doi.org/10.1016/J.PROCS.2022.01.077

Markina, I. S. Defining the Dimensions of National Security, Financial Security and Food Supply Chain in Ukraine. *Int. J. Supply Chain Manag.* **2018,** (7), 608–620.

Mozaffarian, D.; Angell, S. Y.; Lang, T.; Rivera, J. A. Role of Government Policy in Nutrition—Barriers to and Opportunities for Healthier Eating. *BMJ* **2018,** *361*. https://doi.org/10.1136/BMJ.K2426

Müller, P.; Schmid, M. Intelligent Packaging in the Food Sector: A Brief Overview. *Foods* **2019,** *8* (1). https://doi.org/10.3390/foods8010016

Naik, G.; Suresh, D. N. Challenges of Creating Sustainable Agri-retail Supply Chains. *IIMB Manag. Rev.* **2018,** *30* (3), 270–282. https://doi.org/10.1016/J.IIMB.2018.04.001

Nchanji, E. B.; Lutomia, C. K. Sustainability of the Agri-food Supply Chain Amidst the Pandemic: Diversification, Local Input Production, and Consumer Behavior. *Adv. Food Secur. Sustain.* **2021,** *6*, 211. https://doi.org/10.1016/BS.AF2S.2021.07.003

Paam, P.; Berretta, R.; Heydar, M.; Middleton, R. H.; García-Flores, R.; Juliano, P. Planning Models to Optimize the Agri-Fresh Food Supply Chain for Loss Minimization: A Review. In *Reference Module in Food Science*, 2016. https://doi.org/10.1016/B978-0-08-100596-5.21069-X

Pal, M. Nanotechnology: A New Approach in Food Packaging. *J. Food Microbiol. Saf. Hyg.* **2017,** *02* (02). https://doi.org/10.4172/2476-2059.1000121

Pang, Z.; Chen, Q.; Han, W.; Zheng, L. Value-centric Design of the Internet-of-Things Solution for Food Supply Chain: Value Creation, Sensor Portfolio and Information Fusion. *Inf. Syst. Front.* **2015,** *17* (2), 289–319. https://doi.org/10.1007/S10796-012-9374-9/TABLES/7

Pérez-Escamilla, R. Food Security and the 2015–2030 Sustainable Development Goals: From Human to Planetary Health. *Curr. Dev. Nutr.* **2017,** *1* (7). https://doi.org/10.3945/CDN.117.000513/4259862

Poyatos-Racionero, E.; Ros-Lis, J. V.; Vivancos, J. L.; Martínez-Máñez, R. Recent Advances on Intelligent Packaging as Tools to Reduce Food Waste. *J. Clean. Prod.* **2018,** *172,* 3398–3409. https://doi.org/10.1016/j.jclepro.2017.11.075

Ray, P. P. Internet of Things for Smart Agriculture: Technologies, Practices and Future Direction. *J. Ambient Intell. Smart Environ.* **2017,** *9* (4), 395–420. https://doi.org/10.3233/AIS-170440

Rokugawa, H.; Fujikawa, H. Evaluation of a New Maillard Reaction Type Time-Temperature Integrator at Various Temperatures. *Food Control* **2015,** *57,* 355–361. https://doi.org/10.1016/J.FOODCONT.2015.05.010

Simelane, K. S.; Worth, S. Food and Nutrition Security Theory. *Food Nutr. Bull.* **2020,** *41* (3), 367–379. https://doi.org/10.1177/0379572120925341

Ssennoga, F.; Mugurusi, G.; Oluka, P. N. Food Insecurity as a Supply Chain Problem. Evidence and Lessons from the Production and Supply of Bananas in Uganda. *Sci. Afr.* **2019,** *3,* e00076. https://doi.org/10.1016/J.SCIAF.2019.E00076

Sufiyan, M.; Haleem, A.; Khan, S.; Khan, M. I. Analysing Attributes of Food Supply Chain Management: A Comparative Study. *Lect. Notes Mech. Eng.* **2019,** 515–523. https://doi.org/10.1007/978-981-13-6412-9_50

Thow, A. M.; Greenberg, S.; Hara, M.; Friel, S.; duToit, A.; Sanders, D. Improving Policy Coherence for Food Security and Nutrition in South Africa: A Qualitative Policy Analysis. *Food Secur.* **2018,** *10* (4), 1105–1130. https://doi.org/10.1007/S12571-018-0813-4/TABLES/5

Urbano, O.; Perles, A.; Pedraza, C.; Rubio-Arraez, S.; Castelló, M. L.; Ortola, M. D.; Mercado, R. Cost-Effective Implementation of a Temperature Traceability System Based on Smart RFID Tags and IoT Services. *Sensors* **2020,** *20* (4), 1163. https://doi.org/10.3390/S20041163

Vågsholm, I.; Arzoomand, N. S.; Boqvist, S. Food Security, Safety, and Sustainability—Getting the Trade-Offs Right. *Front. Sustain. Food Syst.* **2020,** *4,* 16. https://doi.org/10.3389/FSUFS.2020.00016/BIBTEX

Verghese, K.; Lewis, H.; Lockrey, S.; Williams, H. Packaging's Role in Minimizing Food Loss and Waste Across the Supply Chain. *Packag. Technol. Sci.* **2015,** *28* (7), 603–620. https://doi.org/10.1002/PTS.2127

Vilariño, M. V.; Franco, C.; Quarrington, C. Food Loss and Waste Reduction as an Integral Part of a Circular Economy. *Front. Environ. Sci.* **2017,** *5,* 21. https://doi.org/10.3389/FENVS.2017.00021/BIBTEX

Wicaksono, T.; Illés, C. B. From Resilience to Satisfaction: Defining Supply Chain Solutions for Agri-food SMEs Through Quality Approach. *PLoS One* **2022,** *17* (2), e0263393. https://doi.org/10.1371/JOURNAL.PONE.0263393

Wu, D.; Zhang, M.; Chen, H.; Bhandari, B. Freshness Monitoring Technology of Fish Products in Intelligent Packaging. **2020,** *61* (8), 1279–1292. https://doi.org/10.1080/10408398.2020.1757615

Xie, S.; Tan, H. Z. An Anti-Counterfeiting Architecture for Traceability System Based on Modified Two-Level Quick Response Codes. *Electronics* **2021,** *10* (3), 320. https://doi.org/10.3390/ELECTRONICS10030320

Xiong, H.; Dalhaus, T.; Wang, P.; Huang, J. Blockchain Technology for Agriculture: Applications and Rationale. *Front. Blockchain* **2020,** *0,* 7. https://doi.org/10.3389/FBLOC.2020.00007

Zerbian, T.; de Luis Romero, E. The Role of Cities in Good Governance for Food Security: Lessons from Madrid's Urban Food Strategy, 2021. https://doi.org/10.1080/21622671.2021.1873174

Zhong, R.; Xu, X.; Wang, L. Food Supply Chain Management: Systems, Implementations, and Future Research. *Ind. Manag. Data Syst.* **2017,** *117* (9), 2085–2114. https://doi.org/10.1108/IMDS-09-2016-0391/FULL/PDF

INDEX

A

Active oxygen barrier, 127
Active packaging, 96, 171, 176
  application, 184, 186–188
  bioactive components, 178
  challenges, 177
    compatibility, 110
    extraction method, 110
    food components, 109
    mechanism of action, 108
    techno-economic feasibility analysis, 110–111
    in vivo studies, 108
  Commission Regulation (EC) No. 450/2009, 111
  European Food Safety Authority (EFSA), 112
  European Union's framework regulation (EC) No. 1935/2004, 111
  fruits and vegetables
    Food and Drug Administration (FDA), 107
    microbes isolated from fresh produce, 107
    microbiota of microflora, 106
  headspace gas, 177
  legal regulations
    Commission Regulation (EC) No. 450/2009, 111
    European Food Safety Authority (EFSA), 112
    European Union's framework regulation (EC) No. 1935/2004, 111
    Regulation (EC) No. 1935/2004, 111
  natural antimicrobials
    and antimicrobial mechanisms, 99–101
    bioactive compounds, 99–101
    direct addition, 98, 104
    microbial control, 102
    shelf life increment, 102, 103
    sources, 99–101
  packaging systems, incorporation
    chemical immobilization, 105–106
    coating methods, 105
    polylactic acid (PLA), 105–106
    polymeric films, 104–105
  postharvest treatments, 178
  Regulation (EC) No. 1935/2004, 111
Advance packaging
  anticounterfeiting, 307
    foolproof identifiers, 309–310
    identification issues, 308–309
  food safety and security, 292–293
    agri produce supply chain, 299
    packaging, 299–300
    policy making, 293–294
    supply chain management, 295–296
  FSSAI regulations and guidelines, 310
    agri produces, 311–312
    digital integration, 313
    general requirements, 311
    source and traceability, 312
    transportation, 312
  perishable food, 291
  supply chain management
    agricultural produces, 297–298
    packaging importance, 300–302
    safe agri produce, 296–297
    smart packaging, 303–307
Agricultural product supply chains, 290
Animal source based
  casein, 11–12
  chitosan, 8–9
  dielectric barrier discharge (DBD), 12
  gelatin, 10
  whey protein concentrates (WPC), 12
  whey protein isolates (WPI), 12
Antioxidant packaging

application, 184, 186–188
artificial antioxidants, 179
challenges, 185, 189
natural antioxidant compounds, 179
types, 179–180
Artificial antioxidants, 172
Ascorbic acid (AA)-based scavengers, 126

B

Bacterial cellulose (BC), 14–15
Biopolymers, 2
agricultural products, 34
films, 58
SPI nanocomposite film, 59
animal source based
casein, 11–12
chitosan, 8–9
dielectric barrier discharge (DBD), 12
gelatin, 10
whey protein concentrates (WPC), 12
whey protein isolates (WPI), 12
whey proteins (WP), 12–13
classifications
coatings, 38
films, 37
synthetic biopolymers, 38
European Bioplastics, 35
fruits and vegetables
antimicrobial packaging, 23
antioxidant packaging, 22
edible coatings (EC), 23–24
edible films (EF), 23–24
intelligent packaging, 24
legislations and safety issues, 24–25
microorganisms produced
bacterial cellulose (BC), 14–15
low-density polyethylene (LDPE), 14
poly(3-hydroxybutyrate) (PHB), 14
poly(3-hydroxybutyrate-co-3-hydroxyvalerate) (PHBV), 14
polyhydroxyalkanoates (PHAS), 14
packaging materials, 40–42
mechanical properties, 46–48
multi-layer films, 45
physical properties, 43–46
water vapor permeability (WVP), 44
PBAT (polybutylene adipate terephthalate), 36
PBS (polybutylene succinate), 36
physical and mechanical properties
cellulose, 49, 51
chitosan, 51–52
polylactic acid (PLA), 52–53
starch, 48
plant source based
cellulose, 3, 5
gums, 6–7
pectin, 7
plant protein, 7–8
starch, 5, 6
plasticizers, 36
preparation methods
coating, 18–20
electrospinning, 20–21
extrusion, 17–18
solvent casting, 15–17
properties, 39
recommendations, 25–26
synthetically produced
polyethylene terephthalate (PET), 13
polylactic acid, 13
polyvinyl alcohol (PVA), 13–14
trends & innovations
blending, 53–54
crosslinking, 55–56
nanofiller, 56–58
plasticizers, 54–55
Brown paper bags, 279

C

Challenges
in active packaging
compatibility, 110
extraction method, 110
food components, 109
mechanism of action, 108
techno-economic feasibility analysis, 110–111
in vivo studies, 108
Climacteric fruits/non-climacteric, 267
Code of Federal Regulations (CFR), 256
Commission Regulation (EC) No. 450/2009, 111

Compositions
edible coatings and films
composite-based, 224
lipids, 223–224
polysaccharides, 222–223
protein, 223
Controlled atmosphere packaging (CAP), 171
Controlled modified atmospheric packaging (CMAP), 72
Corrugated fiber board (CFB), 247, 281
Cushioning, 249–250
manufactured materials, 251–253
natural cushions, 251
tests and standards, 257

D

Data carriers
barcodes, 209–210
radio frequency identification tags (RFID), 210
Dehydroascorbic acid (DHAA), 126
Dielectric barrier discharge (DBD), 12
Dipping method, 227–228

E

Edible coatings and films, 219
compositions of
composite-based, 224
lipids, 223–224
polysaccharides, 222–223
protein, 223
fabrication, 224
dipping method, 227–228
panning method, 229
spraying method, 228–229
film-forming process
hot melt extrusion, 226–227
solvent casting (wet process), 225–226
functions of, 229
antimicrobial agents in, 232, 234
antioxidants and antibrowning agents, 231–232
moisture and gas barriers, 230–231
nutraceuticals, 234, 236
nanotechnology
biocomposites, 237
edible nanolaminates, 236–237
montmorillonite (MMT), 237
vegetables and fruits, applications, 237
antioxidant agents, 239
fresh-cut vegetables and fruits, 239–240
microorganism contamination, 238
novel food packaging, 238
whole vegetables and fruits, 240–241
Edible polymers, 221
Ethics and safety issues
environmental impact, 161
labelling, 161
migration component, 161
user manual, 160
Ethylene, 146
Ethylene biosynthesis and role, 148
climacteric fruits, 148
European Food Safety Authority (EFSA), 112
European Union's framework regulation (EC) No. 1935/2004, 111

F

Film-forming process
hot melt extrusion, 226–227
solvent casting (wet process), 225–226
Food and Drug Administration (FDA), 107
Food contact materials (FCM), 256
Food contact substance (FCS), 159
Food industry, 98
Food preservation, 96
Food supply chain, 291
Foodborne illness, 290
Fresh agricultural
advancement, 210–212
bio preservation technology, 199
climacteric fruits, 204
cold storage, 199
data carriers, 209–210
degradation, 200
deterioration mechanisms, 199
ease in communication, 203
freshness indicators (FI), 200–201
integrity indicators, 206–207

natural chromogenic indicators, 203
sensorial and nutritional quality, 198
sensors, 207–209
time-temperature indicators (TTI), 204–206
Fresh produce, 267
Fresh-cut fruits and vegetables (FCF&V)
active packaging, 171, 176
bioactive components, 178
challenges, 177
headspace gas, 177
postharvest treatments, 178
antioxidant packaging
application, 184, 186–188
artificial antioxidants, 179
challenges, 185, 189
natural antioxidant compounds, 179
types, 179–180
artificial antioxidants, 172
controlled atmosphere packaging (CAP), 171
minimal processing, quality
microbiological changes, 175
nutritional changes, 175–176
off flavor and discoloration, 176
physiological and biochemical changes, 173–174
modified atmosphere packaging (MAP), 171
natural antioxidant, 181
adsorption or coating, 183
covalent immobilization, 184
multilayer packaging, 183–184
non-package bound system, 182
package bound system, 182–183
natural-derived antioxidant, 172
pathogens, 170
processing, 169
Freshness indicators (FI), 200–201
Fruit ripening, 147–148
Fruits and vegetables
antimicrobial packaging, 23
antioxidant packaging, 22
edible coatings (EC), 23–24
edible films (EF), 23–24
Food and Drug Administration (FDA), 107
intelligent packaging, 24
microbes isolated from fresh produce, 107
microbiota of microflora, 106
FSSAI regulations and guidelines, 310
agri produces, 311–312
digital integration, 313
general requirements, 311
source and traceability, 312
transportation, 312

G

Generally regarded as a safe (GRAS), 77

H

Halloysite nanotube
ethylene adsorption, mechanism, 155
structure and classification, 153–155
Hot melt extrusion, 226–227
Hot water treatment, 260
Hydrocooling, 259
Hydroxyethyl cellulose (HEC), 129

I

Incorporation technologies
blend films, 132
closures, 134
coatings, 131–132
films, 131
immobilization, 133–134
multilayer scavenging films, 132–133
sachets, 130–131
Inorganic oxygen scavengers
active oxygen barrier, 127
Intelligent packaging
advancements in, 210
indicator systems, 3D system, 211–212
nano biosensors, 212–213
sol-gel method, 211
challenges, 213–214
data carriers
barcodes, 209–210
radio frequency identification tags (RFID), 210
fresh agricultural
advancement, 210–212

bio preservation technology, 199
climacteric fruits, 204
cold storage, 199
data carriers, 209–210
degradation, 200
deterioration mechanisms, 199
ease in communication, 203
integrity indicators, 206–207
natural chromogenic indicators, 203
sensorial and nutritional quality, 198
sensors, 207–209
time-temperature indicators (TTI), 204–206
sensors
biosensors, 208–209
gas sensors, 207–208

J

Japanese Oya stone, 158

L

Legal regulations
active packaging
Commission Regulation (EC) No. 450/2009, 111
European Food Safety Authority (EFSA), 112
European Union's framework regulation (EC) No. 1935/2004, 111
Regulation (EC) No. 1935/2004, 111
Listeria inocua, 77
Low-density polyethylene (LDPE), 14

M

Marketing of fruits, 266
agricultural produce
respiration, 270
transpiration, 271
food waste, 267–269
packaging design, 271
fresh produce supply, 275
primary packaging, 273
rigidity, 275
secondary packaging, 273–274
supply chain, 275–276
tertiary packaging, 274
terminologies
climacteric fruits/non-climacteric, 267
fresh produce, 267
shelf-life, 267
traditional package design, 277
brown paper bags, 279
corrugated fiber board, 281
plastic bags/polyethylene film, 278
plastic netting, 280
rigid plastic packages/clamshells, 279–280
shrink wrapping, 278–279
wood packaging, 280–281
trend in packaging, 281
active packaging, 283
antimicrobial packaging, 283–284
edible coatings and films, 284–285
intelligent packaging, 284
modified atmosphere packaging (MAP), 282
Metallic oxygen scavengers, 125–126
Micro-organisms-based oxygen scavengers
hydroxyethyl cellulose (HEC), 129
natural or bio-oxygen scavengers, 130
polyvinyl alcohol (PVA) films, 129
Modified atmospheric packaging (MAP), 67, 69, 70, 171
developments in films, 71
controlled modified atmospheric packaging (CMAP), 72
generally regarded as a safe (GRAS), 77
intelligent packaging, 74, 75
irradiation techniques, 77
Listeria inocua, 77
microperforated, 78
montmorillonite (MMT), 76
Penicillium digitatum, 77
polyphenol oxidase (PPO), 75
silicon gum-based films, 73
ultraviolet light C (UV-C) irradiation, 72
UV-C treatment, 77
fresh foods, 88–89
gas permeability, 83–84
gas research, developments , 84
nitric oxide (NO), 86

nitrous oxide (N_2O), 85
nobel gases, 86–88
sulfur dioxide, 85–86
microperforated films, 78
mathematical models, 79–83
polymer film, 70
respiration rate, 68
Montmorillonite (MMT), 76, 237
Multilayer scavenging films, 132–133

N

Natural antimicrobials
active packaging
and antimicrobial mechanisms, 99–101
bioactive compounds, 99–101
direct addition, 98, 104
shelf life increment, 102, 103
sources, 99–101
Natural antioxidant, 181
adsorption or coating, 183
covalent immobilization, 184
multilayer packaging, 183–184
non-package bound system, 182
package bound system, 182–183
Natural clay-based ethylene scavengers
consumer perspective, 158
ethylene molecules, competitors, 159
examination for, 160
food contact substance (FCS), 159
limited capacity, 159
packaging material, 159
regulatory compliances, 160
reusability, 159
Natural or bio-oxygen scavengers, 130
Nitric oxide (NO), 86
Nitrous oxide (N_2O), 85
Nobel gases, 86–88

O

Octahedral sheet, 147
Oil-derived plastics, 1
On-the-go (OTG), 117
Oxygen scavengers
ascorbic acid (AA)-based scavengers, 126
defined, 118
degradation rate, 117
dehydroascorbic acid (DHAA), 126
edible oxygen-scavenging films, 127
enzyme-based, 128–129
fruits and vegetables, 116, 119
challenges, 135, 137–138
discoloration, prevention, 123–124
future scope, 139
prevent oxidative deterioration, 120–122
preventing mold growth, 122–123
protecting from insects, 122
regulatory framework, 138–139
saturation level, 138
shelf life and quality, 135
global statistics, 117
incorporation technologies
blend films, 132
closures, 134
coatings, 131–132
films, 131
immobilization, 133–134
multilayer scavenging films, 132–133
sachets, 130–131
inorganic oxygen scavengers
active oxygen barrier, 127
micro-organisms-based oxygen scavengers
hydroxyethyl cellulose (HEC), 129
natural or bio-oxygen scavengers, 130
polyvinyl alcohol (PVA) films, 129
on-the-go (OTG), 117
oxygen scavenging capacity (OSC), 120
and rate, 119
oxygen-scavenging rate (OSR), 119, 120
polymer-based, 128
ready-to-eat food (RTE), 117
systems, types, 124
metallic oxygen scavengers, 125–126
Oxygen scavenging capacity (OSC), 120
and rate, 119
Oxygen-scavenging rate (OSR), 119, 120

P

Packaging design, 271
fresh produce supply, 275
primary packaging, 273
rigidity, 275
secondary packaging, 273–274

supply chain, 275–276
tertiary packaging, 274
Panning method, 229
PBAT (polybutylene adipate terephthalate), 36
PBS (polybutylene succinate), 36
Penicillium digitatum, 77
Plastic netting, 280
Poly(3-hydroxybutyrate) (PHB), 14
Poly(3-hydroxybutyrate-co-3-hydroxyvalerate) (PHBV), 14
Polyethylene terephthalate (PET), 13
Polyhydroxyalkanoates (PHAS), 14
Polylactic acid (PLA), 52–53, 105–106
Polyphenol oxidase (PPO), 75
Polyvinyl alcohol (PVA), 13–14
Polyvinyl alcohol (PVA) films, 129
Precooling, 258
Protective packaging, 248
corrugated fiber board (CFB), 253–256
current trends, 257
hot water treatment, 260
hydrocooling, 259
precooling, 258
vacuum heat treatment, 260
cushioning, 249–250
manufactured materials, 251–253
natural cushions, 251
tests and standards, 257
harvest products' packaging for transportation
challenges in, 261
tests and standards
Code of Federal Regulations (CFR), 256
food contact materials (FCM), 256

R

Radio frequency identification tags (RFID), 210
Ready-to-eat food (RTE), 117
Regulation (EC) No. 1935/2004, 111
Rigid plastic packages/clamshells, 279–280
Ripening
defined, 146
ethics and safety issues
environmental impact, 161
labelling, 161
migration component, 161
user manual, 160
ethylene biosynthesis and role, 148
climacteric fruits, 148
ethylene scavengers, 148
zeolite, 149
halloysite nanotube
ethylene adsorption, mechanism, 155
structure and classification, 153–155
Japanese Oya stone, 158
natural clay-based ethylene scavengers
consumer perspective, 158
ethylene molecules, competitors, 159
examination for, 160
food contact substance (FCS), 159
limited capacity, 159
packaging material, 159
regulatory compliances, 160
reusability, 159
sepiolite, 155, 157
smectite clay, 157–158
zeolite
mechanism, 152–153
properties, 150–151

S

Scavenging of ethylene, 147
Sensors
biosensors, 208–209
gas sensors, 207–208
Shelf-life, 267
Shrink wrapping, 278–279
Silicon gum-based films, 73
Smart packaging technologies
classification, 198
commercial intelligent packaging systems, 202–203
countries, 198
European Commission, 196
European Food Safety Authority (EFSA), 197
fresh agricultural
advancement, 210–212
bio preservation technology, 199
climacteric fruits, 204

cold storage, 199
data carriers, 209–210
degradation, 200
deterioration mechanisms, 199
ease in communication, 203
freshness indicators (FI), 200–201
integrity indicators, 206–207
natural chromogenic indicators, 203
sensorial and nutritional quality, 198
sensors, 207–209
time-temperature indicators (TTI), 204–206
Spraying method, 228–229
Sulfur dioxide, 85–86
Supply chain management
agricultural produces, 297–298
packaging importance, 300–302
safe agri produce, 296–297
smart packaging, 303–307

T

Techno-economic feasibility analysis, 110–111
Tetrahedral sheet, 147
Time-temperature indicators (TTI), 204–206
Traditional package design, 277
brown paper bags, 279
corrugated fiber board, 281
plastic bags/polyethylene film, 278
plastic netting, 280
rigid plastic packages/clamshells, 279–280
shrink wrapping, 278–279
wood packaging, 280–281
Trend in packaging, 281
active packaging, 283
antimicrobial packaging, 283–284
edible coatings and films, 284–285
intelligent packaging, 284
modified atmosphere packaging (MAP), 282

U

Ultraviolet light C (UV-C) irradiation, 72

V

Vacuum heat treatment, 260
Vegetables and fruits
applications, 237
antioxidant agents, 239
fresh-cut vegetables and fruits, 239–240
microorganism contamination, 238
novel food packaging, 238
whole vegetables and fruits, 240–241

W

Water vapor permeability (WVP), 44
Whey protein concentrates (WPC), 12
Wood packaging, 280–281

Z

Zeolite, 149
mechanism, 152–153
properties, 150–151